T0213464

Human-Technology Interaction

Carsten Röcker · Sebastian Büttner
Editors

Human-Technology Interaction

Shaping the Future of Industrial
User Interfaces

 Springer

Editors
Carsten Röcker
Institute Industrial IT
OWL University of Applied
Sciences and Arts
Lemgo, Germany

Sebastian Büttner
Institute Industrial IT
OWL University of Applied
Sciences and Arts
Lemgo, Germany

Human-Centered Information Systems
Clausthal University of Technology
Clausthal-Zellerfeld, Germany

ISBN 978-3-030-99237-8   ISBN 978-3-030-99235-4   (eBook)
https://doi.org/10.1007/978-3-030-99235-4

This Springer imprint is published by the registered company Springer Nature Switzerland AG
The registered company address is: Gewerbestrasse 11, 6330 Cham, Switzerland

Contents

Contributors

Anu-Hanna Anttila Finnish Industrial Union, Helsinki, Finland

Daniel Beicht LMU Munich, Munich, Germany

Dirk Berndt Fraunhofer Institute for Factory Operation and Automation IFF, Fraunhofer-Gesellschaft, Magdeburg, Germany

Andreas Besginow Institute Industrial IT, OWL University of Applied Sciences and Arts, Lemgo, Germany

Tim Bosch Sustainable Productivity and Employability, TNO, Leiden, The Netherlands

Philipp Brauner Chair for Communication Science, RWTH Aachen University, Aachen, Germany

Marc Brünninghaus Industrial Automation Branch INA of Fraunhofer IOSB, Fraunhofer-Gesellschaft, Lemgo, Germany

Jendrik Bulk Faculty 4, Bremen City University of Applied Sciences, Bremen, Germany

Sebastian Büttner Institute Industrial IT, OWL University of Applied Sciences and Arts, Lemgo, Germany

Human-Centered Information Systems, Clausthal University of Technology, Clausthal-Zellerfeld, Germany

Sahar Deppe Industrial Automation Branch INA of Fraunhofer IOSB, Fraunhofer-Gesellschaft, Lemgo, Germany

Milena Frahm Fraunhofer Institute for Manufacturing Engineering and Automation IPA, Fraunhofer-Gesellschaft, Stuttgart, Germany

Matthias Freundel Fraunhofer Institute for Manufacturing Engineering and Automation IPA, Fraunhofer-Gesellschaft, Stuttgart, Germany

Tina Haase Fraunhofer Institute for Factory Operation and Automation IFF, Fraunhofer-Gesellschaft, Magdeburg, Germany

Päivi Heikkilä VTT Technical Research Centre of Finland Ltd, Tampere, Finland

Mario Heinz-Jakobs Institute Industrial IT, OWL University of Applied Sciences and Arts, Lemgo, Germany

Eija Kaasinen VTT Technical Research Centre of Finland Ltd, Tampere, Finland

Alinde Keller Fraunhofer Institute for Factory Operation and Automation IFF, Fraunhofer-Gesellschaft, Magdeburg, Germany

Pascal Knierim Bundeswehr University Munich, Munich, Germany

Oliver Korn Affective & Cognitive Institute, Offenburg University, Offenburg, Germany

Thomas Kosch Utrecht University, Utrecht, The Netherlands

HU Berlin, Berlin, Germany

Frank Krause Sustainable Productivity and Employability, TNO, Leiden, The Netherlands

Mareike Kritzler Siemens Corporate Technology, Berkeley, CA, USA

Christopher A. Le Dantec School of Interactive Computing, Georgia Institute of Technology, Atlanta, GA, USA

Florian Michahelles TU Vienna, Vienna, Austria

Karsten Nebe Faculty of Communication and Environment, Rhein-Waal University of Applied Sciences, Kamp-Lintfort, Germany

Hendrik Oestreich Research Institute for Cognition and Robotics, Bielefeld University, Bielefeld, Germany

Volker Paelke Faculty 4, Bremen City University of Applied Sciences, Bremen, Germany

Jörg Papenkordt Organizational Behavior, Paderborn University, Paderborn, Germany

Sarah Polzer Faculty of Communication and Environment, Rhein-Waal University of Applied Sciences, Kamp-Lintfort, Germany

Carsten Röcker Institute Industrial IT, OWL University of Applied Sciences and Arts, Lemgo, Germany

Alyssa Rumsey School of Interactive Computing, Georgia Institute of Technology, Atlanta, GA, USA

Steffen Sauer Fraunhofer Institute for Factory Operation and Automation IFF, Fraunhofer-Gesellschaft, Magdeburg, Germany

Anne Kathrin Schaar Chair for Communication Science, RWTH Aachen University, Aachen, Germany

Philip Sehr Institute Industrial IT, OWL University of Applied Sciences and Arts, Lemgo, Germany

Kirsten Thommes Organizational Behavior, Paderborn University, Paderborn, Germany

Norimichi Ukita Intelligent Information Media Lab, Toyota Technological Institute, Nagoya, Japan

Gu van Rhijn Sustainable Productivity and Employability, TNO, Leiden, The Netherlands

Florian Warschewske Fraunhofer Institute for Factory Operation and Automation IFF, Fraunhofer-Gesellschaft, Magdeburg, Germany

Ellen Wilschut Sustainable Productivity and Employability, TNO, Leiden, The Netherlands

Martin Woitag Fraunhofer Institute for Factory Operation and Automation IFF, Fraunhofer-Gesellschaft, Magdeburg, Germany

Sebastian Wrede Research Institute for Cognition and Robotics, Bielefeld University, Bielefeld, Germany

Martina Ziefle Chair for Communication Science, RWTH Aachen University, Aachen, Germany

Human-Technology Interaction in the Context of Industry 4.0: Current Trends and Challenges

Sebastian Büttner ⓘ and Carsten Röcker ⓘ

Abstract

The industrial landscape changes at a tremendous pace. With new industrial technologies as well as new ways of interacting with technical systems, the work of humans in industry is transformed. This article should outline the current development of human-technology interaction in the context of Industry 4.0. It outlines the recent technical developments toward Industry 4.0 and the technical and societal trends that play a role for the new industrial revolution. Finally, it derives research trends and challenges and gives an outline of the research topics addressed in this collection.

Keywords

Human-technology interaction · Human-computer interaction · Industry 4.0 · Smart manufacturing · Smart industry · Trends · Challenges

S. Büttner
Institute Industrial IT, OWL University of Applied Sciences and Arts, Lemgo, Germany

Human-Centered Information Systems, Clausthal University of Technology, Clausthal-Zellerfeld, Germany
e-mail: sebastian.thomas.buettner@tu-clausthal.de

C. Röcker (✉)
Institute Industrial IT, OWL University of Applied Sciences and Arts, Lemgo, Germany
e-mail: carsten.roecker@th-owl.de

1.1   Introduction

Currently, the industrial landscape changes at a tremendous pace. In order to produce faster, more efficiently and more flexibly, industry is relying not only on automation but also on strong digitalization and networking. In the German-speaking countries, this development is often summarized under the term *Industry 4.0*, which refers to the idea of a fourth industrial revolution that is taking place [1]. There are also corresponding international initiatives for the digitalization of industry, for example under the terms *Advanced Manufacturing* (USA) [2], *Industrie du Futur* (France) [3], *Smart Industry* or *Smart Manufacturing*, which can be understood synonymously [4].

Simultaneously, interaction technologies have made a huge development in the last decades. The use of mobile devices and touchscreens is ubiquitous, augmented and virtual reality technologies have made their way into the market, new interaction concepts such as gesture control have become established in many areas, and last but not least, the usability of many software products has improved significantly. These developments have also led to increased user expectations regarding the usability and user experience of technical systems.

This article aims at analyzing the interplay between the two aforementioned developments to sketch the resulting trends and research challenges. For this purpose, we provide an overview of the recent technical development toward Industry 4.0, followed by a brief presentation of various technical and societal trends that play a role for the new industrial revolution. Subsequently, we outline the resulting future research challenges for the field of Human-Technology Interaction. Finally, this article summarizes the contributions that are part of this collection.

1.2   Toward Industry 4.0

This section gives a short overview of the historic development of industrial production toward the ongoing changes in industry that are subsumed under the term *Industry 4.0*.

Figure 1.1 shows the idea that the development of industrial production has undergone multiple industrial revolutions: the first revolution contained the introduction of mechanical production equipment using water and steam power, starting in the late 18th and accelerating in the nineteenth century. The technological changes brought along economical changes and a transformation of society as such [6]. Friedmann [7] spoke of a second industrial revolution that took place in the beginning of the twentieth century, which can be characterized from a technical perspective by the introduction of division of labor and mass production that was partly enabled by means of technology. The third revolution started in the 1970s and relates to the introduction of electronics and IT into the production process, which facilitated a high degree of automation. This revolution can be seen as part of the digital revolution, which is still ongoing [8]. The fourth industrial revolution that is referred to here as Industry 4.0 can be seen as a second wave of the digital revolution [9]. While

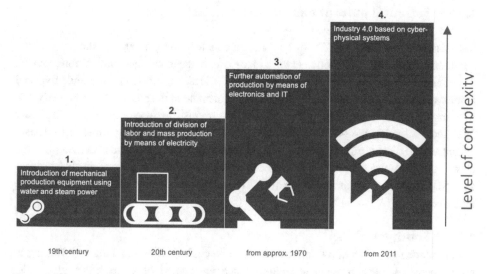

Fig. 1.1 The four industrial revolutions (based on the graphical representation in [5])

digitalization was used in the third industrial revolution to automize single machines, processes or tasks, the vision of the fourth industrial revolution aims at digitalizing and integrating machines throughout organizations and beyond to have an integration along the complete value chain [10].

According to Geissbauer et al. [10] Industry 4.0 is driven by the "digitization and integration of [. . .] value chains", by the "digitization of product and service offerings" and by new "digital business models." Given this transformation of industrial production processes and the introduction of new digital products, services and business models, it becomes obvious that human work also will also change substantially.

1.3 Human-Technology Interaction Perspectives on Industry 4.0

The transition of human work can be viewed from two perspectives. First, new technologies offer new possibilities for organizing or executing work. Technology can help to easily access work-related information [11], to train people on the job [12, 13], to support people during work [14–16] or to improve communication and collaboration [17]. Second, changes in technology, in work processes or in products entail a change in work practices. The interplay between new technologies and new work practices creates challenges for the research in human-technology interaction. Here, we sketch the technological innovations that could extend the existing design space of human-technology interaction in an industrial context, followed by a classification of possible application areas.

1.3.1 Technical Innovations

In the context of Industry 4.0, various interaction technologies are mentioned that are intended to optimize the work between people and technical systems. Among others, Geissbauer [10] mentions mobile and wearable devices, augmented reality and advanced human-machine interfaces as technologies that contribute to the realization of Industry 4.0. In a survey on the topic, Krupitzer et al. [18] distinguish between classic graphical user interfaces following the WIMP (windows, icons, menus, pointer) principle, natural user interfaces (NUI) and virtual and augmented reality (AR) environments. Leaving out the classic graphical user interfaces, we want to provide a brief overview of touch interfaces and NUI, followed by extended reality (XR) interfaces, which subsume AR and VR interfaces.

1.3.1.1 Touch Interfaces

Touch interfaces, which allow for computers to be controlled by touching or gesturing at screen contents, have become ubiquitous since the spread of smartphones and tablet computers at the latest. Consequently, touch panels are established in the industrial environment, e.g., for data input in machine controls and for process monitoring in production, energy and environment, building automation, water management as well as infrastructure and logistics [19]. From a technical perspective, resistive touch screens and capacitive touch screens can be distinguished. Resistive touch screens use two transparent foils on top of the display that are connected by means of pressure to detect touch points. Capacitive touch screens detect conductive objects at the touch-point through capacitive coupling. The different technologies have an impact on the possible use-cases: while resistive touchscreens can only detect a single point, capacitive displays can recognize multiple touch points and therefore allow multitouch gestures. On the other hand, capacitive displays only work if they are touched with electrically conductive material. Consequently, they can cause problems in an industrial context, e.g., if users wear protection gloves [19]. Still, multitouch is considered as an important interaction form for future cyber-physical systems, as this form of interaction is known from personal experience of workers outside the industrial sector [20].

1.3.1.2 Natural User Interfaces (NUI)

The term Natural User Interface (NUI) relates to all interface technologies in which the actual technology of the interface is invisible [21]. Examples for NUI include a broad range of hand or full-body gesture control, gaze control, voice control, proximity sensing and implicit interactions on the input and the full spectrum of human sensing and experience on the output channel [21]. Apart from invisible interfaces, gesture-controlled touch interfaces are also sometimes classified as NUIs (e.g. [22, 23]). These types of interface technologies are not wide-spread within the industrial sector; however, multiple research projects have dealt with the industrial application of NUIs. Gesture control has been used in industry to control robots [24, 25] or for machine control [26], gaze control has been evaluated as a technology for improving process safety of control room workers [27], voice-control has been evaluated in multiple settings, e. g. machine [28] or robot control [29, 30] including work on the question on how to realize voice control in noisy industrial environments [31]. Huge chances are seen in the industrial use of implicit interactions [32] and proximity or location sensing. Proposed industrial systems use the recognition of task process to display context-related information [12] or sense the position of humans or work-related objects as basis for a context-sensitive interaction [33]. Given this broad area of NUI, there is certainly an extensive design space for future applications to be discovered.

1.3.1.3 Extended Reality (XR) User Interfaces

The term extended reality (XR) relates to a range of user interfaces that combine the physical world with the digital world up to interfaces that allow a complete immersion within the digital world. It subsumes all Mixed Reality (MR) variants on the reality-virtuality continuum of Paul Milgram [34] (see Fig. 1.2) and also contains complete virtual environments. Hereinafter, two major variants of XR are described that play a role in the industrial domain: augmented reality (AR) and virtual reality (VR).

AR refers to a specific type of virtual environment that allows people to see a real world augmented with virtual objects. Within the AR environment, people perceive a composite world consisting of existing and virtual objects [35]. According to the definition of Azuma [35], there are three prerequisites that need to be fulfilled to speak of an AR environment: first, the real and the virtual world need to be combined; second, there has to be a real-time interactivity with this integrated virtual and physical world; and third, this environment, which integrates virtual objects, needs to be three-dimensional. There are numerous examples of popular AR applications outside the industrial applications. AR, e.g., is used for games [36], AR can help in the decision process for certain products such as furniture [37] or glasses [38] and AR can help people finding around in other countries [39]. The presentation of Microsoft's Hololens raised many hopes for the industrial use of AR. However, at this time, not many AR systems for industrial applications have been brought to product maturity, so AR can still be considered an important topic for research.

VR subsumes technologies that create virtual worlds in which people can "immersively experience a world beyond reality" [40]. Contrary to AR, VR aims on replacing the perception of the real world with digital content. Looking at the different human senses, different VR technologies have been presented that mainly create visual, auditive and haptic experiences, while olfactory and gustatory experiences have received less attention [40]. While the concept of VR and the required technologies have been available for more than 50 years [41], VR has recently gained traction due to affordable VR headsets that have mainly been developed for gaming and entertainment purposes [42]. Consequently, we expect a rise of commercial VR systems also in industrial application, e. g. for training of new employees.

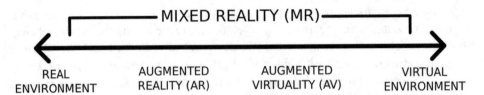

Fig. 1.2 Reality-virtuality continuum. Image source: Wikimedia Commons, https://commons. wikimedia.org/wiki/File:Reality-Virtuality_Continuum.svg, author: Giovanni Vincenti, licensed under CC BY 3.0. Image based on the illustration of Milgram et al. [34]

1.3.2 Application Areas

Based on our previous work [43], we consider the following application areas relevant for the research on human-technology interaction in the industrial domain: manufacturing, logistics, maintenance and repair as well as training, while the latter can be considered as a cross-cutting area that potentially overlaps with the other areas. The application areas are briefly described hereinafter.

1.3.2.1 Manufacturing

Despite the ongoing automation of production facilities, manufacturing is still an area that involves a lot of manual tasks [11]. Consequently, previous research has dealt with the question on how to increase productivity by introducing new interaction technologies in the workplace. To state some examples, Caudell and Mizell [44] presented an AR system that supported manufacturing workers in the aircraft industry. In the field of stationary manual assembly, Boud et al. [45] presented an AR system that is intended to introduce employees to an assembly process by displaying pictograms. Tang et al. [46] developed a similar system to support an assembly process. Multiple stationary assembly support systems have been presented that make use of projections to guide users through an assembly process [12, 14, 15]. And the use of multi-touch displays was investigated to increase efficiency for the control of production facilities [19].

1.3.2.2 Logistics

New interaction technologies have also been investigated in the logistics sector, e.g. by improving warehouse logistics with new interaction technologies. A typical labor-intensive activity in logistics is warehouse picking, where humans collect a set of items from a warehouse. As an alternative to paper-based picking-lists, nowadays pick-by-light systems are often used, which consist of lights or displays attached directly to the compartments in which the items are located. Various works have investigated new interaction technologies to improve the efficiency of this picking task: Schwerdtfeger and Klinker [47] and Günthner et al. [48] used head-mounted displays (HMDs) to help people navigate through warehouses by marking the objects to be picked within an AR environment. The use of HMDs is also referred to as pick-by-vision. Other works have made use of interactive picking carts, e. g. Funk et al. [49] used a picking cart equipped with projectors to enable pick-by-vision without wearing HMDs.

1.3.2.3 Maintenance and Repair

Another broad area of application for new interaction technology is the field of industrial maintenance and repair. This manual task usually requires specific knowledge of workers. Previous work has dealt with the question on how to support human workers with knowledge in the workplace. For this purpose, HMDs, wearables and mobile devices have been used [50]. For instance, Heinz et al. [33] describe an assistance system that supports the maintenance or repair process of complex industrial machines by means of mobile devices or HMDs. Similar maintenance scenarios have also been presented in the automotive sector. Hoffmann et al. [51] investigate the use of AR for repair procedures taking into account the existing prior knowledge about the task. In addition to systems that support individual users, AR systems have been presented that provide remote assistance between two people with different skills levels for supporting collaborative maintenance or repair. For example, Gurevich et al. [52] present the system TeleAdvisor, which supports maintenance or repair tasks through AR projections on site. The projections are controlled by a connected expert that can guide the on-site user through the process. Due to the complexity of human maintenance and repair tasks, we assume that there is huge potential for future research on the use of new interaction technologies in this context.

1.3.2.4 Training

As mentioned, training can be considered as a cross-cutting concern that overlaps with the other areas. On the one hand, there are specific learning systems; on the other hand, there are interactive assistive systems designed for guiding users through the execution of a task, which in the end also facilitates learning. Examples for specific learning systems can be found in the area of VR, where it often is used to train people by simulating a task. This has the advantage that potentially dangerous tasks can be practically trained or that no material is required for the learning phase. As an example, VR has been used to learn welding [53]. Furthermore, AR and VR have been applied to achieve a better understanding of complex machines by creating the possibility of viewing the inner parts of a machine [13]. Interactive assistive systems that facilitate learning have been presented for the familiarization with assembly tasks, e.g. Boud et al. [45] presented an AR application for learning the assembly process of water pumps. Aehnelt and Wegner [54] also presented an assembly assistance system that displays assembly information on mobile devices. The system, which is designed for on-the-job training, aims to combine work with learning experiences. This improves the understanding of the of assembly work, resulting in a positive effect on the independent planning and execution of future work tasks [54]. The field of technologically enhanced learning is broad and there are many more examples for using new interaction technology for learning in the industrial sector. Future research will explore this area in more depth.

1.4   Research Challenges and Contributions in this Collection

Having presented new interaction technologies and application areas in the industrial sector, we would like to sketch the major research challenges on the intersection of interaction technologies and industry that are tackled in this book. By doing so, we also provide an overview on the subsequent chapters in this collection.

1.4.1   How Can Assistance Systems Be Implemented and Integrated into the Work Process?

As the research field of human-technology interaction is concerned with the design of interactions between humans and technological systems, a research challenge that needs to be addressed is the question on how to create methods to design and implement interactive assistive systems in the industrial domain. The following papers focus on supporting companies with methods and tools for taking decisions on the introduction or integration of assistance systems and technologies into the work process. Focusing on digital assembly assistance, Haase et al. [55] present such methods and strategies with a sociotechnical design approach. They develop success criteria, present specific technologies and provide practical examples of assistance systems in industry. Krause et al. [56] consider the case of operator support and present three tools that were developed for small and medium-sized enterprises (SMEs) to reflect on their requirements and in order to make technology choices. Oestereich et al. [57] present design alternatives and a reference architecture for the implementation of adaptive assistance systems as well as a selection of algorithms to realize adaptivity. To demonstrate the feasibility of adaptive assistance systems, they include three exemplary scenarios that make use of the adaptivity. Besginow et al. [58] take a new approach to integrating deep-learning based action detection into stationary assistance systems. Using action recognition, they were able to introduce on-the-fly quality control and inform users whenever errors occurred. Finally, Brünninghaus et al. [59] present the current state of assistance systems for vocational training within a manufacturing context and propose a new system called "XTEND for education."

1.4.2   How Can XR Technology Support Future Work?

Even though past research has presented various concepts and prototypes of systems that make use of XR technology within the industrial domain, we expect that the exploration of use-cases for XR technology has only just begun. While XR headsets are becoming more mature and cheaper, new application areas are simultaneously becoming more feasible. Consequently, we expect more XR concepts and applications within the industry to emerge. Again, from a practical perspective, the question is how the systems will be implemented and rolled out within the industry. Focusing on the design process of XR applications, Paelke et al. [60] present different forms of user-guidance technologies within XR environments. Kosch et al. [61] present a new programming-by-demonstration approach that helps to create AR content for assembly assistance systems. Finally, Rumsey et al. [62] analyze how the introduction of VR technology affects organizational structures. In an empirical study, they investigate organizational challenges coming from the impact of VR on human performance and skills acquisition.

1.4.3 Will Work Become more Human-Centered due to New Technology?

The introduction of new technologies will not only make the current way of working more efficient; rather work practices themselves will be transformed due to new possibilities. Our vision is that work will become more human-centered. Future industry workers might encounter a more human work environment and experience work to be of higher hedonistic and eudaimonistic value. Korn [63] shows how gamification can be used in industrial environments to increase productivity and to motivate staff. Korn not only presents how gamification can be integrated into existing workflows and how user acceptance of gamification can be reached, but also gives specific design recommendations for the implementation of gamification within the industrial domain. Kaasinen et al. [64] studied how personalized job roles and smooth teamwork can improve the well-being of factory floor workers. They envision a close collaboration of human-machine teams in the future of Industry 4.0.

1.4.4 How Can Technology Acceptance and Trust Within Industry 4.0 Systems Be Achieved?

Technology acceptance and trust are essential factors for the successful implementation and integration of new technology within Industry 4.0. A lack of technology acceptance or trust can result in reduced job satisfaction and in performance losses [65]. Focusing on laboratory work, Polzer et al. [66] investigate which factors influence the technology acceptance of technical systems. For this purpose, they apply the technology acceptance model (TAM) in the industrial or more specific laboratory domain. Papenkordt et al. [67] look at factors of smart technologies that promote trust in the systems from a theoretical and empirical perspective. Their results show that immaterial benefits, in particular, affect trust in a strong manner. Providing a final overview on the domain of human-technology interaction, Brauner et al. [68] investigate workers' requirements to achieve human-centered digital production technology that is accepted by its users.

1.5   Conclusion

This article sketched the current development in the area of human-technology interaction in the context of Industry 4.0. It demonstrated the interplay between technological development and the transformation of work. Future research will deal with various questions. First, from an industrial perspective, the question of how to implement and integrate various new systems into the work process, remains a major challenge. In this collection, this is addressed in the context of assistive systems in particular. Nevertheless, this question is of importance for all new interaction concepts and systems in the industrial domain. Second, XR technologies have made a great leap in technical development. Much hope lies in this technology to help deliver information to workers in various contexts. This collection presents concepts for the integration of XR technologies, but of course, future work will further investigate and explore the design space of XR technology. Third, while future work will be informed by new technology, the question will be how more human-centered workplaces can be created. Fourth, new technology requires the acceptance of its users and trust in the technology in order to be successful. Consequently, this topic is addressed here as well. The list of research trends and challenges that are addressed in this collection derived from it is not exhaustive. The development of Industry 4.0 systems has just begun. As researchers and practitioners, let's make sure that Industry 4.0 systems are designed in a human-centered way, so that the lives of future industrial employees can be improved!

References

1. Bundesministerium für Bildung und Forschung: Zukunft der Arbeit. (2021). *Innovationen für die Arbeit von morgen* [German]. Accessed December 01, 2021, from https://www.bmbf.de/upload_filestore/pub/Zukunft_der_Arbeit.pdf
2. President's Council of Advisors on Science and Technology: Accelerating U.S. Advanced Manufacturing: Report to the President. (2014). Accessed January 12, 2021, from https://www.manufacturingusa.com/sites/prod/files/amp20_report_final.pdf
3. Plattform Industrie 4.0: Plattform Industrie 4.0 & Alliance Industrie du Futur. (2018). *Common list of scenarios*. Accessed January 12, 2021, from https://www.plattform-i40.de/PI40/Redaktion/DE/Downloads/Publikation/plattform-i40-und-industrie-dufutur-scenarios.pdf?__blob=publicationFile&v=5
4. Dais, S. (2017). Industrie 4.0 – Anstoß, Vision, Vorgehen [German]. In B. Vogel-Heuser, T. Bauernhansl, & M. ten Hompel (Eds.), *Handbuch Industrie 4.0 Bd. 4: Allgemeine Grundlagen* (pp. 261–277). Springer.
5. Kagermann, H., Wahlster, W., & Helbig, J. (2013). *Umsetzungsempfehlungen für das Zukunftsprojekt Industrie 4.0–Abschlussbericht des Arbeitskreises Industrie 4.0.*
6. McCloskey, D. N. (1981). *The industrial revolution. The economic history of Britain since 1700.*
7. Friedmann, G. (1936). *La crise du progrès. Esquisse d'histoire des idées 1895–1935* [French].
8. Schoenherr, E. (2018). *The digital revolution.* Accessed from https://web.archive.org/web/20180220162425, http://history.sandiego.edu:80/gen/recording/digital.html

9. Kagermann, H., Winter, J. (2018). *Die zweite Welle der Digitalisierung. Deutschlands Chance.* Accessed February 23, 2022, from https://www.plattform-lernende-systeme.de/reden-und-beitraege-newsreader/die-zweite-welle-der-digitalisierung-deutschlands-chance.html

10. Geissbauer, R., Vedso, J., & Schrauf, S. (2016). *Industry 4.0: Building the digital enterprise.* Accessed February 23, 2022, from https://www.pwc.com/gx/en/industries/industries-4.0/landing-page/industry-4.0-building-your-digital-enterprise-april-2016.pdf

11. Fellmann, M., Robert, S., Büttner, S., Mucha, H., & Röcker, C. (2017, August). Towards a framework for assistance systems to support work processes in smart factories. In *International cross-domain conference for machine learning and knowledge extraction* (pp. 59–68). Springer.

12. Büttner, S., Prilla, M., & Röcker, C. (2020, April). Augmented reality training for industrial assembly work-are projection-based AR assistive systems an appropriate tool for assembly training?. In *Proceedings of the 2020 CHI conference on human factors in computing systems* (pp. 1–12).

13. Heinz, M., Büttner, S., & Röcker, C. (2019, June). Exploring training modes for industrial augmented reality learning. In *Proceedings of the 12th ACM international conference on PErvasive technologies related to assistive environments* (pp. 398–401).

14. Büttner, S., Funk, M., Sand, O., & Röcker, C. (2016, June). Using head-mounted displays and in-situ projection for assistive systems: A comparison. In *Proceedings of the 9th ACM international conference on pervasive technologies related to assistive environments* (pp. 1–8).

15. Heinz, M., Büttner, S., Jenderny, S., & Röcker, C. (2021). Dynamic task allocation based on individual abilities-experiences from developing and operating an inclusive assembly line for workers with and without disabilities. *Proceedings of the ACM on Human-Computer Interaction, 5,* 1–19.

16. Sand, O., Büttner, S., Paelke, V., & Röcker, C. (2016, July). Smart assembly–projection-based augmented reality for supporting assembly workers. In *International conference on virtual, augmented and mixed reality* (pp. 643–652). Springer.

17. Adcock, M., & Gunn, C. (2015). Using projected light for mobile remote guidance. *Computer Supported Cooperative Work (CSCW), 24*(6), 591–611.

18. Krupitzer, C., Müller, S., Lesch, V., Züfle, M., Edinger, J., Lemken, A., Schäfer, D., Kounev, S., & Becker, C. (2020). A survey on human machine interaction in industry 4.0. *arXiv preprint arXiv:2002.01025.*

19. Behlen, M., Büttner, S., Schmidt, S., Pyritz, S., & Röcker, C. (2016). *User study on multitouch in the industrial environment.*

20. Gorecky, D., Schmitt, M., Loskyll, M., & Zühlke, D. (2014, July). Human-machine-interaction in the Industry 4.0 era. In *2014 12th IEEE international conference on industrial informatics (INDIN)* (pp. 289–294). IEEE.

21. Jain, J., Lund, A., & Wixon, D. (2011). The future of natural user interfaces. In *CHI'11 extended abstracts on human factors in computing systems* (pp. 211–214).

22. Hofmeester, K., Wixon, D. (2010). Using metaphors to create a natural user interface for Microsoft Surface. In *CHI'10 extended abstracts on human factors in computing systems* (pp. 4629–4644).

23. Preim, B., & Dachselt, R. (2015). Interaktive Systeme: Band 2: User interface engineering, 3D-interaktion. In *Natural user interfaces [German].* Springer.

24. Roda-Sanchez, L., Olivares, T., Garrido-Hidalgo, C., & Fernández-Caballero, A. (2019, June). Gesture control wearables for human-machine interaction in Industry 4.0. In *International work-conference on the interplay between natural and artificial computation* (pp. 99–108). Springer.

25. Sadik, A. R., Urban, B., & Adel, O. (2017, February). Using hand gestures to interact with an industrial robot in a cooperative flexible manufacturing scenario. In *Proceedings of the 3rd International Conference on Mechatronics and Robotics Engineering* (pp. 11–16).

26. Chaudhary, A., Raheja, J. L., Das, K., & Raheja, S. (2013). Intelligent approaches to interact with machines using hand gesture recognition in natural way: A survey. *arXiv preprint arXiv:1303.2292.*
27. Bhavsar, P., Srinivasan, B., & Srinivasan, R. (2017). Quantifying situation awareness of control room operators using eye-gaze behavior. *Computers & Chemical Engineering, 106,* 191–201.
28. Fedosov, Y. & Katridi, A. (2021, May). Concept of implementing computer voice control for CNC machines using natural language processing. In *2021 29th Conference of Open Innovations Association (FRUCT)* (pp. 125–131). IEEE.
29. Janíček, M., Ružarovský, R., Velíšek, K., & Holubek, R. (2021, February). Analysis of voice control of a collaborative robot. *Journal of Physics: Conference Series, 1781*(1), 012025.
30. Rogowski, A. (2012). Industrially oriented voice control system. *Robotics and Computer-Integrated Manufacturing, 28*(3), 303–315.
31. Bartholomew, J. C., & Miller, G. E. (1988, November). Voice control for noisy industrial environments. In *Proceedings of the Annual International Conference of the IEEE Engineering in Medicine and Biology Society* (pp. 1509–1510). IEEE
32. Schmidt, A. (2000). Implicit human computer interaction through context. *Personal Technologies, 4*(2), 191–199.
33. Heinz, M., Büttner, S., Wegerich, M., Marek, F., & Röcker, C. (2018, July). A multi-level localization system for intelligent user interfaces. In *International conference on distributed, ambient, and pervasive interactions* (pp. 38–47). Springer.
34. Milgram, P., & Kishino, F. (1994). A taxonomy of mixed reality visual displays. *IEICE Transactions on Information and Systems, 77*(12), 1321–1329.
35. Azuma, R. T. (1997). A survey of augmented reality. *Presence: Teleoperators & Virtual Environments, 6*(4), 355–385.
36. Tan, C. T. & Soh, D. (2010). Augmented reality games: A review. *Proceedings of Gameon-Arabia, Eurosis.*
37. Raska, K. & Richter, T. (2017). *Influence of augmented reality on purchase intention: The IKEA case.*
38. Rese, A., Baier, D., Geyer-Schulz, A., & Schreiber, S. (2017). How augmented reality apps are accepted by consumers: A comparative analysis using scales and opinions. *Technological Forecasting and Social Change, 124,* 306–319.
39. Tatwany, L., & Ouertani, H. C. (2017). A review on using augmented reality in text translation. In *2017 6th International Conference on Information and Communication Technology and Accessibility (ICTA).* IEEE.
40. Berg, L. P., & Vance, J. M. (2017). Industry use of virtual reality in product design and manufacturing: A survey. *Virtual Reality, 21*(1), 1–17.
41. Sutherland, I. (1965). *The ultimate display.*
42. Wohlgenannt, I., Simons, A., & Stieglitz, S. (2020). Virtual reality. *Business & Information Systems Engineering, 62*(5), 455–461.
43. Büttner, S., Mucha, H., Funk, M., Kosch, T., Aehnelt, M., Robert, S., & Röcker, C. (2017, June). The design space of augmented and virtual reality applications for assistive environments in manufacturing: A visual approach. In *Proceedings of the 10th International Conference on PErvasive Technologies Related to Assistive Environments* (pp. 433–440).
44. Caudell, T., & Mizell, D. (1992, January). Augmented reality: An application of heads-up display technology to manual manufacturing processes. In *Hawaii international conference on system sciences* (Vol. 2). ACM SIGCHI Bulletin.
45. Boud, A. C., Haniff, D. J., Baber, C., & Steiner, S. J. (1999, July). Virtual reality and augmented reality as a training tool for assembly tasks. In 1999 IEEE International Conference on Information Visualization (Cat. No. PR00210) (pp. 32–36). IEEE.

46. Tang, A., Owen, C., Biocca, F., & Mou, W. (2003, April). Comparative effectiveness of augmented reality in object assembly. In *Proceedings of the SIGCHI conference on Human factors in computing systems* (pp. 73–80).
47. Schwerdtfeger, B. & Klinker, G. (2008, September). Supporting order picking with augmented reality. In *2008 7th IEEE/ACM International Symposium on Mixed and Augmented Reality* (pp. 91–94). IEEE.
48. Günthner, W. A., Blomeyer, N., Reif, R., & Schedlbauer, M. (2009). *Pick-by-vision: Augmented reality unterstützte Kommissionierung.*
49. Funk, M., Shirazi, A. S., Mayer, S., Lischke, L., & Schmidt, A. (2015, September). Pick from here! An interactive mobile cart using in-situ projection for order picking. In *Proceedings of the 2015 ACM International Joint Conference on Pervasive and Ubiquitous Computing* (pp. 601–609).
50. Paelke, V., Röcker, C., Koch, N., Flatt, H., & Büttner, S. (2015). User interfaces for cyber-physical systems. *Automatisierungstechnik, 63*(10), 833–843.
51. Hoffmann, C., Büttner, S., Prilla, M., & Wundram, K. (2020, September). Impact of augmented reality guidance for car repairs on novice users of AR: A field experiment on familiar and unfamiliar tasks. In *Proceedings of the Conference on Mensch und Computer* (pp. 279–289).
52. Gurevich, P., Lanir, J., Cohen, B., & Stone, R. (2012, May). TeleAdvisor: A versatile augmented reality tool for remote assistance. In *Proceedings of the SIGCHI conference on human factors in computing systems* (pp. 619–622).
53. Mavrikios, D., Karabatsou, V., Fragos, D., & Chryssolouris, G. (2006). A prototype virtual reality-based demonstrator for immersive and interactive simulation of welding processes. *International Journal of Computer Integrated Manufacturing, 19*(03), 294–300.
54. Aehnelt, M., & Wegner, K. (2015, October). Learn but work! Towards self-directed learning at mobile assembly workplaces. In *Proceedings of the 15th international conference on knowledge technologies and data-driven business* (pp. 1–7)
55. Haase, T., Keller, A., Warschewske, F., Woitag, M., Sauer, S., & Berndt, D. (2022). Digital assembly assistance systems: Methods, technologies and implementation strategies. In C. Röcker & S. Büttner (Eds.), *Human-technology interaction – Shaping the future of industrial user interfaces*. Springer.
56. Krause, F., Bosch, T., Wilschut, E., & Van Rhijn, G. (2022). Cognitive operator support in the manufacturing industry – Three tools to help SMEs select, test and evaluate operator support technology. In C. Röcker & S. Büttner (Eds.), *Human-technology interaction – Shaping the future of industrial user interfaces*. Springer.
57. Oestreich, H., Heinz-Jakobs, M., Sehr, P., & Wrede, S. (2022). Human-centered adaptive assistance systems for the shop floor. In C. Röcker & S. Büttner (Eds.), *Human-technology interaction – Shaping the future of industrial user interfaces*. Springer.
58. Besginow, A., Büttner, S., Ukita, N., & Röcker, C. (2022). Deep learning-based action detection for continuous quality control in interactive assistance systems. In C. Röcker & S. Büttner (Eds.), *Human-technology interaction – Shaping the future of industrial user interfaces*. Springer.
59. Brünninghaus, M., & Deppe, S. (2022). Advancements in vocational training through mobile assistance systems. In C. Röcker & S. Büttner (Eds.), *Human-technology interaction – Shaping the future of industrial user interfaces*. Springer.
60. Paelke, V., & Bulk, J. (2022). Designing user-guidance for extended reality interfaces in industrial environments. In C. Röcker & S. Büttner (Eds.), *Human-technology interaction – Shaping the future of industrial user interfaces*. Springer.
61. Kosch, T., Knierim, P., Kritzler, M., Beicht, D., & Michahelles, F. (2022). Lenssembly: Authoring assembly instructions in augmented reality using programming-by-demonstration. In

C. Röcker & S. Büttner (Eds.), *Human-technology interaction – Shaping the future of industrial user interfaces*. Springer.

62. Rumsey, A., & Le Dantec, C. (2022). Escaping the Holodeck: Designing virtual environments for real organizations. In C. Röcker & S. Büttner (Eds.), *Human-technology interaction – Shaping the future of industrial user interfaces*. Springer.

63. Korn, O. (2022). Gamification in industrial production: An overview, best practices, and design recommendations. In C. Röcker & S. Büttner (Eds.), *Human-technology interaction – Shaping the future of industrial user interfaces*. Springer.

64. Kaasinen, E., Anttila, A., & Heikkilä, P. (2022). New industrial work – Personalised job roles, smooth human-machine teamwork and support for well-being at work. In C. Röcker & S. Büttner (Eds.), *Human-technology interaction – Shaping the future of industrial user interfaces*. Springer.

65. Mlekus, L., Bentler, D., Paruzel, A., Kato-Beiderwieden, A. L., & Maier, G. W. (2020). How to raise technology acceptance: User experience characteristics as technology-inherent determinants. *Gruppe. Interaktion. Organisation. Zeitschrift für Angewandte Organisationspsychologie (GIO), 51*(3), 273–283.

66. Polzer, S., Frahm, M., Freundel, M., & Nebe, K. (2022). Which factors influence laboratory employees' acceptance of laboratory 4.0 systems? In C. Röcker & S. Büttner (Eds.), *Human-technology interaction – Shaping the future of industrial user interfaces*. Springer.

67. Papenkordt, J., & Thommes, K. (2022). Determinants of trust in smart technologies. In C. Röcker & S. Büttner (Eds.), *Human-technology interaction – Shaping the future of industrial user interfaces*. Springer.

68. Brauner, P., Schaar, A., & Ziefle, M. (2022). Interfaces, interactions, and Industry 4.0 – Why and how to design human-centered industrial user interfaces in the Internet of production. In C. Röcker & S. Büttner (Eds.), *Human-technology interaction – Shaping the future of industrial user interfaces*. Springer.

Digital Assembly Assistance Systems: Methods, Technologies and Implementation Strategies

2

Tina Haase, Alinde Keller, Florian Warschewske, Martin Woitag, Steffen Sauer, and Dirk Berndt

Abstract

Flexible manufacturing processes, diversified products and small lot sizes are imposing new demands on assemblers by confronting them with constantly changing work parameters. Innovative HCI technologies and systems help assemblers perform their jobs with digital assembly assistance systems integrated in their workstations. This chapter surveys current technological building blocks, methods and implementation strategies for the design and development of such solutions. First, the authors describe specific implementation and development strategies with a sociotechnical design approach. The goal of such strategies is to develop effective solutions that employees accept and use on a sustained basis. One prerequisite for this is that selected technologies are designed to be conducive to learning and comply with human factors standards. To this end, criteria are developed, which have an impact on success and therefore ought to be ascertained and incorporated in design. Finally, specific assistive inspection and learning technologies are presented, which facilitate the implementation of assembly assistance systems compliant with human factors standards. Specific examples from industrial practice serve to explain the technologies.

Keywords

Digital assembly assistance systems · Implementation process · Conduciveness to learning · Assistive inspection and learning technologies

T. Haase (✉) · A. Keller · F. Warschewske · M. Woitag · S. Sauer · D. Berndt
Fraunhofer Institute for Factory Operation and Automation IFF, Magdeburg, Germany
e-mail: tina.haase@iff.fraunhofer.de; alinde.keller@iff.fraunhofer.de;
florian.warschewske@iff.fraunhofer.de; martin.woitag@iff.fraunhofer.de;
steffen.sauer@iff.fraunhofer.de; dirk.berndt@iff.fraunhofer.de

2.1   Background

Digital assembly assistance systems [1] are being integrated in workstations with growing frequency to enable responding to increasingly flexible manufacturing processes, diversified products and small lot sizes in manufacturing. They help assemblers do their jobs by providing requisite information situationally, validating the assembly process and performing quality inspections.

The authors predicate the design of assembly assistance systems on a fundamental understanding of human and machine collaboration, which situates decisions and responsibilities among individuals. They view assistance systems as skill-reinforcing systems designed to enable learning while working, to increase assemblers' latitude, to adapt to users' needs and work situations, and to assuring the quality of assembled components. The authors therefore refer to them as learning and assistance systems or assistance systems conducive to learning.

This requires an integrated interdisciplinary approach equally geared toward the sociotechnical system and humans and focused on technology and organization. "The analysis and design of an engineered work process in keeping with this approach is not an issue of either technology or humans but rather a striving for a complementary design of individual system elements in one coordinated, unified sociotechnical system" (e.g., [2, 3]). The design of new technologies consequently also includes designing the interactions between technology, work and organization (Ibid.). The capabilities of "skills-driven design" (Ibid.) are a leading criterion of the design process.

This chapter describes the process of designing and developing assistance systems conducive to learning from the perspectives presented in Fig. 2.1.

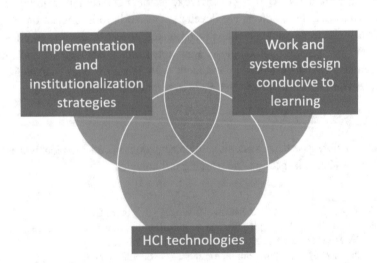

Fig. 2.1 The process of designing and developing assistance systems conducive to learning based on a sociotechnical design perspective [2]

Section 2.2 describes strategies for implementing and institutionalizing digital learning and assistance systems, the organizational and personal factors relevant to embedding of solution in a company permanently and sustainably being the primary focus. The objectives pursued by successful implementation strategies, the challenges that arise in the process, and the specific actions that render the objectives achievable are identified. It is essential to develop a shared vision of the technical solution, its embedding in the existing work processes and its expected impact early on with the employees directly or indirectly affected by the implementation of a digital assistance system.

Section 2.3 describes the human factors design of an assistance solution and the adjacent work processes. An assistance solution conducive to learning, which is perceived to be effective, helpful and motivating by its users and enables achieving the objectives developed conjointly beforehand, requires human factors design throughout the sociotechnical approach. The work system's personal and organizational parameters, which affect success and therefore have to be ascertained and factored into the design process, are described. A methodology for selecting and designing technology is presented. Work design actions conducive to learning are also described.

Key technology design requirements ensue from the fundamental understanding of a user-friendly and user-centered assembly process described at the outset. Section 2.4 therefore describes technologies that enable user and context awareness (Sect. 2.4.1) and human factors engineering optimized for requirements (Sect. 2.4.2) and provides assemblers assistance with quality assurance and process validation (Sect. 2.4.3).

2.2   Strategies for the Successful Implementation and Institutionalization of Digital Assistance and Learning Systems

Targeted actions are employed during the entire process from job clarification through rollout to embed assistance systems in companies effectively.[1] Strategies for implementing and institutionalizing cognitive assistance systems are being systematically developed and studied in the EVerAssist research project.[2] The project's primary objective is to integrate a learning and assistance system into work processes effectively. In the process, strategies are being developed, which respond to didactic and socio-technical challenges during

---

[1] This chapter includes translations of expanded and revised excerpts of A. Keller, S. M. Weber, F. Rentzsch, and T. Haase, Lern- und Assistenzsysteme partizipativ integrieren. Entwicklung einer Systematik zur Prozessgestaltung auf Basis organisationspädagogischer Ansätze. Zeitschrift für Arbeitswissenschaft 79 (4) (2021).

[2] The research and development project *EVerAssist – Einführung und Verstetigung technologiebasierter Assistenzsysteme in KMU* is being funded by the Federal Ministry of Education and Research (BMBF) and the European Social Fund (ESF) in the *Zukunft der Arbeit* program (funding code: 02L19A000ff) and managed by Projektträger Karlsruhe (PTKA).

implementation and institutionalization. Although these strategies are being developed for and applied to maintenance assistance system implementation, the key challenges addressed are similar to those of assembly assistance system implementation. In both cases, this entails designing complex participatory communication processes among different user groups, system developers and HR managers.

This section first presents the multi-level objectives pursued by such strategies (Sect. 2.2.1) and then outlines the challenges that have to be tackled to achieve the aforementioned objectives (Sect. 2.2.2). Finally, it presents the solution options being developed in the EVerAssist project (sect. 2.2.3).

2.2.1 Objectives of a Successful Implementation Process

Successful implementation of learning and assistance systems is manifested on several levels ([4], manuscript in preparation). The following features are indicative of a successful implementation process on the user level (1):

- Users are proficient in the use of the systems.
- The systems are accepted.
- The systems have an educational effect.

The latter implies, for instance, that the systems foster independent action or problem-solving skills, rather than taking their place. Work processes modified for the use of digital assistance and learning systems are indicative of the implementation process's effectiveness on the work systems level (2). Examples of this are altered workflows, time resources freed up to input assistive contents and cognitive ergonomic concepts for the use of assistance systems. Permanently altered informal work routines are also indicative of an assistance system's effectiveness on this level. These features are a prerequisite for assistance systems on the organizational or corporate level (3) that generate select benefits. Objectives could be skills acquisition and retention in the work process, rapid training and retraining, error reduction, quality assurance or material waste reduction, see [5]. Moreover, the integration of an assistance system in a company's primary digital strategy is indicative of a successful implementation process. Since technologies for connected and smart production systems (e.g., sensor systems or image processing) are developing rapidly, effective implementation on the level of digital transformation (4) manifests itself in early incorporation of interfaces to other technical designs and systems, see [6]. This ensures assistive technologies' compatibility with other technical designs and systems, relevant at the present and in the future.

2.2.2 Challenges in the Implementation Process

Since one of the biggest challenges to effective use of an assistance system is its impingement on familiar work habits, one of the most important objectives is to identify the benefits together with employees from their perspective at an early stage and to communicate and present them to identified target groups. An assistance system frequently requires the use and combination of various technological building blocks (e.g. sensor systems or vision systems), which are selected and combined for the specifications of an assembly process. These usually generate several implementation options that must be assessed in terms of the concomitant objectives and ramifications, e.g., expenditures, maintenance costs or suitability for the existing IT systems landscape, with a company's decision-makers. This normally produces custom solutions that meet a company's demands and its employees' needs. The operational specifications have to be drafted early in the implementation process.

Another challenge is the frequent need to integrate several user groups in the process of compiling and using assistive contents (e.g., work scheduling, engineering, quality assurance, management) ([4], manuscript in preparation). This necessitates expectation management to prioritize objectives, such as work documentation, skills acquisition during the work process or error reduction, since they require different technology. What is more, developing processes to update assistive contents and maintain assistive technology is essential to long-term use of an assistance system. Moreover, the implementation of an assistance system is part of a change process in which it is important to create employee awareness of new developments and involve them in design decisions early on. Transparency and data ownership (Who can view my data?!) are particularly issues open for discussion.

2.2.3 Strategies for Implementation and Institutionalization

This section presents participatory design strategies and a methodological framework that includes guiding questions for participatory process design and appropriately selected methods.

2.2.3.1 Overview: Participatory Design Strategies

Participatory process design is ideal for recognizing and systematically addressing potential impediments, e.g., obstacles to acceptance, at an early stage. Participation is understood to be the active involvement of individuals in organizational decision and control processes and is considered the prerequisite for self-organized action by members of and sustainable change in an organization [7].

This extends from the early phase of job clarification through the evaluation of assistance systems in use in work processes. A participatory process ensures that the technical solution is configured so that value is maximized or nothing changes for the worse for employees, organizational units or the company as a whole. Overall, participation proves to be a key design dimension in the context of Industry 4.0 [1, 2, 8].

Whenever the development and integration of an assistive solution additionally requires a variety of technical expertise (e.g., on the technologies employed or human factors engineering), systematic integration of such interdisciplinary knowledge in design decisions becomes another factor for success [9]. Process designers, whose job is to coordinate and implement the participatory design and implementation process, play a key role. They identify the user groups, experts and parties to be involved, activate them, and ensure that the systems are designed for integrated use. For instance, they facilitate the process of working out at a common understanding of technological implementation options and ensure that aspects of corporate culture, e.g., knowledge sharing practices, receive attention and enter into technology selection and design. They also maintain communication between the parties involved, ensuring that expectations are realistic and practical obstacles are identified early on. A methodology for designing an implementation and institutionalization process, which was developed in the EVerAssist project based on organizational education theory [10], serves as a guide for process designers (see Fig. 2.2) ([11], manuscript in preparation). It describes the specifics of the phases of the process (1)–(4). The methodology also includes guiding questions for self-review and contextual review (A)–(C). Their clarification is a prerequisite for situation- and context-aware implementation strategies that can be developed based on the guiding questions (D)-(G).

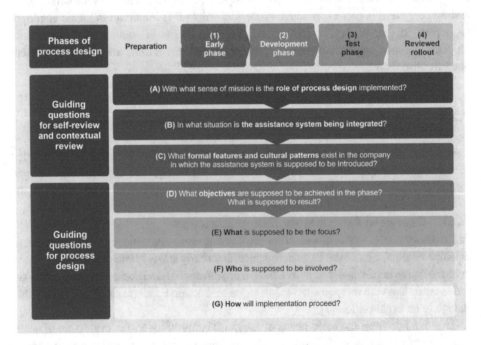

Fig. 2.2 Methodology for participatory integration of cognitive assistance systems, see ([11], manuscript in preparation)

2.2.3.2 Process Design Phases

The methodology first identifies the four phases of the implementation process, expounded below:

1. The **early phase** comprises participatory vision development and job clarification. This entails scrutinizing implementation options and their implications. Unlike upstream phases, such as customer acquisition, the early phase is implemented in the company specifically as part of a change process and every relevant user group is therefore involved.

   Strategic objectives are also prioritized in terms of their feasibility within a defined period, among other things, in this phase. Company management, quality assurance, work scheduling, manufacturing management and assembly specifications, for instance, all frequently differ. Rendering them transparent from the outset and working them out collectively is essential here. The instrument of systematic technology selection and design presented in Sect. 2.3 is used, among others, to concretize job clarification, which is intended to lay the foundation for a technically feasible and educationally expedient solution supported by decision makers and user groups.

2. A technical specification is compiled or software is designed in the **development phase** based on job clarification. Prototypes are designed, evaluated in a participatory manner, and refined iteratively. Both technical prototypes and process prototypes have to be developed. The latter are designs, for instance, of the work and roles in editing processes for assistive contents. Terminal equipment is also selected and tested in this phase. Pilot organizational units in which the assistance system will be tested are selected as well and evaluation criteria are defined. This is guided by the following questions, among others: What objectives are supposed to be achieved for which user group? What indicators can we collect? Which are indicators of achievement of the objectives?

3. The **test phase** is aimed at monitoring and evaluating the use of the technical assistive solution in the daily work routine in a pilot organizational unit and, if necessary, modifying the technical solution. It begins concurrently with phase 2 with the technical integration of the system in the company's IT landscape, its installation in workstations, and the training of end users. The test phase concludes with an evaluation based on the strategic objectives of the assistance system selected and prioritized in the early phase.

4. The experiences from the test phase and the evaluation of the potential benefits are reviewed and assessed in a participatory and interdisciplinary manner in the **reviewed rollout phase**. These assessments serve as the basis for making decisions about the use of the assistance system in other organizational units.

2.2.3.3 Guiding Questions for Self-Review and Contextual Review

Since it additionally explores the situational agency of the individuals who coordinate and design the participatory process, the process design methodology is considered an instrument that helps these individuals introduce the implementation process agilely and situation- and context-aware throughout specific phases.

(A) **With what sense of mission is the role of process design implemented?**

It is based on a review of individual understanding in the activity of participatory process design. Assuming that the suggestions and thoughts contributed by various disciplines and user groups are essential to effective integration, an "expert consulting" mode is presumably less useful. This mode is based on the assumption that established solutions for businesses already exist and the challenge lies in disseminating them. An "integrated design" mode, on the other hand, seems opportune: This mode is based on the presupposition that the path to finding a solution runs circularly and the conviction that process designers' own effectiveness is more the product of inspiration than of drastic action (see [11], manuscript in preparation).

(B) **In what situation is the assistance system being integrated?**

Furthermore, the situation in which an assistance system is integrated has to be reviewed. The project's architecture is one of the things examined here. Is it a research project or a development contract, for instance? Prior history is also reviewed. Among other things, the value propositions negotiated during project initiation are identified here.

(C) **What formal features and cultural patterns exist in the company in which the assistance system is supposed to be introduced?**

What is more, the corporate context has to be analyzed systematically. Criteria, such as company size and sector, are one thing examined. Cultural patterns in companies also have to be reviewed. This raises questions about the prevalent management culture, the forms of employee involvement in decisions to date and their experience with dialogue on equal footing across hierarchies, for instance.

2.2.3.4 Guiding Questions for Process Design

(D) **What objectives are supposed to be achieved in the phase? What is supposed to result?**

The phases of process design are characterized by specific objectives, such as vision development and participatory job clarification in the early phase.

(E) **What is supposed to be the focus?**

Process design additionally poses questions and integrates topics that can be used for an integrated analysis of assistive solutions. Technological questions, for instance, are linked with questions about work process design or changes in the company as a whole.

(F) **Who is supposed to be involved?**

   Process design is typified by well-considered selection of the parties integrated in the individual phases. Already consulting with assemblers during the clarification of strategic issues in the early phase is not always advisable, for instance. Discussion is usually conducted on a level of abstraction with little relation to an assembler's job and far removed from an imaginable solution. The experts in select technologies, work design or educational design also ought to be consulted intermittently or concomitantly to the process when selecting parties for participatory formats.

(G) **How will implementation proceed?**

   Formats are designed and methods are selected in keeping with already formulated objectives, the topics being worked on and the identified group of participants. A distinction must be made here between the level of participation that should be applied, e.g., information vs. participatory design, see [12, 13] or between a more formally or more informally participatory approach, among other things. The former could be information at a general meeting (in the "information" mode), the latter a film self-shot by the developer team with a message to users.

The following table provides an overview of methods and approaches suitable for use in participatory process design (Table 2.1).

2.3   Human Factors Design of the Technological Solution and the Adjacent Work Processes

Designing an integrated learning and assistive solution that is perceived to be effective, supportive and motivating by employees and achieves the operational objectives requires an integrated design consonant with the sociotechnical approach described at the outset.

   A key activity of the design process is the systematic selection and design of technology. The individual objectives and parameters were developed in a participatory manner in the early phase, as described in Sect. 2.2. This serves as the basis for identifying the design decisions that render these objectives achievable. Guidelines that assists companies with this process was developed for this purpose. These guidelines supply important individual and organizational parameters that ought to be ascertained (see Sect. 2.3.1), e.g., by observation or interviews. Technology selection and design options are presented in Sect. 2.3.2.

2.3.1   Individual and Organizational Parameters

Interactive action guidelines that assist operational practitioners during the design process are developed. In the first step, they provide an overview of parameters and individual

Table 2.1 Select methods and approaches for use in process design

Method	Objectives	Implementation	Operational phase and duration
Appreciative inquiry [14]	Gathering prior experiences with assistive technologies or similar projects, making them transparent and referencing them; opening eyes for multiperspectivity	Workshop and interviews, plotted experience curves	Early phase, reviewed rollout
Future search conference [15]	Developing a shared vision	Workshop (critique, future and implementation phase)	Early phase
Systemic impact monitoring [16, 17]	Creating awareness of complexity and interactions when achieving potential benefits; concretizing the shared vision; establishing realistic expectation	Workshop	Cross-phase, beginning in the early phase
Visualizations as translation and mediation tools (see [18])	Producing shared mind maps and mental images, mediating between different mindscapes; sorting views and positions into one big picture	Visualizations of technical design options, e.g., "development blueprint", or scientifically grounded graphics (see [19])	Early phase/ development phase
Transformative evaluation approach [20]	Facilitating concentration on complex effects in practice; empowering user groups; identifying unforeseen obstacles (to acceptance)	Both qualitative and quantitative formats, procedural and iterative design	Test phase, reviewed rollout
Iterative evaluation	Eliminating unforeseen obstacles (to acceptance)	Both qualitative and quantitative formats	Test phase

conditions, which influence success and therefore ought to be ascertained and factored into the design process.

Following the human factors engineering perspective, a technical solution's interaction with an employee is analyzed to draft a defined job assignment subject to various parameters (environmental factors). The relevant job assignment, employee and environmental factors are presented below.

1. Job Assignment

Job assignment (see Fig. 2.3) generates crucial parameters pertinent to the regulation and organization of the work, which play an important role when designing assistance systems that support training and work. The methodology proposed here is used to recorded work-related features of cognitive learning objectives, task, purpose of assistance and type of requirement for the as-is and to-be states (with and without assistance system).

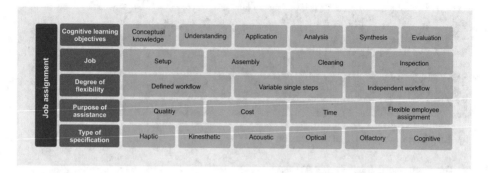

Cognitive learning objectives	Conceptual knowledge	Understanding	Application		Analysis		Synthesis		Evaluation
Job	Setup		Assembly		Cleaning			Inspection	
Degree of flexibility	Defined workflow		Variable single steps			Independent workflow			
Purpose of assistance	Qualitiy		Cost		Time		Flexible employee assignment		
Type of specification	Haptic	Kinesthetic	Acoustic		Optical		Olfactory		Cognitive

Job assignment

Fig. 2.3 Job assignment dimension

Skill level	Novice		Advanced beginner		Competent		Proficient		Expert
Motivation	Intrinsic			Extrinsic			Both		
Motive	Does the job: Meets requirements		Is fulfilled by the work (flow)			Develops personally: career			
Age structure	< 20 years		20 – 35 years		35 – 50 years		> 50 years		
Physical requirements	Visual impairment		Hearing impairment			Limited mobility			
Language	Multilingualism				Monolingualism				
Education	High school diploma		College degree		Professional degree		Graduate degree		
Technical savviness	Little savviness		Mixed			Much experience/savvy			

Employee

Fig. 2.4 Employee dimension

2. The Employee

In addition to job assignment, user-specific criteria also have to be factored into the user-adaptive design of assistance systems conducive to learning (see Fig. 2.4). Skill level, motivation, age structure, physical requirements, language, education and the employee's technical savviness were identified as relevant categories. Not all factors affect the technical solution directly. They can also affect work organization, for instance, by requiring concomitant technology training actions when employees have little prior technical savviness.

The variety of categories is indicative of the complexity of designing a personally effective assistance system conducive to learning. Consequently, it should be possible to adapt a solution throughout the period of its use, e.g., as employees gain experience using the system and develop skills.

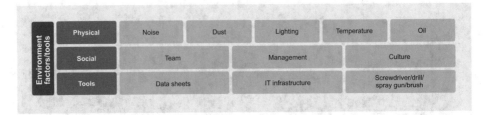

Fig. 2.5 Environmental factors dimension

3. Environmental Factors and Tools

Environmental factors (see Fig. 2.5) are parameters that affect a work system. They can be social/emotional, organizational/communicative, physical/organismic or chemical/material [21, p. 21]. Social environmental factors primarily affect the implementation process and employees' acceptance and use of assistance systems. Physicochemical environmental factors, on the other hand, influence technology selection directly and are therefore given priority in this methodology. Ambient noise and light as well as stress factors, e.g., high ambient temperature, are analyzed.

Existing tools, i.e., technical equipment and established processes, ought to be factored in as well. These include both standard equipment and documents, data and the existing IT infrastructure.

2.3.2 Technological and Educational Design Dimensions

The manifold design decisions for a digital learning and assistance system include both the systematic selection and educational design of technologies. The categories identified are highly application-specific. Maintenance work requires different technologies than assembly does. Some fundamental decisions, e.g., the selection of input and output devices, are generalizable, though.

Figure 2.6 provides an overview of the design dimensions. The selection of the appropriate form is usually contingent on the parameters identified beforehand. Work done with oily hands requires contactless input options, for instance.

Although the choice of media when designing interactive contents is wide, there is a dearth of specific information on their suitability in practice. The methodology processes and links experiences from operational practice with specific design options, thus making it possible to transfer practical experience to practice.

2.3.3 Using the Methodology

The methodology described is intended to provide companies designing learning and assistance systems operational assistance, raise their awareness of relevant parameters

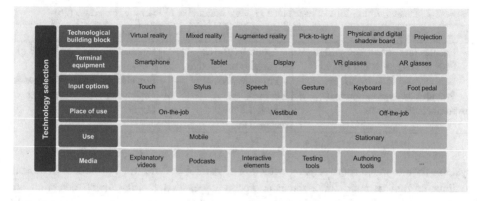

Fig. 2.6 Design decisions during technology selection

during design, and facilitate technology selection with stored experience stories about the design dimensions (in an interactive online version of the methodology). Given the specifics of companies and jobs, the methodology cannot and will not be able to provide design proposal algorithm. Instead, it ought to be treated as a methodology that contextualizes and correlates both positive and negative employee experience stories.

The methodology was initially developed to draft specifications for assembly and advanced maintenance [8], something particularly reflected in the category of task content. The core educational criteria conducive to learning are assumed to be transferrable to other sectors.

The authors expect that adoption of the methodology will prompt a review of corporate strategy and learning culture. Clarification of the objectives pursued with the assistance system (e.g., quality, flexibility, learning effects) already necessitates dialogue with employees from different organizational units, thus initiating a review process crucial to selection and design. This form of visualizing correlations can help structure dialogue while simplifying communication between developers and organization about design conducive to learning. A structured and participatory approach, as described in Sect. 2.2, which ultimately contributes to all participants' acceptance during implementation is recommended here.

2.3.4 Designing Digital Assistance Systems Conducive to Learning

One particular focus is the technological solution's educational design. Designing assistive solutions that are conducive to learning and integrated in workstations requires modifying them for the employees that use them to support their work. Whether a system ought to adapt to its operators automatically (adaptive) or operators ought to modify it themselves (adaptable) is one of the things analyzed here. Adaptation options comprise both the selection of the assistive contents and their level of detail and presentation [22]. Adaptive

Table 2.2 Design dimensions of assistance systems conducive to learning, based on [26]

Personal development	Intrinsic motivation	Social integration
– Enables increase in knowledge – Provides technical information – Is embedded in the work process – Provides an exploration and experimentation mode	– Arouses curiosity – Sustains user activity – Provides personal feedback – Includes reward mechanisms	– Emphasizes the importance of the task – Elevates the status of the work – Establishes interdependence – Creates dialogue

systems require much more sensing detection of employees and work processes to be able to infer current demand for assistance. Technologies that facilitate such detection are described in Sect. 2.4.1. There is, however, a risk that such an automated approach will be less transparent to employees and meet with little acceptance. The authors assume that a hybrid solution in which certain parameters (environmental factors) are derived automatically and operators themselves select content parameters (amount of assistive information) will become more established.

Rather than being limited to the technical system's design, work design conducive to learning ought to analyze the organizational level too. Design dimensions that support assistance systems' conduciveness to learning and thus assist employees with development are being developed in the research project LeARn4Assembly (FKZ: 01PV18007A). They are based on criteria of work design conducive to learning and skills development, see [23–25] and augmented by the dimension of the system's intrinsic motivational incentives.

In keeping with Table 2.2, an assistance system is conducive to learning when it fosters personal development, contributes to intrinsic motivation and enables social integration. Every dimension is operationalized by descriptive criteria. Personal development, for instance, is enabled by permitting employees to acquire knowledge independently and to try out action options or functional mechanisms. Virtual reality modules, for instance, can be integrated to allow safe experimentation.

The following actions influence a digital assistance system's level of conduciveness to learning:

1. Selection of contents

   Contents must be relevant and interesting to the target group so that they are used. One particular challenge is the currency of contents. Only current and regularly updated contents motivate use.

2. Educational editing of contents

   The educational design establishes how learners can learn new content, e.g., by reading texts, watching videos or completing interactive assignments. Along with selecting the appropriate medium for the contents taught, editing content is also relevant. This can be process-driven editing of assignments, for instance.

3. Technical implementation of contents

   Technical implementation determines which terminal equipment, interaction techniques and software solutions deliver contents to employees. Technical implementation boosts conduciveness to learning by integrating communication options or implementing learning by experimenting in game form, for instance.

4. Embedding of the assistance system in the organization

   Embedding an assistance system in an organization is essential to preventing it from becoming a local solution. This includes its integration in the IT infrastructure, work processes and the corporate culture.

Technologies that enable the implementation of assistive solutions with the features described are presented in the following section.

2.4   Implementing Innovative Assistance, Inspection and Learning Technologies

This section describes select technologies that enable implementation of assembly assistance systems consonant with the fundamental understanding described at the outset.

Technologies that collect, interpret and preprocess data and sensor information are introduced, which enable context-aware and user-adaptive assistance (Sect. 2.4.1). Such sensors can also deliver ergonomically relevant information. Ways human factors data can be collected during the work process and supplied to workers as assistive information so they can adjust their performance independently and thus assume responsibility for their own work process are described in Sect. 2.4.2. A quality assurance assistance system is presented in Sect. 2.4.3. Assembly progress is tracked optically using a model-based approach and compared with a reference model. The assistive information is displayed to workers as superimposed AR information on a monitor.

2.4.1   HCI Technologies for User and Context Awareness

Assistance systems are considered particularly good when they are capable of adapting to the particular situation in the work system or the worker's actions. This capability is based on continuous processing of contextually relevant information. Systems and applications with this feature are therefore termed context-aware. Data- and sensor-driven status tracking of all of the system's (contextually relevant) elements collects the requisite information. On the one hand, this can pertain to work equipment, e.g., tool availability, current machine sequencing or required parts inventories. On the other hand, it also includes environment data or workers' workloads and actions during assembly operations. Suitable data processing can extract the current situation in the work system from the totality of individual information.

This serves as the basis for designing an assistance system that responds to any given situation. This can enhance the user experience significantly, thus contributing to effective implementation of an assistance system in a company. Examples of specific options for employing context-aware applications are:

- needs-driven provision of assistive contents (e.g., by recognizing the current step),
- troubleshooting support (e.g., for missing material, tool malfunctions or errors in the assembly sequence),
- predictive planning of steps as a function of planning data and work equipment availability, and
- the assistance system's response to a worker's workload (e.g., precautions or traffic light rating systems).

This section provides an overview of a potential approach to acquiring data from contextually relevant information at assembly workstations. The focus is on tracking the status of work equipment and the work environment with low-cost sensors and recognizing human activity with inertial sensors. The authors follow the principle of introducing sensor technology in the work system as noninvasively as possible, i.e., keeping the technology or interaction with it from impeding existing operations and avoiding modifications of existing systems or the IT infrastructure.

2.4.1.1 Data Acquisition Pipeline for Contextually Relevant Information

The proposed approach follows the data acquisition pipeline frequently found in industrial IoT applications (see Fig. 2.7). It pursues the principle of having distributed devices collect heterogeneous data in the work system and transmit them to a central location (server)

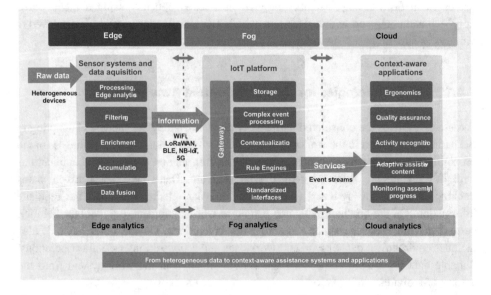

Fig. 2.7 Data acquisition pipeline

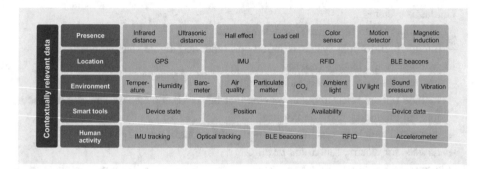

Presence	Infrared distance	Ultrasonic distance	Hall effect	Load cell	Color sensor	Motion detector	Magnetic induction
Location	GPS		IMU		RFID		BLE beacons
Environment	Temper-ature	Humidity	Baro-meter	Air quality	Particulate matter	CO_2	Ambient light	UV light	Sound pressure	Vibration
Smart tools	Device state		Position		Availability		Device data
Human activity	IMU tracking		Optical tracking		BLE beacons		RFID		Accelerometer

Contextually relevant data

Fig. 2.8 Overview of sensors that collect contextually relevant data

where the data are standardized and context information is generated and provided cumulatively. This approach scales easily with the number of additional data sources that can be introduced into the system. Only new devices have to be configured and integrated in the pipeline, whereas the central services remain largely unaffected when the respective protocols are supported.

The acquisition of environment data, e.g., on noise, dust and temperature, and the tracking of work item status, e.g., components or shipping equipment, require different principles of sensing. Figure 2.8 provides an overview of commonly used sensors and scanning principles. Low-cost commercial sensors accessed by edge gateways are often suitable for such use cases. The gateways supply sensors with power, retrieve relevant measurements through the respective interfaces, and transmit data through the network. Gateways based on low-cost hardware have been implemented in manual assembly workstation use cases [27]. Gateways that have even more computational power and perform preprocessing tasks are steadily growing in importance (edge computing). Typical operations that can be performed right at the edge include smoothing, aggregation, normalization and more complex application-specific functions. They are intended to process data where it is produced, ideally to eliminate reliance on a network (entirely local applications in the workplace) or to minimize network data throughput.

Smart tools constitute another category of potential contextually relevant data sources. These tools or systems with data interfaces are often used to validate processes. Either they are integrated directly in the sensor network and transmit process data to the server or they must communicate through an additional controller. The following smart tools are found in industrial assembly workstations and elsewhere:

- Configurable tools for process validation
  - are primarily employed for screwdriving systems,
  - can be enabled and disabled by signals, and
  - monitor the tightening process by transmitting torque, angle of rotation, screwing time, etc.

- Guidance systems
  - are employed to minimize search times and
  - include pick-to-light systems, digital bit holder and digital shadow boards
- Measurement and testing systems
  - can process measurements automatically and
  - include optical measurement systems that inspect completeness and correctness in manual mounding tasks and/or use
  - automatic weighing systems for completeness control and
  - devices for electric functional testing.

People can be considered another data source in the work system. Their movements can be recorded by motion capture systems and subsequently analyzed. Wearable sensors or vision systems can collect data to model a digital body model. Assigning movements to a specific activity during assembly falls into the domain of human activity recognition, something employed in many other domains and often based on methods of machine learning and artificial neural networks. It is unsuitable for industrial practice since a tremendous amount of labor is required to teach such systems. Interpreting motion data to yield information on the human factors of work without using machine learning methods is described in Sect. 2.4.2.

Data collected using different sensing principles can be transmitted to the server. Lightweight IP transmission protocols that follow the publish-subscribe pattern are particularly well suited, depending on the data source. OASIS's standardized, open Message Queuing Telemetry Transport (MQTT) protocol is a prominent representative. The protocol is easily built, numerous client libraries exist and it performs well in unreliable networks. OPC UA is another widespread mechanism for transporting data between clients and servers. It is often employed wherever machine manufacturers want to provide a data interface, the machine itself functioning as an OP UA server. The key is the well-designed information model that can be used to describe every device semantically. Numerous domain-specific specifications upon which new applications can build exist already. Data protocols and interfaces for smart tools and motion capture systems are dependent on the manufacturer and often have to be integrated in sensor networks manually. At times, this necessitates handling large quantities of data.

Using an IoT platform is expedient for server-side data processing. It standardizes and persists data from different protocols in databases. It also simplifies device management and the implementation of security guidelines. Devices are enriched with metadata and sensor values are mapped on the platform to physically existing objects or assets. The API platform is the starting point for building context-aware systems and enables standardized retrieval and processing of information from individual applications. Open source stacks are alternatives to IoT platforms. They comprise different services that represent the requisite functions singly. Essential components include message brokers, data aggregators, databases and visualizations, which have to be interconnected manually, however.

Visualizing information in a standardized form once it is retrievable from the server is an expedient next step. Visualizations of information collected during the assembly process can provide useful insights on processes and facilitate the selection of specific use cases and the definition of context-aware applications' objectives (see Sect. 2.2.3). Web-based visualization tools in which user-definable dashboards can be assembled out of various widgets, are suitable for this.

2.4.1.2 Complex Event Processing as Central Building Block

Once the data from all integrated sources are available in a standardized form, the actual information processing can begin. The intention is to process information so that assistance systems can retrieve contextually relevant information whenever it is needed or certain events occur in the work system. Standard approaches that process data in a downstream analysis process (batch processing) reach their limits here. For one thing, occasional storage of high frequency data in a database is very resource-intensive. Moreover, the assistance system must respond to state changes at maximum speed, making downstream interpretation with a time delay inexpedient. Stream processing methods resolve this problem. They process and analyze data virtually in real time. Rather than storing and analyzing incoming data across-the-board in due time like batch processing, stream processing interprets input data as continuous streams of information analyzed directly upon receipt. Only information with value for the respective applications is stored or transmitted to other systems. Constantly recording workplace concentrations of particulates is not particularly expedient when the assistance system's job is to warn workers in the event the concentration grows too large. Defining and comparing incoming data with limits is more expedient. As soon as the particulate concentration exceeds the defined value, an appropriate event is triggered (e.g., "particulate concentration reaches critical value"), which other systems can store or consume.

Interpreting contextually relevant information as discrete-time events enables employing complex event processing (CEP), a specific method of analyzing different information streams to deliver contextual information. CEP combines different information streams or single events to produce higher-level events with more information content. Complex events are defined using user-defined queries and continuously interpreted by a CEP engine. The engine compares information streams with given queries and decides whether the currently available data match the defined patterns. Typical queries include temporal rules indicating that events must occur in a particular time frame or sequence. An examples applicable to contextually relevant assistance systems illustrates the CEP method.

When a certain particulate concentration is exceeded in the workplace, the system generates an alert that makes the worker aware of this. The defined limit concentration is $100 \ \mu g/m^3$ for a 24 h mean value. A sensor with a sensing range of $0–1000 \ \mu g/m^3$ is selected to detect particulates.

```
@source(receiver.url='http://shopfloor-sensors/pm')
define stream PMStream(value double);

define stream AvgStream(average double);

@sink(publisher.url='http://alert-receiver/pm-alert')
define stream AlertStream(average double);

@info(name='Aggregation')
from PMStream#window.time(24h)
select avg(value) as average
insert into AvgStream;

@info(name='RangeFilter')
from AvgStream[average>100]
select *
insert into AlertStream;
```

First, the input stream "PMStream", supplied by an HTTP interface in this case, is defined. An intermediate stream, "AvgStream", is created, which contains the moving average. The alert is published in the output stream "AlertStream". The query "Aggregation" first selects the particulate concentration reading in a sliding 24 h window and aggregates the values in a mean that it pushes into the AvgStream. Then, the average in the query "RangeFilter" is analyzed for the limit and, if necessary, an alert is triggered. The advantage of CEP is the powerfulness of the query languages employed. Their range of operations and the reusability of event definitions permit great flexibility. Moreover, existing solutions are designed so that experts can define events themselves without losing time programming. The technique can also be combined excellently with assembly flowcharts. Individual stages of assembly flowcharts are described by predefined complex events and can thus be used effectively to monitor progress.

2.4.2   HCI for Ergonomic Assistance

In addition to supplying contextually relevant information to employees, smart sensors can also capture data about the employees themselves or their completed actions and thus modify work processes for employees' personal requirements. This enables responding to demographic change as the number of older employees grows in companies, for instance. As mentioned at the start of this chapter, the standardization and targeted improvement of workflows in manual manufacturing and assembly have become crucial objectives for small and medium-sized businesses to boost productivity and corporate value creation in the industrial process chain. The personalization of products, shrinking lot sizes, and growing complexity and variety of product models is confronting employers with the challenge of continuously redesigning and reassessing their workplace ergonomics.

This section presents a method of detecting and assessing good and poor ergonomic movements in the workplace with the aid of instrumentation. A worker's movements during an assembly process are captured by inertial sensors (see Sect. 2.4.1) and evaluated based on four ergonomic factors. By providing feedback on their current ergonomic stress during work, this tool enables employees to assess the ergonomics of their work and design their work themselves.

2.4.2.1 Method of Detecting Poor Ergonomic Posture

Section 2.4.1 describes the use of MEMS technology, an effective means of generating information about parts and tools in a work system. Miniaturized inertial sensors enable employees to analyze workplace ergonomics themselves (human activity recognition). Eleven Xsens (xsens.com/products/mtw-awinda) MTw Awinda motion analysis system sensors are used in the method of detecting poor ergonomic posture. The 16-g sensors attached to work clothes with Velcro straps deliver measurements at a scan rate of 60 Hz. A body model (see Fig. 2.9) and various fusion algorithms provide locations, orientations and joint angles to ascertain ergonomic parameters.

The four factors of *working zone, working posture, working angle* and *working position*, which are relevant to manual assembly, were defined for the method. They comprise information about working posture and work execution and thus information about physical stresses in the workplace. These four factors are expounded below.

Working Zone

Information about individuals' workstations is essential to an assessment of their physical stresses during manual labor. Drilling is more strenuous at head level than at waist level, for instance. Does an individual have to drill close to or far from their body? This affects the stress on the musculoskeletal system. The *working zone* factor includes working height and working distance to detect this stress.

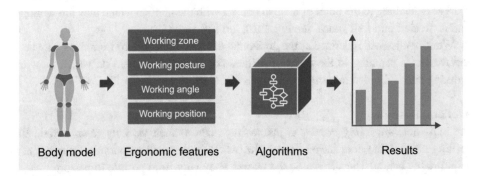

Fig. 2.9 Overview of the method's poor posture detection step

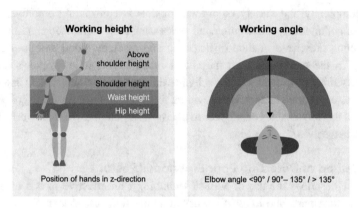

Fig. 2.10 Working zone (working height and working distance parameters)

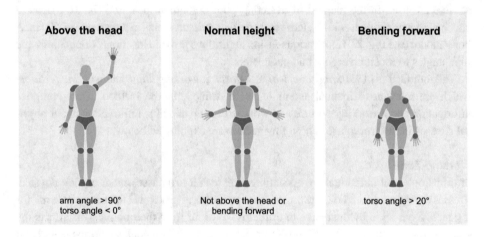

Fig. 2.11 Working posture (arm angle and torso angle parameters)

Working height is defined as the vertical distance (z direction) from the spot where the worker is standing to the point of manual contact with the item on which they are working and is divided into four zones, see Fig. 2.10, left [28, p. 23].

Working distance is defined as the horizontal distance (x direction) from the spot where the worker is standing to the point of manual contact with the item on which they are working and is divided into three zones, see Fig. 2.10, right [28, pp. 23–24].

Working Posture

An individual's *working posture* is just as important as their *working zone*. Individuals drilling close to the floor have to bend low, whereas they drill ergonomically better at a comfortable height. The stresses on the human body vary from posture to posture.

The ergonomic factor of *working posture* is defined as the stance in which work is done and is divided into three classes (see Fig. 2.11).

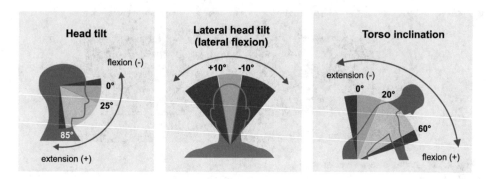

Fig. 2.12 Detail of the 11 working angles defined

Working Angle

Attaching components to a hard-to-reach location forces the body into an unnatural position through turning and bending. A limited view or cramped workspace necessitates tilting one's head very far to the side and extending one's elbow or bending one's torso.

The factor of *working angle* determines whether respective body angles are good (green), fair (yellow) or poor (red).

Figure 2.12 shows the head tilt angle, lateral head tilt (lateral flexion) and torso inclination. The assistance system draws on altogether eight more angles to assess ergonomics, namely head rotation, adduction/abduction and anteversion/retroversion and pronation/supination of the shoulder, flexion/extension of the elbow, pronation/supination of the forearm, and flexion/extension and radial movement/ulnar movement of the hand.

The recommended values for the angles are based on values from the Institut für Arbeitsschutz, see [29, pp. 2–6].

Working Position

An individual installing cable ducts in a ceiling, for instance, assumes an unnatural position. Unnatural positions arise because the work constrains movement. Muscles fatigue quickly. Conversely, high biomechanical loads also occur in the joint areas when dynamic work is done in a position with extreme angular joint positions. The ergonomic factor of *working position* determines whether the respective work is done statically or dynamically (see Fig. 2.13). Every degree of the *working angle* is examined.

A segment's respective angles of static work may not simultaneously change more than 5° positively or negatively for at least 4 s. If they do, the work done in this period is dynamic. The recommended value for the time frame is based on the value from the Deutsches Institut für Normung, see [30, p. 12].

2.4.2.2 Ergonomic Feedback for Employees

The method for detecting poor posture is primarily intended to estimate the overall stress level and distribution. The four factors' calculated parameters are weighted and rated with a

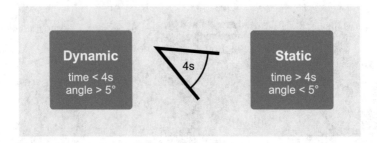

Fig. 2.13 Overall stress level of work

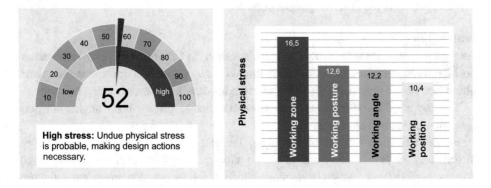

Fig. 2.14 Overall stress level of work (left) and stress distribution of the four factors (right)

score. Higher scores are given to poor parameters and high percentages times relative to the total time. Work above head level, for instance, receives a higher score than work at waist level because the stress on the musculoskeletal system is higher. The higher the score is, the higher the musculoskeletal stress is too. Each factor is a percentage of the overall stress assessed. Ultimately, the overall stress is assigned a score and an informational text is added (see Fig. 2.14). The following result data are based on manual piping assembly completed by a subject bending, at normal working height and overhead.

The overall stress level categorized in four risk ranges of low, medium, elevated and high stress serves as a guide. The parameters are assigned to the risk ranges and assessed based on the established EAWS and LMM methods [31, 32]. The risk range's limits are fluid because of personal methods of working and performance requirements. One subject scored 52 in the example, indicating high stress. Excessive stress on subjects is therefore probable and workstation design actions are necessary. The stress distribution is additionally presented in Fig. 2.14 (right) to estimate the individual factors' influence on the overall stress level. The *working zone* with a score of 16.5 accounts for the largest share of the overall stress level. The *working position* with a score of 10.4 has the least influence. The overall stress level and distribution can be used for an initial ergonomic assessment of the upper body during the subject's manual assembly work.

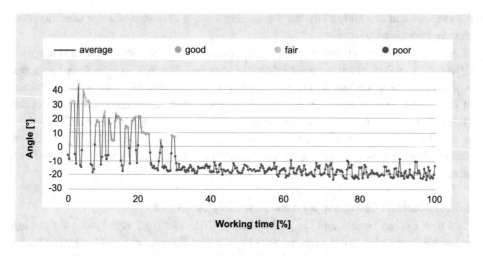

Fig. 2.15 Percentage frequencies of the flexion/extension *working angle*

Another of the stress assessment's purposes is to visualize scores by factor (partial stresses). This enables anyone, even without expertise in human factors, to quickly conclude whether one factor influences overall stress more than another does. The system delivers the results of the four factors of *working zone, working posture, working angle* and *working position* in real time at a scan rate of 60 Hz. The single factor of *working angle* is assessed in the following. The remaining three factors of *working zone, working posture* and *working position* are assessed similarly.

The *working angle* is used to ascertain whether respective body angles are good, fair or poor during work. Information can be provided on 18 body angles during manual work. The assessment of the torso angle is presented in this section as an example.

Figure 2.15 shows the subject's torso angle during piping assembly. Data points in the negative range are indicative of torso extension by the worker. Data points greater than zero are indicative of torso flexion. The first 30% of time working was spent in flexed and extended. Body angles were good, fair and poor ergonomically. Posture after approximately 30% of time working was solely extended. Such angles are ergonomically poor.

A visualization in line or pie charts or a traffic light rating system can provide employees a visual assistance system taking the form of a real-time ergonomics monitoring system. The method can be used both to compare workstations by ergonomic factors and to initiate design actions to prevent poor posture.

2.4.3 HCI for Quality Assurance

Custom equipment manufacturers generally manufacture one-offs and short runs. Since the requisite parts vary widely at times while lot sizes are comparatively smaller, the manufacture and assembly of the parts constitute a major challenge because conditions are changing constantly.

An important step in manufacturing is the machining of blanks into machined parts with CNC machines. Machined parts are secured in so-called clamping fixtures so that they can be automatically milled and drilled in machining centers. Modular systems with a multitude of parts, which can be manually assembled into such clamping fixtures, exist in short run manufacturing. It is tremendously important here that clamping fixtures are built precisely as specified. If the components used are the wrong ones or are assembled in the incorrect position, the machine tool spindle can inadvertently collide with the clamping fixture or an incorrectly clamped part.

2.4.3.1 Current Clamping System Assembly Situation

Pilot user Kolbus GmbH und Co. KG creates a CAD model of every part. A clamping fixture and a machine tool path are additionally designed whenever a part is supposed to be machined by a CNC machine. Until now, trained professionals have manually assembled the clamping systems required in manufacturing based on de-rived 2D drawings, design program screenshots, photos of previously assembled clamping systems, manually stored memos and mechanical templates and jigs, which are used to comply with specified clearances and angles. A clamping system was inserted in the CNC machine for machining and the first blank of a short run, the so-called "original part", was secured. Since it was impossible to ensure that the clamping system was error-free, the machine's machining speed was greatly reduced and the machine operator retracted it point by point. Whenever an assembly error was detected, the machining program had to be terminated by emergency stop and the clamping system had to be removed from the machine, corrected and retested. This approach is unreliable in complex and unclear assembly situations, though, since it requires the operator's unbroken concentration.

The digital disruption arising during manual assembly leaves a major gap, which can increase stresses on employees, time spent working and, in the event of an error, costs caused by breakdowns.

2.4.3.2 Clamping System Assembly Solution

We therefore propose an assembly assistance system that uses intuitive visualizations to help users complete assembly correctly [33]. This solution's key feature is an AR visualization that shows workers the next parts that have to be installed in a clamping system assembly in the correct location at the correct time. Predesigned CAD models of clamping systems and blanks serve as the basis of this visualization. Once the clamping systems have been designed during job planning, an appropriate assembly sequence is defined once in a linear assembly flowchart in which one (or more) assembled part is assigned to each assembly step. The assembly flowchart can be loaded into the assistance system together with the CAD data as often as desired. Augmented reality presents assembled parts in step-by-step instructions that users follow (see Fig. 2.16). Since problems that were unforeseen during design can arise during assembly, e.g., sequencing problems or clamping problems

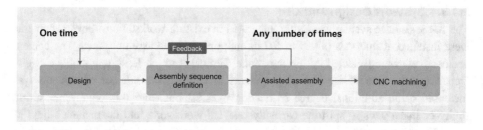

Fig. 2.16 The data and information stream during assisted assembly

that are only detected by assemblers with experiential knowledge, an additional feedback function is integrated to relay such information to the design unit. Depending on the severity of the problem, assemblers may continue assembling after receiving feedback or have to wait until the assembly and CNC machining program have been updated. Once the clamping system has been assembled with the aid of the assistance application, workpieces can be clamped and machined with the preprogrammed CNC program.

2.4.3.3 AR Technology as the Outcome of Systematic Technology Selection

The methodology introduced in Sect. 2.3.1 defines work processes such as this one as assembly processes characterized by a set procedure in lone work. The primary objective of assistance is to enhance quality. Visual assistance is the most intuitive tool for industrial assembly jobs. The solution developed is intended to address the entire range of staff from novices to experts, regardless of age. It can be assumed that employees are more extrinsically motivated and intent on completing their job assignments as well as possible. Errors that occur can be traced directly to the employee making them. The assembly job provides little room for creative fulfillment, though. An employee survey revealed a high level of technical savviness overall, the complexity of the solution having to be a concomitant of the harsh demands in industrial manufacturing.

Augmented reality is generally a suitable solution for assembly jobs since it presents needed information at the right time and in the right place. Workspace monitors were selected as the output devices for the following reasons:

- AR glasses suffer from technically inadequate display accuracy. No AR glasses currently attain continuous deviations of <1 mm. Moreover, staff and management rejected constant use because of the anticipated mental and physical stress.
- When used, smartphones and tablets constrain at least one hand needed for assembly.
- Projected images are suitable for planar applications with rather diffuse surfaces but not for complex assemblies with any shiny metal and greasy surfaces.

Visualizing the assembly steps required customizable complexity corresponding to the user's training level.

2.4.3.4 Systems Design and Use

The AR assistance system was easily integrated in existing workstations. Several cameras were installed at intervals of 1.5 to 5.0 m and aimed at the assembly area. The camera perspectives were selected to produce one vertical and at least one oblique view. Depending the size of the clamping systems assembled, the number of cameras was increased or an additional degree of freedom was established by absolutely measuring rotary or linear axes to shift the currently relevant work zone into the camera's field of view. Touchscreens that display the camera images in real time with the AR visualizations were mounted at an ergonomic distance as input and output devices, (see Fig. 2.17).

A user assembling a clamping system loads the appropriate CAD model with the pertinent assembly flowcharts. Then, assembly is completed following the predefined sequence, the appropriate parts being installed in each step. The user independently decides when an assembly step is finished and then moves to the next step manually by way of a touchscreen. VR or AR views can be selected on the display. The VR view displays a 3D model of the completed assembly, either as a complete model that provides the user an overview of the assembly or as a submodel of the current assembly step. Users can navigate freely in the VR view, viewing details more easily.

The contours of the assembled parts in the AR view are derived from the CAD model and projected onto the camera image, producing an intuitive, topographical display of the parts' shapes and locations. The object edges in the camera image only coincide with the superimposed CAD contours when the right element is installed in the correct location, (see Fig. 2.18). Ambiguities in one camera view's image are clarified by a different view of the assembly.

Users can independently choose between novice or expert view mode. Novice mode presents every step of the assembly flowchart separately, whereas expert mode combines

Fig. 2.17 Assembly assistance by stationary augmented reality: left, schematic of an assembly station with three monitors and five cameras; right, a worker assembling a clamping system using the proposed assistance system

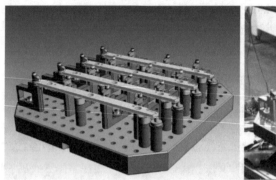

Fig. 2.18 Example of a clamping assembly: left, CAD model with base plate (green), clamping elements (gray) and machined workpiece (orange); right, visualization of a step in the assistance application

several steps, thus increasing the number of parts per assembly step. If problems with the instructions are detected or things important for subsequent assembly are noticed during assembly, a memo function can be used to store information on assembly steps. Recorded problems can be analyzed during design and job planning, resulting in modifications of assembly or the assembly flowchart. Assembly instructions are automatically overlaid on the right step during subsequent assembly.

2.4.3.5 Findings

In all, four workstations with differently sized assembly areas were equipped with the assistance system. Employees' responses to this form of assembly assistance have invariably been positive because of its simplicity and intuitive use. The system is constantly being used by experienced users and to train new staff.

All employees with experience with conventional assembly and the newly introduced assisted assembly were surveyed after 6 months of training. They saw the greatest advantage over the previous procedure in the way instructions were presented and the lessened cognitive load. The AR visualization enables employees to position and align components much more easily. They no longer have to take measurements with tools or create templates and other fixtures. They consistently noted that this assistance system significantly reduces assembly errors known to cause high resultant costs. The greatest advantage for job scheduling is the continuous digitalization of data, thus enabling the transmission of all necessary information without format incompatibility und significantly simplifying change management.

This also coincides with the primary assessment. Only gross negligence can still cause assembly errors. Even though the system itself is passive and does not perform any active inspection, employees have already displayed enough personal responsibility that the machine collisions described at the outset have no longer occurred since the systems were introduced. This has made it possible to machine original assemblies continuously in the CNC machine since time-consuming point-by-point verification is unnecessary. Repeat designs can immediately be run at full machine speed. This translates into a time gain at the CNC machine and thus faster job changes.

Although it was not a focus of development, an additional time gain was measured in assembly. It is attributable to the intuitive assembly visualization and the problem feedback function that optimize and thus expedite assembly sequences once problems have been solved.

Using this system in all four assembly stations can save 1700 machine hours at the CNC machining centers and 2900 person hours per year when clamping systems are assembled in-three-shift operation.

2.5 Conclusion and Outlook

2.5.1 Conclusion

An integrated method of introducing, designing and technically implementing digital assistance systems for manual assembly has been presented. The prerequisite for an effective system is acceptance. This enables achieving a range of operational objectives derived from the attributes of good work as well as more traditional operational objectives, such as time and quality.

One of this chapter's main findings was that the selection and combination of such objectives for the use of assistance systems vary depending on the assembly process and the company. This necessitates a deliberation process in which technical and work design implementation options and operational objectives are coordinated. This chapter additionally described specific actions that enable and facilitate the early process of collectively creating a vision and formulating objectives.

This chapter presented important personal and organizational parameters that influence technology selection and design and therefore must be ascertained and incorporated in the process. This is done in educational and work organization design. Since technologies and interactions that enable context-aware and user-adaptive assistance directly in the work process are required to reproduce these specifics in the assistance system, this chapter also presented such technological building blocks, e.g. an IoT platform and sensor data acquisition, fusion and interpretation. Select examples illustrated the selection and design process as well as the technologies' use in operational practice. Such assistance solutions can be used to retain and develop assemblers' skills, improve ergonomic conditions or support quality assurance, for instance.

An interdisciplinary team that integrates the perspectives of a company's various hierarchies and units as well as scientific expertise is clearly needed to achieve the stated objectives. Work designers and human factors engineers that select and design technology, education experts, technology experts and corporate knowledge facilitators who convey company-specific professional requirements and parameters are all needed. This chapter provides insights on the work of process design, which, among other things, systematically develops the deliberation process in the early phase.

2.5.2 Outlook

Digital assistance systems are being employed for manual assembly processes with increasing frequency. Along with providing direct assistance to perform jobs, assistance systems will also contribute to humane work in the future. Current research studies demonstrate that contents conducive to learning can help enhance individuals' appreciation of their own jobs, thus strengthening their identification with a company, a product and a job. Potential contents, e.g., an assembled part's function, costs incurred or personal contributions to their company's sustainability, enable employees to contextualize their own jobs in the corporate process. Moreover, the risk of disqualification can be countered and monotonous workflows can be transformed.

This necessitates a design process that has its origins in relevant operational questions, is followed by a systematic process of technology selection and design, and involves employees in the process.

Corporate adaptability is one target category that was not addressed in this chapter and will grow in significance in the future. Assistance systems can be an important building block that strengthens corporate adaptability since, among other things, they cut training times drastically. This requires stable processes for rapidly creating and reviewing new assistive contents, though. This target dimension and its influence on the design of assistance solutions will have to be studied in other research projects.

Acknowledgments The authors would like to thank Krister G. E. Johnson (Fraunhofer Institute for Factory Operation and Automation IFF) for his translation of the German manuscript into English and KolbusGmbH & Co. KG for their support during the development of worker assistance systems.

- Avilus+: Part of this study was completed in the AVILUSplus research project funded by the German Ministry of Education and Research (BMBF) under the funding code 01IM08002.
- 3D-Montageassistent: BMBF: 03220441 G
- 3D-KOSYMA: BMBF: 03ZZ0446H
- LeARn4Assembly: BMBF: 01PV18007A
- EVerAssist: BMBF: 02L19A000
- MORPH-IT: BMBF: 01PJ21003A

References

1. Apt, W., Bovenschulte, M., Priesack, K., Weiss, C. & Hartmann, E. A. (2018). *Einsatz von digitalen Assistenzsystemen im Betrieb (Forschungsbericht Nr. 502)*. Accessed March 25, 2021, from https://www.bmas.de/SharedDocs/Downloads/DE/Publikationen/Forschungsberichte/fb502-einsatz-von-digitalen-assistenzsystemen-im-betrieb.pdf; jsessionid=BAE9C68A66386F5555A9C6159F329EAC.delivery1-replication?__blob=publicationFile&v=1
2. Hirsch-Kreinsen, H. (2019). Entwicklung und Gestaltung digitaler Arbeit. In M. Becker, M. Frenz, & K. Jenewein (Eds.), *Digitalisierung und Fachkräftesicherung: Herausforderung für die gewerblich-technischen Wissenschaften und ihre Didaktiken* (pp. 17–30). Bertelsmann.
3. Grote, G. (2018). Gestaltungsansätze für das komplementäre Zusammenwirken von Mensch und Technik in Industrie 4.0. In H. Hirsch-Kreinsen, P. Ittermann, J. Niehaus (Eds.), *Digitalisierung industrieller Arbeit. Die Vision Industrie 4.0 und ihre sozialen Herausforderungen*, 2. aktualisierte und überarbeitete Aufl.: Nomos (pp. 215–232).
4. Keller, A., & Fischer, E. (2021). Cognitive assistance systems as boundary objects. Theorizing and analyzing digitally networked communication practice. In S. M. Weber & J. Elven (Eds.), *Beratung als symbolische Ordnung. Organisationspädagogische Analysen sozialer Beratungspraxis*. VS.
5. McKinsey Global Institute. (2015). *The Internet of Things: Mapping the value beyond the hype*. Accessed March 25, 2021, from https://www.mckinsey.com/~/media/McKinsey/Industries/Technology%20Media%20and%20Telecommunications/High%20Tech/Our%20Insights/The%20Internet%20of%20Things%20The%20value%20of%20digitizing%20the%20physical%20world/Unlocking_the_potential_of_the_Internet_of_Things_Executive_summary.ashx
6. Wichmann, R. L., Eisenbart, B., & Gericke, K. (2019). The direction of industry: A literature review on Industry 4.0. *Proceedings of the Design Society: International Conference on Engineering Design, 1*(1), 2129–2138. https://doi.org/10.1017/dsi.2019.219
7. Weber, S. M., Göhlich, M., Fahrenwald, C., & Macha, H. (Eds.). (2013). *Organisation und Partizipation. Beiträge der Kommission Organisationspädagogik*. VS.
8. Haase, T. (2017). *Industrie 4.0. Technologiebasierte Lern- und Assistenzsysteme für die Instandhaltung. wbv*. Bielefeld.
9. Keller, A., & Weber, S. M. (2020). Trans-epistemic design(-research). Theorizing design within Industry 4.0 and cognitive assistive systems. *Proceedings of the Design Society: DESIGN Conference, 1*, 627–636. https://doi.org/10.1017/dsd.2020.173
10. Göhlich, M., Novotny, P., Ravensbek, L., et al. (2018). *European Research Memorandum of Organizational Education*. Accessed December 14, 2020, from https://digilib.phil.muni.cz/bitstream/handle/11222.digilib/138258/1_StudiaPaedagogica_23-2018-2_13.pdf
11. Keller, A., Weber, S. M., Rentzsch, F., & Haase, T. (2021). Lern- und Assistenzsysteme partizipativ integrieren. Entwicklung einer Systematik zur Prozessgestaltung auf Basis organisationspädagogischer Ansätze. *Zeitschrift für Arbeitswissenschaft, 79*(4). https://doi.org/10.1007/s41449-021-00279-2
12. Arnstein, S. R. (1969). A ladder of citizen participation. *Journal of the American Planning Association, 25*(4), 216–224.
13. Wright, M. T., von Unger, H., & Block, M. (2010). Partizipation der Zielgruppe in der Gesundheitsförderung und Prävention. In M. T. Wright (Ed.), *Partizipative Qualitätsentwicklung in der Gesundheitsförderung und Prävention* (1st ed., pp. 35–52). Huber.
14. Cooperrider, D. L., Whitney, D., & Stavros, J. M. (Eds.). (2004). *Appreciative inquiry handbook. For leaders of change*. Berrett-Koehler.

15. Weisbord, M., & Janoff, S. (1995). *Future search. An action guide to finding common ground in organizations and communities*. Berrett-Kohler.
16. Baumfeld, L., Hummelbrunner, R., & Lukesch, R. (2012). *Instrumente systemischen Handelns. Eine Erkundungstour*. Rosenberger Fachverlag.
17. Keller, A. (2018). Kognitive Assistenzsysteme in der Prozessindustrie – Mitarbeiter werden zu Mitgestaltern. In Schenk M (Ed.), 20. *Forschungskolloquium am Fraunhofer IFF. Intelligente Produktionssysteme*. Accessed April 10, 2021, from https://www.iff.fraunhofer.de/content/dam/iff/de/dokumente/publikationen/forschungskolloquium-2018-fraunhofer-iff.pdf
18. Weber, S. M. (2013). Partizipation und Imagination. In S. M. Weber, M. Göhlich, C. Fahrenwald, & H. Macha (Eds.), *Organisation und Partizipation. Beiträge der Kommission Organisationspädagogik* (pp. 71–82). VS.
19. Haase, T., Radde, J., Keller, A., Berndt, D., & Dick, M. (2020). Integrated learning and assistive systems for manual work in production - proposal for a systematic approach to technology selection and design. In *Advances in usability, user experience, wearable and assistive technology. Proceedings of the AHFE 2020 Virtual Conferences on Usability and User Experience, Human Factors and Assistive Technology, Human Factors and Wearable Technologies, and Virtual Environments and Game Design*, July 16–20, 2020, USA (pp. 853–859).
20. Weber, S. M. (2012). Transformative evaluation. In U. Kuckartz & S. Rädiker (Eds.), *Erziehungswissenschaftliche Evaluationspraxis: Beispiele – Konzepte. Methoden* (pp. 120–141). Beltz Juventa.
21. Schlick, C., Bruder, R., & Luczak, H. (2018). *Arbeitswissenschaft*. Springer Vieweg.
22. Haase, T., Berndt, D., & Herrmann, K. (2019). Anforderungen an die lernförderliche Ge-staltung arbeitsplatzintegrierter Assistenzsysteme. *lernen & lehren, 35*(19).
23. Dehnbostel, P. (2007). *Lernen im Prozess der Arbeit*. Waxmann.
24. Franke, G. (1999). Erfahrung und Kompetenzentwicklung. In P. Dehnbostel et al. (Eds.), *Er-fahrungslernen in der beruflichen Bildung – Beiträge zu einem kontroversen Konzept* (pp. 54–70). Kieser.
25. Franke, G. (1987). *Der Lernort Arbeitsplatz: eine Untersuchung der arbeitsplatzgebundenen Ausbildung in ausgewählten elektrotechnischen Berufen der Industrie und des Handwerks*. Beuth.
26. Fredrich, H., Dick, M., & Haase, T. (2021). Zur Passung von Arbeitsanforderungen und digitalen Assistenztechnologien in handwerklichen und industriellen Montageprozessen. In *ARBEIT HUMAINE gestalten. 67. Frühjahrskongress der Gesellschaft für Arbeitswissenschaft 2021*. GFA Press.
27. Ohannessian, H., Warschewske, F., & Woitag, M. (2019). Online data acquisition and analysis using multi-sensor network system for smart manufacturing. In *Smart SysTech 2019, European Conference on Smart Objects, Systems and Technologies* (pp. 1–7).
28. REFA-Verband für Arbeitsstudien und Betriebsorganisation. *Arbeitplatzgestaltung: Ermittlung zulässiger Körperkräfte. Lehrunterlage (Bestell-Nr. 037301/1)*.
29. IFA Institut für Arbeitsschutz. (2015). Bewertung physischer Belastungen gemäss DGUV - Information 208-033 (BGI/GUV-I 7011). In *Deutsche Gesetzliche Unfallversicherung (Hrsg.) Belastungen für Rücken und Gelenke – was geht mich das an?*
30. *Norm, DIN EN 1005-1: Sicherheit von Maschinen - Menschliche körperliche Leistung - Teil 1. Begriffe*. (2009).
31. baua Bundesanstalt für Arbeitsschutz und Arbeitsmedizin: Gefährdungsbeurteilung mit Hilfe der Leitmerkmalmethode. Accessed August 08, 2019, from https://www.baua.de/DE/Themen/Arbeitsgestaltung-im-Betrieb/Physische-Belastung/Leitmerkmalmethode/Leitmerkmalmethode_node.html

32. Schaub, K., Caragnano, G., Britzke, B., & Bruder, R. (2013). The European assembly worksheet. *Theoretical Issues in Ergonomics Science, 14*(6), 616–639. https://doi.org/10.1080/1463922X. 2012.678283
33. Sauer, S., Berndt, D., Niemann, J., & Böker, J. (2010). Worker assistance and quality inspection for manual mounting tasks. A virtual technology for manufacture. In V. Lohweg (Ed.), *Bildverarbeitung in der Automation. Lemgo, 10.11.2010, 1. Jahreskolloquium Bildverarbeitung in der Automation; BVAu 2010*. Hochschule Ostwestfalen-Lippe.

Cognitive Operator Support in the Manufacturing Industry - Three Tools to Help SMEs Select, Test and Evaluate Operator Support Technology

Frank Krause, Tim Bosch, Ellen Wilschut, and Gu van Rhijn

Abstract

Technology aimed at providing cognitive support to operators in the manufacturing industry, is developing rapidly. Through technologies such as near-eye displays and augmented and mixed reality, the operator receives digital information and instructions to help the operator learn, and work efficiently and accurately. Companies, especially SMEs, often have difficulty deciding whether cognitive operator support technology can help them, which specific technology suits their business best and what is required to implement the technology. This chapter describes three tools that were developed for, and are offered to SMEs to help them with this process. For each tool we describe our approach and provide one or more examples of application in industry. The first tool is a workshop aimed at selecting the best technology match considering the SME's goals, production process and staffing. The second tool is a short test of the selected technology on the shop floor. This allows companies to gain experience with a technology and form a better opinion. The third tool is a business case evaluation, which is described and illustrated by industrial use cases. Finally, lessons learned and suggestions for future developments of the tools are given.

Keywords

Cognitive operator support · Industry 4.0 · Manufacturing · Assembly · Work instructions · Pilot testing · Cost benefit analysis · Digitalization · Human factors · Ergonomics

F. Krause (✉) · T. Bosch · E. Wilschut · G. van Rhijn
TNO, Netherlands Institute for Applied Scientific Research, Leiden, The Netherlands
e-mail: frank.krause@tno.nl

3.1   Introduction: Outline of the Chapter

In manufacturing industry there is a growing interest in assistive technologies, technologies that are specifically aimed at supporting operators at performing the tasks assigned to them. In this chapter we focus on cognitive operator support technologies. These technologies are in essence also work instructions, but the digital nature and type of technologies allow for both better operator support and more control over the operator having the correct set of instructions. For companies, especially SMEs lacking research and development departments or large engineering or IT staff, selecting the right technology as well as implementing the technology of choice can be difficult.

Because of these challenges above, we created three tools to help SMEs select, test and evaluate operator support technology. In this chapter we describe these tools we recently developed together with SME manufacturing companies. The chapter is organized in the following way:

- Section 3.2 introduces the development of, and interest in the assistive technologies inside Industry 4.0. Further we briefly describe similar available tools.
- Section 3.3 describes our first tool, an Operator Support-canvas. It facilitates the discussion about the needs of a manufacturing company, the future users of the technology (e.g., shop floor operators, technicians and engineers), the available technological solutions and, lastly to support them finding the right match. Besides assisting the discussion, the tool was developed to transfer knowledge to companies. It gives understanding of what it means to digitalize assistance and helps to understand support requirements from production processes in combination with strengths and weaknesses of specific technology applications.
- Section 3.4 is about testing on the shop floor. As a second tool, and logical follow-up on the first, we support a small-scale test on the shop floor to pilot the selected technology, during which data is collected on usability, acceptance and performance in a representative company use case.
- Section 3.5 describes our last tool which is used to perform a business case evaluation.
- In Sect. 3.6, in conclusion, we share what we have learnt from applying the tools and using the operator support technology.

Sections 3.2, 3.3 and 3.4 each contain a description of the tool and ends with a subsection with use case descriptions. In this subsection we share results from applying each tool in several industrial use cases. Although the tools can be used consecutively, we have not yet been able to apply all tools to the same use case. Unfortunately, therefore, we cannot not yet evaluate the added value of using all tools as a three-step approach, as we have intended them to be.

3.2   Industry 4.0 and the Augmented Worker

Within the current industrial revolution (Industry 4.0), there is a strong focus on digitalization and robotization to optimize manufacturing processes. Along with Industry 4.0 there is
the development of the augmented worker or operator 4.0, or as Romero et al. [1] put it:
*"The operator 4.0 generation represents the 'operator of the future', a smart and skilled
operator who performs 'work aided' by machines if and as needed"*. Although robotization
on the shop floor is rapidly increasing [2], operators remain an important asset and are not
likely to be replaced altogether by robots and automatization. In fact, while many
companies are trying to get a grip on Industry 4.0, Industry 5.0 has already made its
appearance: a world in which people work alongside robots and smart machines [3]. New
technologies allow for optimal use of the operator's unique qualities such as flexibility,
creativity and fuzzy decision making [4] whereby human limitations are overcome by new
technology. In the physical domain we see exoskeletons providing support to workers to
limit physical loads on the body and hopefully reduce the prevalence of musculoskeletal
disorders (e.g., [5]). Also in the physical domain are the collaborative robots, which can
take over tasks at the workstation, that operators cannot (high precision) or prefer not
(tedious, high repetition) to do. This domain is developing quickly, and more and more
research is done on how to truly collaborate with the robot at task level, as is done for
example in the EU funded Rossini project (https://www.rossini-project.com).

Finally, in the cognitive domain, this involves supporting workers in tasks by means of
digital work instruction platforms, near-eye displays (i.e., smart glasses) and augmented
reality and mixed reality applications. An essential element of these types of cognitive
support is to have a minimal distance between the location of the instruction and the
location where the work needs to be performed and provide only the information needed to
perform the task fast and without making errors.

AR, MR and other cognitive support solutions have become more widely available for
industry. Tablets, smartphones, smart glasses and in-situ AR projections in combination
with industrial software solutions, support a transition from experimental prototypes in
laboratory settings to industrialized applications on the manufacturing shop floor. AR
systems and their application in manufacturing have been reviewed on training by Werrlich
et al. [6], on industrial applications by Bottani and Vignali [7], de Souza Cardoso et al. [8]
and Egger and Massood [9], on head-worn displays by Bal et al. [10] and on maintenance
work by Palmarini et al. [11]. There is a growing interest from companies to implement
operator support solutions in manufacturing, assembly and service and maintenance
settings for training as well as guidance purposes.

3.2.1 Developments Leading to an Interest in Cognitive Operator Support

3.2.1.1 Zero Defect and First Time Right for High-Mix Low-Volume and High-Complexity Manufacturing

To remain competitive and maintain production work in Europe, European manufacturing SMEs have followed the trend of producing highly customized products, leading to demanding manufacturing processes where batches of products are small and often as small as one piece. Quality demands are high and zero-defect production and first-time right is very important. However, this high mix, low volume and high complexity production process is more prone to mistakes because product variants may show only hardly visible differences and because for workers it is impossible to remember all different products and manufacturing procedures by heart. This is an important driver behind SMEs seeking to upgrade their operator production guidance from building from memory, paper instructions or paper on glass to more sophisticated support methods.

3.2.1.2 Travel Restrictions from COVID-19 Pandemic

Most companies exporting machinery have been forced to deal with COVID-19 related travel restrictions leading to strict choices on whether to travel or not, if at all possible [12]. Companies have been forced to delegate installation, servicing and maintenance tasks to local technical staff while providing support through regular mobile technology such as online meetings and WhatsApp video conferencing. At several companies we have seen that this has not only opened their eyes to the technologies' potentials for online work and remote working, but also led to post-corona travel reduction goals.

3.2.1.3 Employment: Personnel Shortages and Inclusiveness

A third important driver for operator support systems is related to employment and can be split into two related areas. First, many companies have trouble finding skilled personnel, especially highly educated technical workers for service and maintenance of technical installations and non-technically educated people for manual tasks such as assembly. Cognitive support or assistive technologies may help to reduce the required job competences and thus enlarge the pool of suitable workers. At the writing of this chapter, we have experiments planned on this topic with results to come available in 2022. The research is aimed at technical staff for service and maintenance of warehouse installations and at relatively simple warehouse tasks. The latter seeks to answer the question to what extent these tasks can be made available to workers with a distance to the labour market, by using assistive technology. This would not only be beneficiary to companies' continuity, it also adds to inclusion, which is the second employment related reason why companies and researchers are interested in operator support technologies.

In 2014 the European Union started its Cohesion Policy[1] of which the nineth priority was aimed at social inclusion. In 2015 the Netherlands adopted the Participation act (Participatiewet). This law is aimed at creating jobs for people with a distance to the labour market. Although there are signs that this law does not have the desired effect [13], companies offering sheltered work have shown a strong interest in both physical and cognitive operator support technology. Early research on AR operator support has already focused on cognitive impairments [14]. TNO has researched the technology in several companies [15, 16] and created a guide, matching technologies to applications in these kind of companies [17]. At the writing of this chapter, several new projects have recently started, so inclusiveness can be seen as important driver behind the development of these technologies.

3.2.1.4 SME's Technology Position

SMEs play a major role in most economies. According to the World Bank,[2] SMEs represent about 90% of businesses and more than 50% of employment worldwide. However, for SMEs with scarce human resources on research and development, it is difficult to follow all the latest developments in automation and digitalization. A recent survey on innovation and digitalization among more than 4000 SMEs in the Netherlands [18] shows that around a third of the companies are currently working to a large or very large extent on improving customer service, acquiring new customers, organising internal processes better or more efficiently, or launching new or improved products. About 40% of manufacturing SMEs indicated that they are working on automation and digitalization of production processes. This often concerns digitalization at a company level, such as engineering, purchasing, inventory management or planning via ERP systems. In the Netherlands, digitalization of the industrial workplace itself is lacking behind. Although many aspects of production and other business operations are digitalized, many manufacturing companies still work with text documents that have no direct connection to CAD or PLM software and to other manufacturing software systems, such as ERP or MES.

Order information, work instructions and manual quality inspection is quite often paper-based. The most frequently mentioned reasons for not being actively involved with digitalization: other priorities within the company and insufficient time available, lack of clarity about costs and benefits of digitalization and a lack of knowledge/skills within the company and the entrepreneur him/herself often play a role as well.

In summary, SMEs recognize the need for digitalization. At the same time, SMEs need practical support and knowledge in the field of digitalization that fits in with their daily operations.

---

[1] See: https://ec.europa.eu/regional_policy/en/policy/how/priorities
[2] See: https://www.worldbank.org/en/topic/smefinance

3.3   Operator Support (OS) Canvas Workshop as a Selection Guide

3.3.1   OS-Canvas in Short

The operator support or OS-canvas (Fig. 3.1) is a custom designed whiteboard with six predefined sections: (1) goals, (2) operators, (3) scope, (4) process description, (5) information needs and (6) context aspects (Figs. 3.4 and 3.5). The board is used to structurally approach a customer's questions regarding operator support technology, structure the process and facilitate an open discussion between participants leading to consensus. This is done in a work session in four steps that will typically be on-site and take half a day. The first step is an introduction in the different types of operator support technology for guiding them through, or training them for, manual work processes, like assembly, service, installation. This is followed by the company showing their product and providing insight in their production process and current way of working. After this, in the third step, all elements of the canvas board are filled in. In the last step, we briefly go over the business case: what investments and recurring costs are involved, what can be expected gains and how long does it take to earn back the investment.

The session is hosted by two experts in the field of operator support technology. Further participants are two to three representatives of the company. We recommend having important stakeholders present: management as investors/decision makers, an engineer as process owner and developer and an operator who has product and process knowledge and could be the future end-user. After the canvas session the company should have more knowledge on the available operator support technologies and their potential to meet the company's needs, have a prioritized list of suitable technologies and have an idea of involved costs versus benefits.

In the following paragraphs we will describe the different elements of the work session and canvas board more in detail.

3.3.2   Technology: What Kind of Technologies Are Available?

Cognitive operator support is aimed at providing support on operations that cannot be remembered, take valuable time, like searching, or are prone to mistakes, like similar components in adjacent bins. Furthermore, OS technology is used for training (complex) procedures to improve learning speed, reduce supervision efforts and enforce a standardized way of working.

The workshop is aimed at charting these operations and matching them to the proper technology solution. The canvas currently identifies the following four different technology groups (Table 3.1) that provide digital working instructions to the operator.

Fig. 3.1 Discussion in front of the OS-canvas board during a session at De Gier

3.3.3   Filling in the Canvas

As mentioned above, the canvas board itself contains six sections that form the basis for discussion and that we identified as being essential to making the right technology choices. In the following we explain how each of these topics relates to obtaining more insight in the operator support questions a company has.

3.3.3.1 Goal: Why Implement a New Way of Providing Work Instructions?

What does the company intend to achieve with new operator support technology? The goals are obviously the starting point: without making the goals explicit, there is no way to decide which technology is best. We try to quantify their goals as much as possible because this facilitates the selection process and can be used as KPI selector when piloting a certain technology. Companies have a wide range of goals they want to achieve with operator support. The canvas organizes goals into four categories: product quality, productivity, staffing and company image. Foremost the company's interest lies in using OS to increase product quality: zero-defect and first-time right production. With respect to productivity, a company may want to reduce the time new operators need to reach target production speeds and to reduce the time and effort needed for supervision from more experienced operators. This simultaneously leads more continuity in the work of experienced operators. In the staffing category companies are seeking to enlarge their staffing pool with the technology by assisting the operator to perform more complex tasks. With employees being able to perform more tasks, flexibility goes up and often personnel costs can be reduced. This can be seen in regular production environments, but the technology has also proven to be very

Table 3.1 Brief overview of technologies identified in the operator support canvas workshop

1.	Displays—static or mobile 	Instructions are offered via electronic working instruction software on a monitor or tablet. It allows for stepwise instructions but is less suitable for tasks requiring the operator to be very mobile, without running the risks of mistakes associated with needing to transfer the information from the display to the place where the work needs to be done. Most of the commercially available EWI software platforms can be connected to external devices such as pick-to-light sensors, torque tool controllers or barcode scanners. Typical manufacturing tasks supported by this kind of technology are manual line or work cell assembly.
2.	Wearable near eye displays—non-context aware 	Highly mobile are the solutions in this second group through which 2D information is presented on near-eye displays, often also referred to as smart glasses. Glasses in this group are not context aware, the information does not adapt to what the user is looking at. Typical manufacturing tasks supported by this kind of technology are logistic activities (e.g., order picking through pick-by-vision), strenuous maintenance tasks (e.g., in or under machinery) or assembly of large machinery or constructions.
3	Projection-based technology 	In-situ spatial AR projections (sAR) often in combination with time-of-flight cameras such as Microsoft's Kinect 2. Wherever possible and needed, work instructions are presented step-by-step on top of or close to the assembly piece thereby keeping information and product within the field of view. Information about a part's assembly location and its orientation can be transferred as well as directions in which to move a piece. Motion detection is used to guard operations such as picks from the proper bin and to automatically advance instructions upon detection of a finished action. Typical manufacturing tasks supported by this kind of technology are assembly and welding.
4	Context aware augmented reality and mixed reality 	Context aware applications of augmented reality (AR) and mixed reality (MR) show information on a head-mounted see-through display (e.g., Microsoft's HoloLens 2) or handheld device, allowing the user to look at the object of work and have an (3D) information layer added to it. In case of a MR see-through display virtual displays and controls may be added. For both devices stepwise instructions are provided to the operator. A technology group that has received extensive interest is remote assistance wherein an expert guides a non-expert operator on site for remote installation, service and maintenance applications. As this technology is not yet part of the canvas methodology it will not be described in the examples in this chapter.

successful in sheltered workplaces where people with for example cognitive impairments can work without high output demands and with sufficient support from coaches [15, 16]. Lastly, and generally a side goal, a company may want to relay the image to their customers that they seek technological advancement and do their utmost to maintain high quality standards.

3.3.3.2 Target Group: Who Is It for?

For discussion purposes it is good to know something about the target user group: which operators will be using the technology, are they experienced or inexperienced and which level of competence and (digital) skills is present and what should be supported? The technology does not have clear advantages for one single target group. Users from all ranges in cognitive capabilities may benefit from cognitive support. However, in many sectors with staffing problems we see an interest in being able to reduce the required cognitive capacities and skills levels to be able to perform the work.

3.3.3.3 Process in Scope: Which Process Steps Are Reviewed in the Canvas Session?

In this step we define which (sub)process will form the basis for the analysis in the following workshop steps. This can be a procedure from start to end or a specific part that is complex or important to get right. Before the session, companies are asked to select a procedure (assembly, inspection or maintenance) that is typical for their production process, so that knowledge from the session is more widely transferable.

3.3.3.4 Process Description: What Are the Process Steps?

In case of an assembly, the (sub)assembly is split in assembly steps. This action is important because for each step, in the next part of the canvas, it is defined what the information requirements are to be able to perform this step. A process breakdown is used to determine all relevant steps. This can be done by creating a commonly shared process map using the 'MAS' methodology. MAS stands for 'Montage Afloop Schema' (assembly process flow, [19], see example in Fig. 3.2). The customer may explain the process steps via work instructions, video or by demonstrating them.

3.3.3.5 Information Needs: What Is Needed for Comfortable, Fast and Zero-Defect Process Execution?

In this step a detailed assessment is made of the previously defined assembly process. For this process we use cards depicting a certain type of information. Each card is connected to a specific question about the process that is asked during the session. The visible process description on the canvas board helps to guide the discussion. Each card's backside shows which technologies are fit for providing this specific type of information. Figure 3.3 shows examples of a few cards. Note that our fourth technology group is split into handheld and head mounted devices.

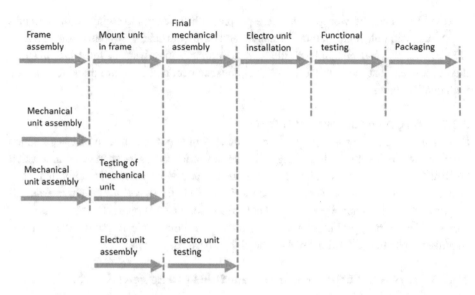

Fig. 3.2 Example of an assembly process scheme (MAS), which visualizes the sequence of the various process steps or assembly tasks

Fig. 3.3 Examples of information cards; last is a picture of a card's backside showing which technologies are fit for this specific purpose (design: B. Lammers, RoboHouse, Delft)

The card questions need to be answered by staff that know the process in detail; in general, this is an experienced operator. Questions that are asked are: to what extent does the assembly require a stepwise approach, what type of tools are used and is tool selection critical, is the parts placement order critical, when placing parts, can an operator make mistakes in their orientation? In total there are 14 cards that create an inventory of the information needs.

The magnetic cards are placed in one of the four information sections (see Fig. 3.4). These sections are defined by two factors: how often in the process is this type of information needed to prevent mistakes or to speed up assembly (low/high frequency), and, if a mistake from missing this information is made, what are the risks (low/high costs)? In connection to the second question, it is important to identify where the mistake is likely to be discovered. In some cases, the operator will discover the risks and be able to correct it himself, in other cases the risk may result in machine failure after delivery to a customer.

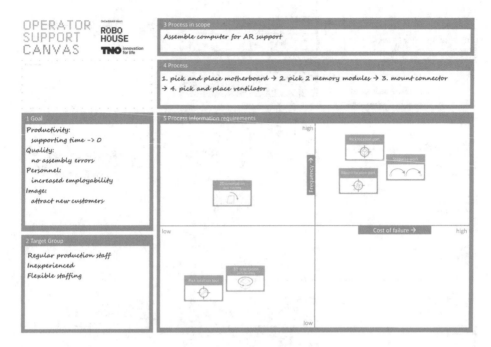

Fig. 3.4 Canvas set-up. Step 5, Process information requirements, is split into four sections by two axes: the frequency that a type of information is needed, and the risk associated with mistakes based on missing this type of information

These factors define the risks of mistakes in which costs can be in time, money but also less quantifiable costs such as reputational damage or reduced customer confidence.

After placement of all relevant cards, they are turned around to reveal which technologies are most suitable to supply the specific type of information (see Fig. 3.3, most right picture). This overview, in connection to their position on the canvas, already provides clues towards the technology that is most promising to meet the company's needs, given their goals, operators and production process. The last part serves as further refinement.

3.3.3.6 Context: What Requirements Come from the Context?

After having identified the information needs, the context in which information is supplied, is reviewed. This too is done via a standard checklist consisting of seven items. These items stem from requirements a certain technology might have, such as being able to fixate the assembly piece in case of projections, or from requirements from the assembly process itself which may or may not match very well with certain types of technologies, such as required mobility or needing to have both hands free for assembly.

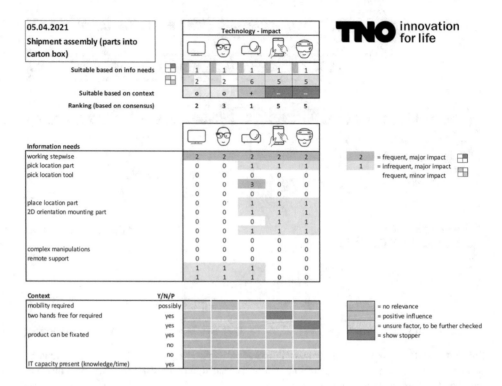

Fig. 3.5 Example of the outcome of a canvas session; the ranking is defined after discussing the results and is based on consensus (see also Sect. 3.3.4)

3.3.3.7 Report: Canvas Summary and Short Business Case Analysis

After having filled in and discussed all elements of the canvas board, the outcome is filled in, in a MS Excel-template. This gives an overview of information and context, and of technology suitability in relation to these factors (see Fig. 3.5). The ranking of technologies is defined after discussing the results and is based on consensus. This overview is part of a short report the companies receive directly after the workshop.

The report also contains the outcome of a first raw estimation of foreseen investments (hardware/software licenses) and monetary and non-monetary benefits for the company. After having gained more experience with the technology, a more elaborate analysis of costs and benefits is made. This is further described in a later section in this chapter.

3.3.4 Use Case Descriptions: Canvas Examples from Two Use Cases

3.3.4.1 Company A: Shipment Assembly

Company A manufactures industrial chairs and large monitor stands. As a use case for the canvas workshop, they had selected the process of getting a shipment ready for an overseas customer. The company's goal was to improve their shipment's quality by more standardization and checks on performed actions. The target group consisted of reasonably experienced personnel. After having observed the use case process, the canvas was filled in. It should be noted that information requirements were relatively low, which already reduces the added value of AR technology. Further, there were few critical steps with respect to task execution. Most important was not to forget any items for shipment, but even then, as this mostly concerned regular assembly material, this could be solved quite easily were it to be discovered at the customer.

Based on the position of the cards and the context, projection technology had a slight advantage over digital work instructions and near-eye displays (see Fig. 3.6). AR and MR technologies were considered unsuitable because of the context: two hands were needed therefore the tablet was not usable, and the technology would be needed all day making a head mounted solution less desirable from a user acceptance perspective and battery life duration. As a consensus outcome the different technologies received a ranking. Based on the results, a short pilot was organized using projection technology. The results are not described in this paper; however, it can be mentioned that although the pilot was successful and delivered the desired outcome, implementation costs were considered out of balance with current business risks.

3.3.4.2 Company B: Assembly of a Smart Wallet Counter Display Model

Company B is a sheltered workplace with diverse assembly work. Their use case was the assembly of a smart wallet's display model. The primary goal was to select a technology suitable of training their operators to learn to assemble the product independently without coaching from a team lead, thus reducing production cost and making team leads available for other work. Additional goals were to improve assembly quality, to increase the skills and independence of their operators and increase their employability within the company and lastly to improve the company's image and attract different customers with products of increased complexity. The group of operators consisted of people with a low level of work experience, workers with an autism spectrum disorder, people with cognitive disorders and people dealing with illiteracy.

The assembly consists of 14 steps with a cycle time of 5–7 min. Information requirements were particularly high with respect to the assembly order and the mounting orientation of several parts as mistakes in this area render the product unusable. Further, information on tool usage and on some complex manipulations were important. Based on the ability to fulfil information requirements, spatial AR and AR/MR technologies were most suitable. However, context factors such as usage intensity and needing to use two hands for assembly, ruled out the latter two. Spatial AR was found to be very interesting

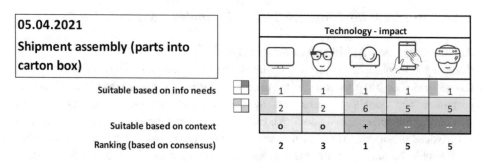

Fig. 3.6 Segment of the canvas results from company B (see also Fig. 3.5)

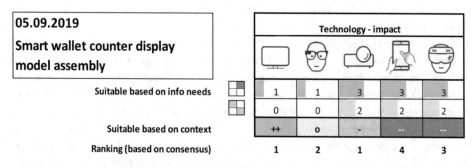

Fig. 3.7 Segment of the canvas results from company A. Figure 3.5 shows an example of a full overview of results

based on guidance potential, however it would require keeping the product in the same place during assembly, which was seen as a challenge. Based on the canvas session the company said it would look at the operator support's potential for other assemblies and, based on the outcome, decide between spatial AR and digital work instructions on a monitor, optionally supported by a physical pick-to-light solution. Figure 3.7 shows a segment of the canvas results with the technology ranking outcome, bases on consensus between all workshop's participants.

3.4 Pilots on the Shop Floor

3.4.1 How Do We Set Up the Small-Scale Pilots?

The pilot set up differs depending on the use case and the technology and of course on the company's wishes. In general, we try to create a pilot that provides data from at least eight operators to have some scientific value. This is not always possible and not always needed by the company, for example if it is their only wish to be able to try out one or more technologies in a company specific use case. Apart from the number of participants, the pilot duration is considered. From experience we have learnt that pilots in assembly processes generally take a few weeks to be able to involve enough workers and collect reliable manufacturing and operator data.

In manual assembly, the pilot is almost always set up in production, where the pilot workstation replaces an existing one. Cycle times vary depending on the product that is assembled. In the use case we describe below (see company C in Sect. 3.4.2), the duration of a single assembly was approximately 1 h. To be able to check for learning effects and production speeds, the aim was to have each operator work a full day at the task.

When OS technology is used to guide installation start-up, servicing or maintenance tasks, a procedure is performed only once for every available technology. A test with one operator takes less time, even if more than one technology is tested by the same operator. For this reason, in these types of pilots, it is often possible to collect usability data from up to four operators in 1 day (see company E in Sect. 3.4.3).

Trying to set up pilots in regular production processes poses extra challenges to the researcher. E.g., in case of the pilot at company C (see next section), small delays in getting the pilot workstation ready, led to a diminishing order size, as production could not be halted while waiting for the pilot. Apart from trying to fit in with regular order processes, staffing the experiments can be a challenge, especially in SMEs, as often not enough staff is available, either being too experienced or too inexperienced. This generally leads to not reaching the desired number of participants or having an inhomogeneous population. For the participating company, this does not have to be a problem, since many of these projects are aimed at transferring knowledge. The companies generally learn enough to be able to decide future actions with respect to the technology.

3.4.2   Use Case Descriptions: Results from Two Shop Floor Pilots with Operator Support Technology

3.4.2.1 Company C: Precision Machining

About
Company C manufactures and assembles precision components and products for several markets such as exoskeletons, rescue equipment and eye surgical instruments. Assembled products typically consist of many parts. The company has a large flexible workforce that requires frequent learning to assemble a product. This accompanied by a substantial amount of guidance time by experienced operators. Company C was interested in testing an operator support system, in this case spatial AR, to see how this would affect guidance time, productivity and error rate. Further they were interested in learning how long it would take for operators to reach target assembly times and how much time programming would cost.

Fig. 3.8 Workstation set-up for the pilot at Company C

The Pilot

A pilot was set up around these questions [15]. Based on hardware requirements and our experience with installation we advised Company C on the required adaptations to the workstation to be able to suspend a projector and time-of-flight camera from an overhead beam. For the pilot the assembly of an exoskeleton component, the so-called Smartjoint, was chosen. The assembly consists of 42 parts, all available in bins on a rack in front of the operator (see Fig. 3.8). The number of parts increases initial searching time and the time to remember all steps, if remembered at all. Further, the assembly consist of several crucial steps such as the application of glue to certain parts. This is guarded by the system's movement tracking feature. Target assembly time is 51 min for a complete assembly.

As mentioned above, testing in regular production processes often means having to compromise with respect to scientific standards. Planning can suddenly change and cherry-picking subjects for the test is seldom possible. This was also the case in the Company C pilots, although the company did their utmost to meet our requirements. In the end seven operators took part in the pilot, two experienced and 5 without experience in assembling this specific product. Six operators assembled 10 products; one operator could only assemble one product with the system. During the test, cycle time was recorded by the projection system, product quality was inspected afterwards by the quality officer and system usability was assessed using the System Usability Scale.

Results

The pilot showed that all participants were able to assemble the products almost flawlessly, with a minimum of guidance by an experienced operator. Participant numbers and production figures were too low to be able to compare quality to that of experienced operators. Nevertheless, error rate was considered to be sufficiently low and partly due to an unnoticed error in material supply. Further, after four assembly cycles, the novice operators reached the target cycle time of 51 min. In the old situation (paper-based instructions) generally more than 20 work cycles were needed to reach the target cycle time. During the pilot, cycle times never reached experienced levels (38 min). From other pilots we have learned that productivity with this specific type of system often remains the same or slightly decreases due to system response times. We expect this to improve with system upgrades and with adaptive systems that reduce instructions with raised experience levels [20].

After a day of working of working with the system, 6 participants were asked to evaluate the system. We used the System Usability Scale (SUS, [21]) and a short questionnaire on their user experience. Five out of six participants reached an SUS-score above 70, which is considered the threshold value for mature and useable systems [22]. The user experience was generally positive. Four out of five inexperienced operator wanted to continue working with the system. Feedback from the operators showed ways to improve the system, such as also mentioning required tools in the projected instructions. This was now done only by highlighting the required tool. Last, it was expected that more experienced operators would become annoyed by the required manual advancement of the system by using a virtual button on the desktop.

The company especially valued the built-in motion tracking feature as this prevents operators to skip critical steps. E.g., if the operator forgets to pick up and apply glue for certain screws, the system does not advance.

3.4.2.2 Company D: Sheltered Workspace

About

Company D is a sheltered workspace in which people work and receive additional coaching or training depending on their abilities and needs because of cognitive, social, physical or a combination of impairments. The work in company D is organized by departments which work on a variety of packing and (sub)assembly work for e.g., sun panel manufacturing. Available work generally varies depending on the actual production contracts. To meet the demand of their contracts and the variety of changing products next to employee turn-over and a daily varying available workforce, there is a continuous demand for training and a desire to make work available for lower skilled personnel. Training is done on the job by either a team lead or an experienced co-worker. The company was interested whether an operator support system could be used to train employees to assemble a product independently and whether more people and lower skilled personnel would be able to perform a subassembly of a sun panel clamp guided by the system.

The Pilot

In the pilot project (see also [15]) we compared on the job training with both a face-to-face training (n = 8) and a training using spatial augmented reality or sAR (n = 9). Both groups received a training session of 45 min learning to assemble the same product. The working

Fig. 3.9 Impression of two workstation setups: AR guided training (top) and face-to-face training (bottom) both with working instructions (in Dutch)

instructions for the projection-based sAR system consisted of short video's, pictures, symbols and text. Projections were used to support pick-to-light functionality, highlight placement locations, show part orientation and accompany it with a short instruction text. Work instructions contained 21 steps: 9 picking steps, 9 placement steps, one step with tooling, one cleaning and one quality check. After the instruction was programmed, the instruction was reviewed with an experienced operator. The face-to-face training was done by an experienced operator who showed the steps and a paper instruction. The two workstations were placed in a quiet corner of the production floor with partition walls (see Fig. 3.9).

Results

For the task completion time both training methods showed significant cycle time improvement. Cycle times at the start of the training were faster for face-to-face training (3.7 min) than sAR training (6.7 min). However, no significant differences between methods were found at the end of the training period (2.7 vs. 2.1 min) or after the training period

(1.6 vs. 1.7 min). After 45 min of training all participants (except one) were able to assemble the product correctly without instructions or technology support. There were no significant differences in product quality between methods during and after the training period. Additional supervisory effort was needed during sAR training because participants had to learn to work with the technology. There was no reduction for the amount of supervision time. For both methods, supervision reduced about 90% after the training period, when participants were building the product without additional support. Mental workload was relatively high at the start, during, and after participating in the training for both methods. Mental workload seemed higher for sAR after the training. All novice lower skilled operators were able to assemble the product successfully, except for one person. This person showed a low ability to rotate the parts correctly despite receiving projected sAR instructions showing the shape of the correctly rotated part and a short video how the part should be placed (see Fig. 3.9, top right). For this person, this may have been caused by low mental rotation capabilities. An interesting observation by the test leader, was that despite the structure and support sAR gave in every assembly step, people still asked for confirmation from the experienced co-worker or the test leader, who were always close-by. This indicates that more emphasis should be given to provide positive feedback to the operator with sAR, even when they perform well. Further observation during this pilot showed that people in the sAR condition actually had to learn two things: (1) the assembly of the new product and (2) how to work together with the technology. This might account for higher cycle times in the sAR condition at the beginning of the training. When they worked without the system, after training the transfer of the learned knowledge was good and operators no longer depended on the technology to execute the assembly. After the pilot, company C see saw potential in several applications: to train new operators and lower skilled operators, to train experienced operators to assemble new products or product variants and to give operators a refresher training after job rotation or prolonged breaks. Also, they highlighted that it was important as a sheltered workplace to keep innovating, to remain competitive and attractive for their clients, and for their employees so they could learn so called twenty-first century skills.

3.4.3 Additional Examples: Two Short Test Descriptions

3.4.3.1 Company E: Various Operator Support Solutions in Maintenance of Sorting Systems

Company E is a turnkey system integrator providing a wide range of logistic systems, including sorting systems and order pick systems for the e-commerce and retail market. Installation, maintenance and service of sorting systems is done by service engineers and assembly operators of company E, the customer's technical service department or a local distributor. One of the challenges the company is facing, is the strong dependency of technical specialists in tight labour market conditions. Furthermore, the company would like to shorten reaction times to avoid unnecessary machine down time and lower cost. As

Fig. 3.10 Work instructions for a machine maintenance task provided on a tablet (left), Iristick smart glass (middle) and the MS HoloLens 2 (right)

the company's business is growing the challenges will become even more urgent in the near future. The company showed interest in operator support technologies to remotely guide workers with step-by-step instructions.

In a small-scale pilot we evaluated three types of operator support technologies (see Fig. 3.10). The tablet and smart glass application showed step-by-step information with pictures and text. A short video was added in case of complex tasks. The mixed reality application also showed text and pictures on virtual displays. Additionally, an overlay with a 3D animation or 3D annotation on the physical machine were used to support complex procedures (see Fig. 3.11).

A typical but rather complex maintenance procedure was executed by four operators. All operators had sufficient technical skills but had no experience with the evaluated maintenance procedure. The procedure included safety procedures, mechanical activities and activities related to the machine control PC. Maintenance procedures were designed with support of the companies engineering department. All four operators worked with all technological solutions in a randomized order. All operators were able to execute the maintenance task without additional expert support. On the Technology Acceptance Model (TAM) questionnaire the operators rated all OS systems as positive. Beforehand we had expected the MR solution to have an added value in complex activities (e.g., loosening an invisible bolt, removing a part following a predefined trajectory in narrow spaces). However, this was not visible during observations. In interviews with operators, using the UX laddering method [23], operators confirmed this.

3.4.3.2 Company F: Manual Electronic Product Assembly Supported by Digital Work Instructions

Company F is an SME manufacturer of medical and life science products. As other manufacturing companies in Western Europe, Company F experiences less predictable market demands, a shorter product life cycle and a need for zero defect and first-time right

Fig. 3.11 Impression of MS Guides virtual display (left) and REFLEKT One 3D animation (right) in the MS HoloLens 2 application

Fig. 3.12 Workstation setup in Company F

production. To face these challenges and support a standardized way of working, unambiguously instruct operators, improve version control of instructions, the company evaluated digital work instruction technology. Furthermore, the company would like to improve traceability of parts, products and production orders by eliminating manual registration of batch- and serial numbers of parts. In a small-scale pilot, step-by-step digital work instructions were evaluated (see Fig. 3.12). The pilot showed the limited added value of work instructions in high volume batch production as operators know what to do and production variation is limited. However, the operator support technologies supported digitalization of the work cell and thereby eliminated all manual registration activities, enforced a standardized way of working, improved version management and thereby alerts the operator to product changes.

3.5   Business Case Analysis

As with any investment in industry, companies would like to know if there is a positive business case for these operator support technologies: what are the costs, what are the benefits and how long does it take to earn back the investment. Part of the toolkit to help SMEs with the selection of technology, is an operator support business case analysis tool, TNO developed for this purpose. The tool is based on previous research on cost-benefit analyses in ergonomics [24, 25] and builds upon previous similar tools to define costs and benefits of ergonomic changes. Key elements of this MS-Excel-tool, and previous ones, are the participatory approach and considering both quantifiable and non-quantifiable outcomes of a change. In the following paragraph we will describe the different elements of the tool and give some examples of analyses for pilot projects in manufacturing companies.

3.5.1   Quantifiable Costs and Benefits

The main drivers behind innovations are cost reductions and productivity gains [26]. There-fore, it is important that a tool gives insight in these factors. It can be quite difficult to obtain all relevant figures. This is one of the reasons a participatory approach is used to fill in the tool. Some figures need to be estimated, especially if a new technology has not been tested in a controlled environment that allows to draw conclusions on e.g., productivity effects or effects on error rates. Only if the estimates are made together with the company, the outcome of the analysis will be credible. Data provided by hardware and software suppliers have limited value, since they cannot take the company's actual use case details and starting situation into account, which strongly affect the business case. The tool becomes more reliable if it is used with data from an experiment.

Effects on *productivity* (relevant parameters in italics) are quantified as the delta between *production time x operator costs* before and after investment. Effects on *error rate* may reflect in many costs such as *rework time, quality inspection time, scrapped material* but also *recovery costs* if an error is left undiscovered until it reaches the customer, or worse if it leads to costly *failures or downtime*. Other effects from using operator support systems can be reduced *instruction time* in assembly and *problem-solving time* in technical systems. The latter two were pursued in the Company H and I use cases, respectively (see Sect. 3.5.3). Using operator support technology to remotely support installation, mainte-nance and repair of exported products, can obviously reduce *travel time and costs* but also support company growth without an expansive grow of the customer service department.

On the investment side, costs are divided into *capital expenditures* (CAPEX) and *operational expenditures* (OPEX). CAPEX are all one-time costs made to get the new technology started, such as *software and hardware costs, installation costs* and *training costs*. In many pilots so far, an important cost element was the *time needed to create or transform instructions* for all products to be assembled with operator support. Because

Table 3.2 Overview of general investments and potential effects, quantifiable and non-quantifiable. This can differ from company to company and may be adjusted dependent on the customer

Investments	Quantifiable effects	Non-quantifiable effects
• Software	• Production time	• Production continuity
• Hardware	• Staffing costs	• Production flexibility
• Training	• Rework	• Risk management
• Infrastructure	• Quality checks	• Image
• Integration	• Scrap	• Attractiveness as employer
	• Travel time	

many companies involved in our pilots had not changed to work instruction software, investments were relatively high and often outweighed the benefits of the operator support technology. The same is true for costs involved in integrating a new technology into a company's existing IT-infrastructure and its software, such as an ERP system. Although this integration with existing data systems is key to create a successful solution, these costs are less predictable as they depend on the company's existing IT-infrastructure and on their demands and wishes. At the writing of this chapter, the latter costs are not yet incorporated in the tool. Considering their size and effect on the business case, we plan to do so in future versions of the tool. In the current version of the tool, the costs of changing to a digital work instructions platform are left outside the equation. We see this as a necessary investment companies will need to make anyway, if they want to digitize their manufacturing shop floor. All CAPEX are calculated into yearly costs based on the depreciation time. OPEX are also calculated as yearly costs and generally consist of software licenses, maintenance fees and costs made to keep work instruction up to date.

All quantifiable costs and benefits (Table 3.2) are totalled into the annual benefits or costs of running the system. Negative outcomes are not necessarily a showstopper because non-quantifiable results from the new technology can be seen as enough to justify a negative outcome of this step.

3.5.2 Non-quantifiable Costs and Benefits

It is important to also acknowledge the non-quantifiable 'soft' effects or non-monetary effects of introducing new technology. Although they may be difficult or impossible to quantify, they can play an important role in the decision to invest, even—as mentioned above—if a cost benefit calculation turns up with a negative figure. The non-quantifiable nature of these effects from technology further stresses the need to follow a participatory approach in applying a business case analysis tool. In general, we divide non-quantifiable effects into three sections in our tool: customers, operations and personnel.

As we have often seen in pilots, the pilots with new technology are followed with great interest by the company's marketers who will not miss the opportunity to show an

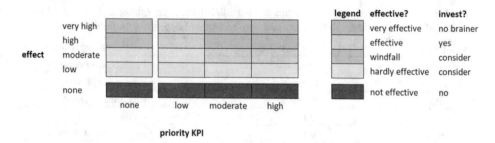

Fig. 3.13 Diagram used to plot non-quantifiable effects and label their effectiveness based on the combination of the effect's magnitude and its priority in the company's list of KPIs

experimental workstation with operator support to visiting customers. The image the company has, to existing and potential customers, is an important aspect to which being innovative and having high quality standards can add.

In operations companies mention production continuity and flexibility as important effects. Being able to increase the pool of operators that can assemble a product is an important asset of the operator support technology. Using operator support technologies forces the company to think about optimal assembly processes, leading to a more standardized way of production where risks are better managed.

For personnel, the technology can increase continuity and reduce disturbances. Personnel may value having to ask for help less frequently and as a result there is less disturbance and stress for more experienced staff that is often called in to provide support.

To help businesses with evaluating these non-quantifiable effects, each effect is scored on how high the aspect is prioritized a one of the company's KPIs, and how large the expected effect will be. Depending on the combination of scores, each benefit (or cost) is labelled 'very effective', 'effective', 'a windfall', 'marginally effective' and 'not effective' (see Fig. 3.13). The investment may have large effects on an aspect that is not high on the list of priorities. This is considered a windfall, which could tip the scale if it coincides with other effects on KPIs that are higher on the company's priorities. If this is the only effect, investing in the technology is hardly worthwhile. All effects are plotted in a diagram similar to Fig. 3.13.

The results from both the quantitative and qualitative analysis are discussed with the company. The business case analysis tends to form a valuable addition to the 'gut-feeling' entrepreneurs often have towards investments of their company. The detailed approach to required investments and the wide view on effects is appreciated and often provides insight in both costs and benefits previously overlooked.

3.5.3   Use Case Descriptions: Was There a Business Case in Our Pilots?

From applying the tool, we have found that there is no general business case for operator support technology. Justifiable investments strongly depend on the company's goals, technology position, products and many more aspects. But obviously also the technology plays an important role as there is a large price range between all available technology, both in CAPEX and OPEX. We have found that yearly software license fees can easily tilt the scale for an SME with interest in operator support in one workstation. In addition, the scalability can be an issue, especially in cases where hardware investments per workstation or operator are high: in these cases, there is hardly any gain from applying the technology to more workstations, CAPEX and OPEX more or less rise incrementally making it harder to make the business case. As technology costs of this relatively new technology are expected to drop, this will obviously positively affect the business case. In the paragraphs below we will provide some examples of the outcome of the business case for a few types of use cases.

3.5.3.1 Company G, Sheltered Workplace2: Moderate to Strong Business Case
Spatial AR, using projection technology in combination with Kinect motion detection, was used in a sheltered workplace in one workstation of a production line of child safety seats (see Fig. 3.14). The analysis showed that there was a strong business case for the system based on multiple benefits, both quantitative and qualitative. On the quantitative side the business case was formed by combining two workstations into one. This was possible through the system's movement tracking feature which made it possible to use two types of torques tools and control their application based on the location of use. Using two torque tools with different settings was not allowed in the regular set-up. This led to better line balancing and reducing the number of employees on the line. Even if sheltered workplaces have a goal to help people to work, working efficiently remains important to be able to be competitive. The perceived reduced guidance time added to the equation. On the qualitative side the improved quality control was highly appreciated considering the high standards that apply to child safety seats. Further, employees who previously were not able to perform the assembly tasks, could now, with the help of the system, perform a more difficult task. This added to the company's staffing flexibility and improved employees' sense of efficacy, the latter being of high priority for sheltered workplaces. This pilot was described in a previous publication [15].

In general, we have found that operator support technology's business case in sheltered workplaces is often viewed differently compared to 'regular' production sites. The companies have a strong mission to help people obtain and keep work, to increase self-esteem and to help them to learn and develop. Companies see great value in operator support technology and have chosen to invest, even—within limits—without a positive quantitative business case.

Fig. 3.14 Application of projection-based technology to support impaired workers in child safety seat assembly in company G

Fig. 3.15 Application of projection-based technology in manual gear box assembly in Company H

3.5.3.2 Company H, Gearbox Assembly: Weak Business Case

Quite different was the business case for a manufacturer of gearboxes for opening large windows. Their business location among a large number of agricultural companies, makes it difficult to find personnel that is proficient in either Dutch or English. Dutch agriculture has many seasonal workers from (south)eastern European countries, and these workers form the largest flexible staffing pool for the company. A flexible pool is needed because orders can have a large influence on the required production capacity. Even though the products do not change a lot, turnover is high, therefore there are not many who are experienced in building the product. Frequent new personnel require a lot a guidance time from experienced staff. Operator support via projection technology (see Fig. 3.15) was tested for 2 weeks to evaluate the system's usability and its effect on guidance time, also considering the system's features to provide guidance without words.

The evaluation showed that productivity was lowered, something we have observed quite often with the system. Guidance time, on the other hand, was strongly reduced making up for a slightly positive business case. The company, however, presented another aspect to the business case which they found to tilt it to the negative side. In their view, because of the relatively simple—and stable by design—product, every investment in operator support was keeping them from investing what was actually required: investing in robotization. This was strongly influenced by the relatively high required investments in software and hardware. Simpler and cheaper technology would substantially reduce the bar for SME to invest in this type of technology.

3.5.3.3 Company I, Step-by-Step Remote Assistance: Strong Business Case

Our last example is of a manufacturer of potato, vegetable and fruit weighing and packing machinery. Their products can be found all over the world. Like many other companies exporting machinery, the company was interested in the potential of step-by-step remote assistance, more specifically in assisting remote technicians in solving the most common malfunctions. On average 10 h a month were spent on telecom platforms solving these malfunctions. Remote support was seen as an opportunity for an additional service, while at the same time reducing the number of—often unpaid—calls for support during malfunctions.

Currently all knowledge on malfunctions was in the heads to two experienced technicians, therefore this knowledge first needed to be extracted and transformed into procedures. This already proved to be very valuable for the company, apart from the technology, as it reduced their vulnerability with respect to knowledge preservation. As further qualitative benefits were mentioned, an image improvement by showing technological advancement and better insight in malfunctions allowing them to optimize product design.

On the quantitative side, considering all costs from software, hardware and instructions building, and the benefits from incoming licensing fees and reduced un-accountable time on calls, there was a positive business case.

3.6   In Conclusion: Lessons Learnt and Future Developments?

The three tools we describe in this chapter suit the needs of SME companies. In applying them, we too have come to insights. These insights are on several levels: at the company side, on implementing technology, on our side concerning our methodology and tools and lastly, at the technology side, about the ins and outs of each technology. In this section we summarize and share our lessons learnt over recent years and conclude with an outlook on required future developments.

3.6.1   Implementing Cognitive Operator Support

Most companies are willing to embrace new technologies but face a number of challenges and implementation barriers. In general, company's lack one or more of the following: sufficient investment funding, knowledge of operator support technology, and skilled programmers to implement solutions on the shop floor. However, most importantly we have found many companies to be lagging with respect to documenting and adhering to standard procedures.

SME manufacturing companies thinking of operator support implementation quite often start off with no or a low level of digitalization. Standard procedures or work instructions for assembly tasks, servicing or maintenance are often missing, outdated or poorly documented. Most training is done by face-to-face on the job training. As a result, knowledge and information of procedures is bound to specific experienced operators on which a company relies. Formalization of procedures, however, is essential for making working instructions for operator support. If this is the case, companies have to make quite a big step from going without formal working instructions or procedures, to a system of well documented structured procedures that can be digitalized.

Operator support is only as good as the content it displays and having adequate digital working instruction is key. In literature some guidelines for designing assembly instructions are mentioned (e.g., [27]). These principles should be applied, but even then, we have learned from our pilots that creating clear, self-explainable and consistent work instructions, is an art in itself, relying as much on information about process steps as on insight in how people perceive and process information. Creating good instruction requires investing effort and time in the content. It requires digital CAD drawings, pictures, videos and text, and a step-by-step structure where information is clustered or separated. Additionally, we advise to have instructions reviewed by both an experienced and an inexperienced operator. For operator support in sheltered workplaces the quality and inclusive design of the working instructions is even more important as often illiteracy or low reading skills need to be considered.

There are large differences in the skills required to work with the technology and create work instructions. To suit the need of the company and to create and update work instructions we believe it is best not to have to rely on third parties. Our advice is to have a product owner (e.g., process engineer) within the company who is responsible for creating and updating the digital working instructions. This, however, poses a problem for many companies: the frequency with which the software is used may be relatively low, making it hard for a product owner or dedicated programmer to maintain sufficient skills and knowledge on using an application.

A standalone application of this kind of technology is usable in a pilot setting. However, to fully optimize the technology's use and to use it at its full potential, integration with other digitalization systems within the organization like ERP or MES is essential, so that changes in the bill of materials, bill of processes and CAD files are automatically transferred to work instructions. This enables work instruction configuration for product variants, user and competence management and it supports the use of existing manufacturing data.

3.6.2 Our Methodology and Tools

Through our experience with applying and evaluating innovative technology, and with collecting data and user experiences, all on the shop floor, we have gained a lot of insights and developed our methods and tools. Below we reflect on our findings thus far and give an outlook.

3.6.2.1 OS-Canvas

Our canvas has proven to be a valuable method to quickly access together with the companies all possible procedures that could benefit from operator support technology within their processes (e.g., assembly processes, service and maintenance and remote assistance) and the technology specific advantages or disadvantages given the contextual factors that are present. As a future development the OS canvas could entail more canvasses, each tailored to the different types of procedures i.e., assembly processes, service and maintenance and remote assistance. Although they all share the need to transfer information about the procedure to follow, their contexts differ largely and with this certain operator support requirements. Assemblies can be frequently recurring procedures at a fixed workstation or engineering machinery to order. A fixed workstation creates completely different options for operator support, than a machine on site at the customer requiring irregular servicing and regular maintenance. Remote support, in the spotlights due to COVID 19 travel restrictions, is even more different from a context point of view. Currently, context is the last aspect that is covered in the canvas workshop. It may be better to create different operator support canvasses, depending on the application situation, thus maybe making the canvas more concise or alternatively more detailed but relevant on all aspects.

3.6.2.2 Evaluating Usability

As was mentioned in a previous section, evidence-based usability research to assess the effect of cognitive operator support in a production environment, is generally very challenging due to the nature of production. There are many factors that can interfere with a scientist's need for experimental control. Availability of operators and products may vary during the planned pilot period for measuring data. Research is hindered by the lack of available operators (low scientific power), by long cycle times per product (e.g., 3 days), requiring long test periods, or by a short time window for piloting the operator support technology, forcing to either reduce the number of participants or the duration per participant. Further, company benchmark or baseline data may be unavailable or of poor quality.

Measuring and finding proof of a new technology's effects, is more difficult than testing its potential. Quantifiable effects of operator support technology on relevant manufacturing KPI's like cycle times and product quality can be measured relatively easily, but data is not always reliable, and disturbance can easily affect collected data. Effects on the operator, i.e., workload, acceptance, perceived usability, job satisfaction and job quality, are harder to establish without relying on self-reporting and questionnaires. Even more challenging is to measure subjective effects when dealing with operators in sheltered workplaces. They often have more difficulties to reflect on their own work and goals and others have trouble understanding the questions asked due to illiteracy or the semantic nature of questionnaires. This shouldn't lead to skipping the opinion of end-users to avoid risk of underutilizing the opportunity to support operators in a manner that will benefit them e.g., in terms of mental workload, usability and job quality, and thus benefit the company. We continue to develop questionnaires and methodologies which give insight into the human factors aspect of the implementation of OS technology within a company's production process.

Doing research within production requires extra resources from the company in staffing, equipment and sometimes materials, e.g., for building a construction to install a projector and time-of-flight camera or for additional quality control. It helps to clearly discuss these requirements with the use case company before the start of a pilot. This leads to a test plan that is more realistic and is more likely to result in the intended data quality and with this, in valid and reliable results.

3.6.2.3 Business Case

An important lesson we learnt from all business cases, is that they stretch further than the operator support software, hardware and workstation where it is applied. One company mentioned that, having inexperienced operators relatively quickly up to speed by the operator support technology, would allow them to reduce production batch sizes, maintain smaller stocks and be able to more quickly adapt to design changes from their customers. This would reduce waste and increase their attractiveness to customers.

In almost all business cases, implementing the operator support technology would require integration with existing production software. This is something we can currently only mention to our customers, but we are not yet able to quantify these costs in our tool. These costs strongly depend on the customer needs, the existing company systems and the properties and possibilities within the operator support system. On this last point there are large differences between technology providers.

Whether, or how, we should try to add these aspects to the tool, remains to be discussed. Trying to create an all-encompassing business case tool, could easily result in a tool that might be correct, but unworkable. A managing director is used to seeing opportunities and taking risks and does not necessarily require an exact and complete dataset. Having seen best practices in other companies, experienced a technology in action in their own company, learnt about involved aspects via the tool, both quantitative and qualitative, and having some approximate figures is often enough to decide on whether there is a business case or not.

3.6.3 Technology

Although the technology is aimed at supporting the operator at work, we find that the human centredness in their design is limited leading to sometimes poor usability. Especially head mounted equipment should be carefully evaluated to decide whether and to what extent (duration and frequency of use) the technology is applied. Due to comfort issue, combination with own prescribed glasses, weight and blockage of (peripheral) vision. Even if it may be beneficial to the production process, it may not be at the expense of the worker needing to wear the equipment.

In this respect, we have found that the added value of head mounted displays, either as near-eye displays or as AR/MR solution, to be limited compared to using a mobile phone or tablet. Only in highly mobile working tasks requiring the use of two hands, such as order picking, such applications have clear added value. But even then, pick by voice could be the more comfortable option. In complex manipulations, we expect the combination of video and AR on a mobile device to perform similarly or only slightly less than AR/MR technology such as Microsoft's HoloLens 2 or Magic Leap's model 1. From a performance point of view, this remains to be researched. However, considering wearability, ruggedness, price and required programming skills, the mobile phone and tablet, currently seem an attractive alternative. On the other hand, considering rapid technology evolution, these issues could quickly be resolved, putting AR/MR technology's distinct benefits in a stronger spotlight.

Similarly, investments for the large-scale roll-out of more complex technologies, currently surely affect a company's decision to implement a technology or not. Using work instruction software and outstanding work instructions on a simple tablet, could prove to have a better business case then any of the more complex technologies. Obviously, this balance also changes with every new version of a technology and with prices dropping rapidly.

3.6.4   Future Developments

In addition to the required and expected developments mentioned above, we would like to conclude with our view on the future of cognitive operator support in the manufacturing industry.

For the near future the gap between SME and technology providers needs to close. Accessible platforms may be easy to use but their functionality may be insufficient to meet a company's most important needs. Complex platforms, on the other hand, have many functionalities but require skills many manufacturing SMEs currently do not have.

A bit further away, we remain convinced that despite robotization, operators will remain to play a role on the shop floor. However, the setting in which they operate will continue to grow in complexity and required flexibility, because of artificial intelligence, collaboration with robots and all of the production system's elements connected and communicating. Cognitive operator support will be necessary to prevent the operator from getting lost in all complexity but also to be able to apply control to an extent that guarantees job satisfaction. For this the technology must have a strong human centred focus, as is advocated by Industry 5.0, leading to highly usable and wearable equipment and to systems being able to adapt to the user's preferences and level of expertise. Usability also applies to the ease with which engineers can create high quality work instructions. Developments with respect to easy creating of work instructions, are on the way: semi-automatic generation of instructions based on data such as CAD and BOM is being researched by several parties [28, 29]. Usability of such work instructions should receive equal attention but is by no means guaranteed. Human factors and ergonomics professionals need to remain alert.

Acknowledgments This work is funded by the Dutch Ministry for Economic Affairs, Smart Manufacturing Industriële Toepassing Zuid-Holland (SMITZH), Interreg VL-NL project Fabriek van de Toekomst (FOKUS) and RAAK SIA project Assemblage 4.0. We thank the people from the manufacturing companies for their collaboration and allowing us to present their use cases in this chapter.

References

1. Romero, D., Bernus, P., Noran, O., Stahre, J., & Fast-Berglund, Å. (2016). The operator 4.0: Human cyber-physical systems & adaptive automation towards human-automation symbiosis work systems. In I. Nääs et al. (Eds.), *Advances in production management systems. Initiatives for a sustainable world. APMS 2016. IFIP Advances in information and communication technology* (Vol. 488). Springer. https://doi.org/10.1007/978-3-319-51133-7_80
2. IFR. (2019). *World Robotics 2019 Industrial Robots*. International Federation of Robotics.
3. Longo, F., Padovano, A., & Umbrello, S. (2020). Value-oriented and ethical technology engineering in Industry 5.0: A human-centric perspective for the design of the factory of the future. *Applied Sciences, 10*(12), 4182.

4. Zolotová, I., Papcun, P., Kajáti, E., Miškuf, M., & Mocnej, J. (2020). Smart and cognitive solutions for operator 4.0: Laboratory H-CPPS case studies. *Computers & Industrial Engineering, 139*, 105471.

5. De Looze, M. P., Bosch, T., Krause, F., Stadler, K. S., & O'Sullivan, L. W. (2016). Exoskeletons for industrial application and their potential effects on physical work load. *Ergonomics, 59*(5), 671–681.

6. Werrlich, S., Eichstetter, E., Nitsche, K., & Notni, G. (2017). An overview of evaluations using augmented reality for assembly training tasks. *World Academy of Science, Engineering and Technology International Journal of Computer and Information Engineering, 11*, 10.

7. Bottani, E., & Vignali, G. (2019). Augmented reality technology in the manufacturing industry: A review of the last decade. *IISE Transactions, 51*(3), 284–310. https://doi.org/10.1080/24725854. 2018.1493244

8. De Souza Cardoso, L. F., Mariano, F. C. M. Q., & Zorzal, E. R. (2020). A survey of industrial augmented reality. *Computers & Industrial Engineering, 139*, 106159.

9. Egger, J., & Masood, T. (2020). Augmented reality in support of intelligent manufacturing: A systematic literature review. *Computers & Industrial Engineering, 140*, 106195.

10. Bal, M., Benders, J., Dhondt, S., & Vermeerbergen, L. (2021). Head-worn displays and job content: A systematic literature review. *Applied Ergonomics, 91*, 103285.

11. Palmarini, R., Erkoyuncu, J. A., Roy, R., & Torabmostaedi, H. (2018). A systematic review of augmented reality applications in maintenance. *Robotics and Computer-Integrated Manufacturing, 49*, 215–228.

12. Rapaccini, M., Saccani, N., Kowalkowski, C., Paiola, M., & Adrodegari, F. (2020). Navigating disruptive crises through service-led growth: The impact of COVID-19 on Italian manufacturing firms. *Industrial Marketing Management, 88*, 225–237.

13. Van Waveren, B. (2020). *Dutch Participation Act not (yet) a success, ESPN Flash Report 2020/ 02, European Social Policy Network (ESPN)*. European Commission.

14. Funk, M., Mayer, S., & Schmidt, A. (2015). Using in-situ projection to support cognitively impaired workers at the workplace. In *Proceedings of the 17th international ACM SIGACCESS conference on computers & accessibility* (pp. 185–192). ACM.

15. Bosch, T., van Rhijn, G., Krause, F., Könemann, R., Wilschut, E. S., & de Looze, M. (2020). Spatial augmented reality: A tool for operator guidance and training evaluated in five industrial case studies. In *Proceedings of the 13th ACM International Conference on PErvasive technologies related to assistive environments* (pp. 1–7). Association for Computing Machinery.

16. Wilschut, E. S., Bosch, T., Könemann, R., Van Rhijn, G. J., Hosseini, Z., & De Looze, M.P. (submitted). *The effect of work capacity of impaired workers on performance and perceived workload in assembly training with augmented reality support*.

17. Kranenborg, D. K., de Looze, M., Wilschut, E., Cremers, A., & Hazelzet, A. (2021). Inclusieve technologie voor mensen met een psychosociale arbeidsbeperking. *Sociaal Bestek, 83*(83), v.

18. KvK. (2019). *Digitalisering in het MKB. Rapportage van onderzoek uitgevoerd in het KVK Ondernemerspanel, maart 2019*.

19. Van Rhijn, J. W., De Looze, M. P., Tuinzaad, G. H., Groenesteijn, L., De Groot, M. D., & Vink, P. (2005). Changing from batch to flow assembly in the production of emergency lighting devices. *International Journal of Production Research, 43*(17), 3687–3701.

20. Westerfield, G., Mitrovic, A., & Billinghurst, M. (2015). Intelligent augmented reality training for motherboard assembly. *International Journal of Artificial Intelligence in Education, 25*(1), 157–172.

21. Brooke, J. (1996). Sus: *A quick and dirty usability. Usability evaluation in industry*, p. 189.

22. Bangor, A., Kortum, P. T., & Miller, J. T. (2008). An empirical evaluation of the system usability scale. *International Journal of Human–Computer Interaction, 24*(6), 574–594.

23. Abeele, V. V., & Zaman, B. (2009). Laddering the user experience. In *User experience evaluation methods in product development (UXEM'09)-workshop*.
24. Koningsveld, E. A. P., Bronkhorst, R. E., & Overbosch, H. (2003). Costs and benefits of design for all. In *Proceedings of the 15th Triennial Congress of the International Ergonomics Association and the 7th Joint Conference of Ergonomics Society of Korea/Japan Ergonomics Society*.
25. Koningsveld, E. A. P., Dul, J., Van Rhijn, G. W., & Vink, P. (2005). Enhancing the impact of ergonomics interventions. *Ergonomics, 48*(5), 559–580.
26. Van Rhee, H. J. (2019). What improves wellbeing and performance simultaneously? A study of measures taken by SMEs. In *10th international scientific conference on the prevention of work-related musculoskeletal disorders*.
27. Haug, A. (2015). Work instruction quality in industrial management. *International Journal of Industrial Ergonomics, 50*, 170–177.
28. Claeys, A., Hoedt, S., Schamp, M., Van De Ginste, L., Verpoorten, G., Aghezzaf, E. H., & Cottyn, J. (2019). Intelligent authoring and management system for assembly instructions. *Procedia Manufacturing, 39*, 1921–1928.
29. Neb, A., Schoenhof, R., & Briki, I. (2020). Automation potential analysis of assembly processes based on 3D product assembly models in CAD systems. *Procedia CIRP, 91*, 237–242.

Human-Centered Adaptive Assistance Systems for the Shop Floor

4

Hendrik Oestreich, Mario Heinz-Jakobs, Philip Sehr, and Sebastian Wrede

Abstract

Most of today's assistance systems for the shop floor do not consistently follow a human-centered design approach. An explicit modeling of concepts for instructions, users and adaptations is required when aiming for a holistic approach. Therefore, we first define a morphological box to capture design alternatives for adaptive functionalities. Afterwards, a reference architecture is presented, consisting of several building blocks that implement an adaptation loop to continuously adapt the provided assistance and the system behavior to the current user. A selection of algorithms to realize adaptive behavior concludes our contribution. Following a human-centered design approach, three exemplary scenarios for assistance on the shop floor demonstrate the application of our approach.

Keywords

Adaptive assistance systems · Human-centered computing · Interactive learning environments · Computer-managed instruction · Software engineering · Shop floor

H. Oestreich (✉) · S. Wrede
Research Institute for Cognition and Robotics (CoR-Lab), Bielefeld University, Bielefeld, Germany
e-mail: hoestreich@cor-lab.de; swrede@cor-lab.de

M. Heinz-Jakobs · P. Sehr
Institute Industrial IT (inIT), OWL University of Applied Sciences and Arts, Lemgo, Germany
e-mail: mario.heinz@th-owl.de; philip.sehr@th-owl.de

4.1   Introduction

Industrial assistance systems are designed to support employees in carrying out complex work processes by providing interactive step-by-step instructions. The systems guide the worker through a work process by presenting text, image, and video instructions for individual work steps. By interacting with the system, the worker can follow the steps until the process is completed. The systems thus particularly serve as cognitive support for workers to reduce their cognitive load and enable them to carry out more complex tasks [1]. In recent years, a wide variety of assistance systems for different fields of activity and user groups have been developed and evaluated. The considered fields of activity include industrial manufacturing, logistics, maintenance, training, and other related domains [2].

Besides regular inexperienced and experienced industrial workers, some systems further covered specific user groups like elderly workers, workers with a physical or mental disability, or workers with a migrant background who do not yet speak a required language [3]. Furthermore, the use of different stationary and mobile technologies such as computer displays, in-situ projections, tablet computers, head-mounted displays, and smartwatches has been investigated for the implementation of assistance systems [2, 3]. Figure 4.1 shows two examples for assistance systems based on in-situ projection and a head-mounted display.

To provide workers with an appropriate level of assistance, systems must be able to adapt to different situations [4]. In the field of human-computer interaction, a situation is defined by the situational context including details of the user, the task to accomplish, the available resources as well as the social and physical environment [5]. Thus, an adaptive system has to know with which user it is currently interacting, what the task is that should be accomplished, and which resources are available. The social environment might also play an important role, if a worker is part of a line and rather new, he/she might be slower and needs different support from an assistance system, than the other colleagues in the same assembly line. The physical environment describes aspects like lighting, acoustics, temperature, and furthermore. If this information is aggregated, it allows a system to detect different situations and to adapt appropriately. The design space of possible adaptations is broad: From adapting the (graphical) user interface, over switching modalities used for interaction, to manipulating the instructions that are shown.

While early assistance systems were only able to capture a rather limited amount of context information, rapid developments during the last years in sensor technology and intelligent algorithms for data processing have enabled assistance systems to collect and process detailed context information. Adaptive assistance systems for industrial applications nowadays already support different tasks, resources, and environments. The use of information on the individual user, on the other hand, has so far only been covered by few systems and there only to a limited extent. While this poses a central problem for the application of assistance systems, it offers an exciting field for research. A key aspect of future systems lies in the individual adaptation to different users. Using the adaptive functionalities in a human-centered way, personalized user experiences can be created

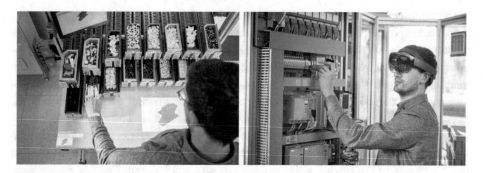

Fig. 4.1 Examples for industrial assistance systems based on in-situ projections (left side) and a head-mounted display (right side)

which are conducive to learning, support individual performance and even enhance the support on an ergonomic level when compared to current assistance systems. Previous studies and feedback from the practical application of assistance systems in industrial settings have shown that without adaptation, the provided assistance and interaction strategies often do not fit the individual experience and performance level of the user [4, 6]. Then, the users no longer benefit from the assistance after a certain training period. Sometimes the working speed is even limited by the system, due to the required interactions, which is counterproductive from an economical perspective.

Adapting the assistance to the experiences, skills, and needs of a user offers to provide individually tailored instructions, interaction strategies, and personalized interfaces. This can not only improve the performance of the user but also enhances the user experience and the acceptance of the user of the systems. This allows users with different levels of experience, from beginners over advanced to experts, to benefit from the use of an assistance system. For instance, beginners are offered comprehensive instructions and an interaction strategy that structures the work process, experts are offered more passive access to instructions and a tool to control the work process without the system restricting the performance. Another group that can especially benefit from this kind of adaptive assistance system are people with cognitive disabilities. The members of this group are characterized by comparatively high heterogeneity in terms of their abilities and needs. A significant proportion of this group also shows problems regarding memory.

Concerning the purpose of assistance systems, a distinction can also be made between an unlimited application for the permanent support of workers and a limited application for the learning or training of work processes. The permanent support requires an adaptation of the assistance to a level necessary for the respective user depending on his performance level and experience level. The support of learning and training processes, on the other hand, poses different demands, like for example a targeted step-by-step reduction of the assistance level to achieve structured learning of the work steps or alternating phases of support and no or limited support.

Our contribution to the research in this area is at first a description of the concept of human-centered adaptivity in the context of assistance systems. Using a morphological box we then explain which design decisions have to be made regarding the development of assistance systems by listing six dimensions and realization options in Sect. 4.2. Afterwards, we provide three exemplary scenarios in Sect. 4.3 to describe what individual adaptations would be chosen in each case and how the different dimensions would be parametrized to support the purpose of the adaptation. Subsequently, in Sect. 4.4, we present a reference architecture for human-centered, adaptive assistance systems and describe which components are needed, how modeling can support the adaptivity, and which methods to implement adaptive functionalities can be used. Finally, we conclude with a summary of our contribution and limitations of the current state of our work (Sect. 4.6).

4.2   Human-Centered Adaptivity

The topic of human-centered design of assistance systems has been addressed in various publications in recent years. Some guidelines for the design [7] and introduction [8] of assembly assistance systems were contrived to give developers and companies an idea of what has to be considered in the implementation of these cyber-physical systems in the workplace. Researchers explored the design possibilities for the assistance systems [9] and enhanced these to reach a certain maturity and thus implementation in various commercial systems. To successfully introduce these systems, the users should be included in the decision-making process [10] and the systems have to offer good usability and user experience [11]. But all these approaches reach a limit when they do not consider the integration of adaptivity to further enhance the interaction between human and the system. Therefore, we looked at the human-centered design process described by the ISO 9241-210:2019 and applied it for the development of adaptive assistance systems:

1. *Analysis:* Gather information about potential users, their tasks and goals, the work context and the technical preconditions.
2. *Definition of Requirements:* Prioritize features and adaptations needed in the assistance system for the desired use case. Not all features and adaptations are useful for every context. Therefore, it is important to choose a focus for the assistance, which kind of support is needed the most, which adaptations help to make it more successful?
3. *Prototyping and Implementation:* An iterative process is crucial to ensure the participation of the users in the process. Starting with prototypes, this approach helps to verify a common understanding and to steer the development process into the right direction.
4. *Evaluation:* Continuous evaluation of the artefact of the current development cycle ensures to test the usability and usefulness of the assistance system itself, but as well to investigate the effects of the adaptations. Are they helpful, do they work as expected or

do they interfere with each other. In a human-centered design process the users stay in the focus since they will be working with the product regularly in the future.

Considering the iterative design process with its four steps and using the morphological box presented and explained in the next two subsections, the development of adaptive assistance systems will be driven by a human-centered perspective and increase the chances of a successful implementation.

4.2.1 Design Space for Adaptable Human-Centered Assistance

Enhancing an assistance system with adaptive features seems quite complex in the first place. That is why we developed a systematic design approach for human-centered assistance systems which can guide the developers through the many questions that have to be asked and answered during the implementation process.

We identified several dimensions that are essential to cover most of the questions that will come up in the design process of the system. Afterwards, a morphological box was developed which structures the dimensions and associated implementation options (see Fig. 4.2). The morphological box was inspired by the model of Streicher and Smeddinck [12] which comes from the domain of serious games but which has similarities with adaptive assistance systems. Combining their model with the distinctions presented by Burgos et al. [13] and our previous work [14], we derived our approach on how adaptation design decisions can be made in a structured way during the development process of adaptive assistance systems.

A team of developers can start at the top of the box, choose one option per dimension and then discuss for the next row which option supports the previously chosen options the best. Be aware that sometimes aspects of more than one option are affected, but still the overall design process starts at the top row and ends at the bottom row with a resulting path through the alternatives of each dimension.

4.2.2 Dimensions of Adaptive Assistance

The following paragraphs will discuss the individual design dimensions of adaptive assistance and explain the options that we identified. Each dimension has an alphabetical letter as an identifier and the options are numbered from left to right to allow direct referencing.

4.2.2.1 Goal of the Adaptation

Since adaptive features tend to make systems more complex and error-prone, because they rely on situation detection and make the system behavior less predictable, the first question to be asked should always be: What is the goal that should be reached through adaptivity

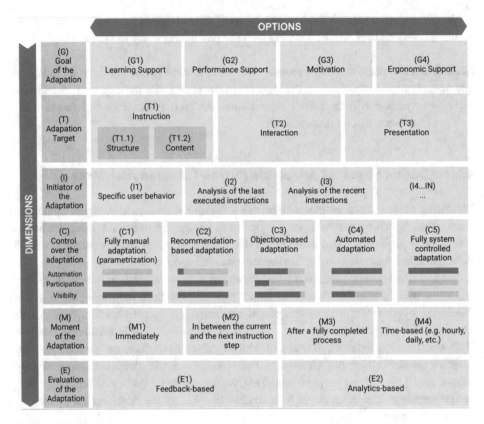

Fig. 4.2 Morphological box for adaptation design of assistance systems, each row and their cells are described in Sect. 4.2.2

and are there alternative actions which are less expensive? However, there are also lots of good reasons for adaptive features which should be explained in the following section.

Browne et al. [15] list different purposes for adaptive user interfaces: Extending a system's lifespan, widening of a system's user base, improving operational speed and accuracy, and also more user-specific: satisfying user wants, enabling user goals, and enhancing user understanding. Not all of these are relevant for the enhancement of assistance systems, but adaptivity on a system level offers even more possibilities than on the user interface level alone.

Similar to the e-learning domain, it seems that demands for more individualization and personalization are becoming more popular [16]. A one-size-fits-all approach does not lead to satisfying solutions when it comes to assistance on the shop floor: "The interface needs to adapt to the user's changing skills and requirements, and the assistance provided by the system needs to be relevant to the tasks the user is performing" [15].

Especially when looking at special user groups like people with disabilities and handicaps assistance systems offer a solution for more flexible and versatile work but demand individualized support depending on the users' capabilities.

Looking at the literature, we found four main purposes when it comes to adaptivity in assembly assistance systems:

Learning Support (G1) Assistance systems are often used to train new workers or to further qualify existing staff [17, 18]. The systems are used to train the assembly of specific products, special tasks or even to learn new skills. Sometimes the training is executed at specific workstations, but this is expensive and the setup of the workstation has to provide the required supplies, materials, and tools for the assemblies to be learned. The opposite approach of integrating working and learning tasks into the working process was called 'learnstruments' by McFarland et al. [19]. This concept offers the advantage of eliminating the transfer from a different learning environment to the workplace and decreases the need for additional workstations. Unfortunately, the concept was not explained from a technical perspective and remained a prototypical implementation.

Normally, if a worker learns to assemble a new product, he/she needs a longer time per product in the beginning and this decreases over time [20]. This is described by several learning curve models, which describe the performance progress over time that happens during learning [21]. Furthermore, the amount of support a worker needs also decreases over time. At first, very comprehensive instructions help to understand what the worker is supposed to do, when he/she is supposed to do it, and also why.

If the worker learns the assembly process over time, more details of the instructions and their sequence are memorized. For instance, the amount of information and the number of steps that have to be confirmed decreases over time, if the support is designed human-centered.

In procedural learning scenarios, a decrease in the perceived workload can be detected [22]. The findings from Letmathe et al. [23] show that a good performance can be established if understanding and skill development (learning through repetition) are combined.

Besides the learning curves, models for forgetting also exist and may play an important role in adaptive assistance systems. Depending on the time that has passed since the last assembly of a product, the instructions may be adapted for the next iteration. Wilschut et al. [24] investigated chunking of instructions for different learning support systems with the result that it should be avoided for novice workers since it increases the mental workload. Still, they advocate for more adaptivity since chunking makes sense for more efficient instruction for experienced workers.

The concept of flow is another argument for the need for adaptive instructions. Baker et al. [25] describe flow as the state between boredom and frustration which is most conducive for learning. To keep the operator in the flow state, the assistance system would have to monitor him/her closely to adapt the instructions to avoid overstraining and underchallenging.

The way people learn best can also be different and depending on a person's characteristics and preferences. Therefore, different learning modes might be desirable as Heinz et al. [26] presented. Smart workbenches can even be used to qualify workers beyond their normal task as [27] demonstrate. They train the workers to learn the MTM codes for their tasks while doing their normal work. This concept can be used to make normal work tasks more interesting and to enhance standard work with new challenges.

But how sustainable is the learning with assistance systems at all, especially compared to personal training? Büttner et al. [28] present a study where they investigated the long-term effects of learning with projected instructions. While their AR training could not beat the personal training regarding speed and recall after 24 h, they could show that the long-term effects do not differ between the different training methods. Hinrichsen and Bornewasser [29] emphasize the need for situational cognitive support the assistance systems have to deliver and point out, that the informational design also plays an important role. The informational design is also crucial for adaptive features since it allows to adapt the information to the user.

Performance Support (G2) Besides learning, the other main reason to deploy assistance systems on the shop floor is to support workers in flexible production with changing products and variants. While the business goals, in this case, are often to improve the product and process quality, from a human-centered perspective, cognitive support and an improved ergonomic support for the workers will be the main goals. If workers should be capable of assembling a wide range of products or many variants of a product, the need for support is different in the particular learning phases. Product variants have many similar parts in the assembly process and just differ in a small amount of the assembly specifics. Still, these differences are important to notice and assistance systems should deliver the right information at the right moment to let the workers pay special attention. Although the goals of improved efficiency, higher quality, and more produced units may be motivating factors for a company to use assistance systems, human-centered goals to deliver cognitive support and to decrease mental workload in an increasingly complex world should be regarded too.

Hold et al. [30] describe assistance systems as cyber-physical-systems and emphasize the importance of sensor integration into these systems. Advanced sensors can facilitate adaptive support by allowing the monitoring of the user and the assembly process. Picking sensors can observe if the right part was picked for the current assembly step. Image processing and gesture recognition can be used to automatically proceed in the instruction. But to provide more comfort and decrease manual interaction, these systems have to work very robustly and should not be too dependent on the surroundings and environmental factors.

The goals of human-centered adaptive performance support should be to provide assistance when needed but to stay in the background if workers are fully aware of the current task and its requirements.

Motivational Support (G3)   Another topic often addressed by adaptive solutions is the preservation and the increase of motivation. Monotonous work processes can lead to boredom and unmotivated staff. Adaptive assistance can help to counteract these effects. Dostert and Müller [31] propose to include motivational theories from the e-learning domain and gamification elements into assistance systems. This concerns the instruction design on the one hand but also the overall system design on the other hand. Schulz et al. [32] investigated the effects of branded gamification in assistance and report a direct impact on the individuals' attitude towards the task. Streicher et al. [12] also mention the Fogg-Behavior-Model which describes the relationship between motivation and ability and in which cases triggers (adaptations) succeed in changing behavior and in which they fail. Transferred to the domain of assistance systems, adaptations might not help to change a specific behavior of the worker, if he/she has a low ability level and a low motivation at the same time.

Ergonomic Support (G4)   Cognitive assistance systems can also improve ergonomic support for their users. This can be established through an adaptive choice of modalities that an assistance system uses to interact with its user. E.g. in assembly situations where workpieces do not fit onto an assembly workstation, comparable to maintenance scenarios of larger machines, the assistance system has to adapt to the current situation. As demonstrated by Josifovska et al. [33] in one situation a user might interact with a tablet as usual by looking at the graphical user interface on the screen and interacting through touch inputs. In another situation the user might need both hands and the system should be able to switch so speech output of the instructions and speech recognition as the input modality. Similar situations could be solved by using AR-Devices for a certain part of an instruction. A human-centered design demands a fluent experience when modalities are switched within an instructional process.

Another assistance system presented by Rönick et al. [34] adapts the height of the working table to the individual user to optimize the operator's working posture. Petzoldt et al. [35] also report about physically adaptive workstations with assistance systems, but criticize that the support in some systems only considers the situation before starting the actual work.

Also, much simpler adaptations such as font size and personalized positioning of information elements can improve the ergonomic support of a system for the individual worker. Other features like volume for speech output and display brightness for graphical user interfaces can also be adapted, but are often already built-in features of devices (like e.g. tablets) that are used.

Adaptation Target   After defining the different purposes of adaptivity in the previous section, now the different possibilities for adaptation targets should be discussed. We argue that these can be grouped into three main categories:

1. *Instruction (T1):* The first category can further be divided into two subcategories:
   (a) *Structure (T1.1):* The specific sequence of instructions is what we call structure. Manipulations can be used to condense the instructions, individualize their sequence or to introduce and remove special steps which should not be part of the long-term instructional model.
   (b) *Content (T1.2):* What is shown for each instructional step depends on the content and further configuration of the system. In some cases, it makes sense to add different kinds of meta-data for each step to satisfy individual learning types. An adaptive system is capable of choosing the right kind of content (e.g. text, images or animations), depending on the users' preferences and needs.
2. *Interaction (T2):* The way the system interacts with the user may also change over time or depending on the current situation. This includes changes in the used modalities of the system, but may also include where the system requires manual confirmation from the user.
3. *Presentation (T3)*: Adaptations on the presentation can include changes of display and font sizes, the layout and the language of the system.

Similarly, Taoum et al. [36] distinguish between five types of pedagogical actions which can be used to enhance procedural learning with an intelligent tutoring system. These actions include manipulation of the virtual environment (highlighting/animating objects), of the interaction (navigation by the system vs. by the user), of the structure of the system (displaying documentation), of the system dynamics (explaining the behavior), of the pedagogical scenario (making an evaluation). While these seem to be tailored towards a specific intelligent tutoring system, our categorization tries to generalize for a broader application.

Initiator of the Adaptation After identifying the main targets of the adaptations, suitable triggers that initiate an adaptation must be found. The main sources that initiate adaptations are either the situation detection and the changes in the user model. The first option example for this dimension (I1) describes specific user behavior that was recognized by the system. This might be for example that a user has picked a wrong part from one of the storage boxes, that he/she looked confused into a camera equipped with emotion recognition algorithms, or that the worker has skipped forward an instruction faster than expected. The next example (I2) describes analytics that take a longer period of actions into account and examines if trends can be discovered. A prominent example of this would be to look into the execution parameters of the last assemblies to see if these indicate a learning curve for the specific user. Comparing the gradient of the curve to the overall average can help to decide whether the support has to be increased or if it can be minimized. The third example (I3) describes the use of interaction data with the system. If a user always maximizes a picture to see a detail, it might be more comfortable for him/her if the picture is displayed larger by default. The dimension of initiators for an adaptation is huge and the most suitable initiator is always very specific for the purpose of the adaptation and further depends on the sensory equipment of the assistance system. Therefore, an arbitrary number of further initiators can be added to the options (I4...IN).

Control over the Adaptation To ensure human-centered adaptivity, it is important to determine the amount of control that should be given to the user. The spectrum of human control is defined by the three (sub-)dimensions of automation, (decision) participation, and visibility. If a system is capable to determine the need for an adaptation and the execution of it all by itself, the automation level is high. The participation dimension describes the amount of decision control the user has in the application of an adaptation.

Lastly, the visibility describes the transparency with which the adaptations are executed and how they are communicated to the user. The different possible manifestations are described in the following: The adaptive features can either be controlled completely by the user which is rather defined as a parametrization of a system (C1), or the control can be completely on the side of the system (C5). Between these two extremes, there can be many nuances.

In many scenarios it makes sense to use a recommendation-based adaptation approach (C2): The system identifies the need for an adaptation, informs the user, and lets him/her decide to accept or decline the adaptation. In other cases, an objection based approach (C3) might be suitable, where the user gets informed and can decline the adaptation in a certain amount of time and if he/she does not make use of the intervention, the system executes the adaptation. On the other end of the spectrum, the system takes control of the adaptations and might at least make it visible to the user (C4) or do it completely in the background (C5), without explicitly notifying the user at all. All cases have to ensure that the adaptations lead to an improved user experience because otherwise, the acceptance of the system might decrease dramatically. Negative feelings of confusion and paternalism should be avoided [37].

Moment of the Adaptation Next, a frequency with which the adaptations are allowed to happen should be defined. This has a huge impact on the user experience and acceptance of the adaptive system. If adaptations are too frequent, the user is distracted and may get annoyed. On the other hand, if adaptations are too infrequent, the moment of need is not matched and the system gives the impression to not work properly. While some adaptations should be triggered immediately (M1), because, for instance, they point out a mistake to the user, other adaptations can happen in between tasks (M2) or even after fully completed processes (M3). For some cases, it might also make sense to use time-based adaptations (M4) because, either the initiator information is not available and only time-based progress can be estimated, or, if analytic functionalities are rather complex, they might require a nightly adaptation computation.

Evaluation of the Adaptation Finally, an important step of adaptivity that should not be omitted is the evaluation of the adaptations. It can either be implemented based on user feedback (E1 - e.g. "How satisfied were you with the adaptation of the system") or it may be based on an analytic feedback loop (E2) which measures the performance progress after the adaptation and compares it to the state before. The feedback can be gathered explicitly or implicitly and the measurements should be integrated directly into the system itself. The analytic loop can compare the individual user data over time, compare current performance against average performance of other users or even against goal values such as MTM execution times [38] for a specific action.

As this section showed, several design choices must be made in the development process of adaptive assistance systems. The following section will motivate three exemplary scenarios and illustrate what concrete choices can look like.

4.3 Three Exemplary Scenarios for Adaptivity

Since the requirements and goals for adaptivity can be very versatile, this section should illustrate three scenarios in the context of adaptive assistance. The goals are different in each scenario and thus the chosen adaptation options are different for each dimension. Depending on the goal, the best fitting adaptation target to support the goal is chosen first, then the initiator and afterwards the degree of control that the user should have are chosen. Finally, the moment and frequency with which the adaptations should be triggered are selected and a suitable evaluation method is picked to close the adaptation loop.

Nelles et al. [39] argue that personas can be used to establish a better understanding of the situation and context. Thus, we use them to describe potential users who would be affected by the proposed adaptations. In this case, these are fictional personas and they roughly reflect users and use cases we experienced in previous industrial projects. While our personas are kept prototypical, the technique of personas can generally be grounded on questionnaires or structured interviews [39].

We visualized the chosen adaptation option for each persona in Fig. 4.3. The yellow path describes scenario 1, the red path scenario 2 and the blue one scenario 3.

4.3.1 Scenario 1

In this scenario (Table 4.1) a contract worker should learn a new assembly task. Since the colleagues and supervisors are busy and cannot help and the training of new workers should be realized in a more standardized procedure, the company decided to integrate assistance systems into the workstations on the shop floor. The main goal to be supported through adaptivity is learning support (G1) in this scenario.

There are two possible ways to design the support during the assembly training: (1) Active learning support: Here the assistance systems actively support the learning process. This can be done by providing additional background information for the

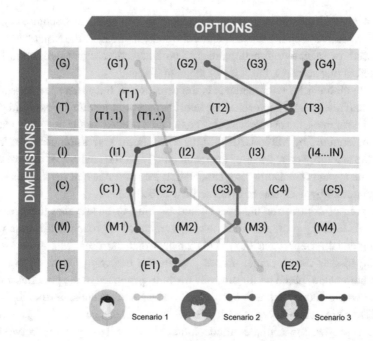

Fig. 4.3 Paths through the adaptation options (for an explanation of the dimensions and their options see Fig. 4.2)

Table 4.1 Persona for scenario 1

 First name: Adrian **Last name:** Hoffmann **Age:** 32 **Mother tongue:** German	**Character:** Adrian Hoffmann is employed as a contract worker in a company which manufactures electrical components. He is quite new to the factory, but he has assembly experience from other companies. Learning new procedures is satisfying for him since he is eager to learn and likes to figure out how things work.

assembly or by introducing quizzes in between repetition to determine the progress in learning. (2) Passive learning support: In passive learning support, the assistance system adapts to the current need for information according to the learn state of the worker. This is done by reducing the provided information as the performance of the user improves. The worker is motivated to memorize the instructions and to progress in learning the assembly.

Thus the main dimension targeted by the adaptivity will be the instruction (T1). As both ways (active/passive support) suggest, either the content (T1.2) or the structure (T1.1) may be addressed by adaptive functionalities. The interaction dimension might also be manipulated in this scenario, but it is not the main focus.

A key requirement for both methods is the detection of the current learner state of the worker. There are several approaches to detect progress in learning. Possibilities are the decrease of error rates, the difference between the time needed for the assembly and a time-threshold determined by MTM-UAS [38], or a non-linear regression of a learning-curve model to the assembly data. Most of the methods rely on the analysis of the last executed usages of the system (I2). During implementation the most relevant parameters should be chosen, supported by user tests to evaluate their effectiveness.

The degree of control for the user should be high to ensure that the user stays in control of his learning progress. Thus, the recommended approach in this case would be the recommendation approach (C2). This ensures that the user feels in control of the system and his/her own learning process. While the detection for an adaptation need should be automated, based on the analysis methods proposed in the previous paragraph, the worker would choose autonomously if he/she wants to accept an adaptation proposal. This leads to a high visibility of the adaptations from a user's perspective. The frequency should be kept low, for example an adaptation should only happen in between whole assemblies (M3). Active learning support that also points the user's attention to mistakes made in the assembly process might also justify more frequent interactions. Since we assume that the training happens based on a learning analytics approach, the evaluation can happen automatically in the background based on the gathered data of the executions (E2).

4.3.2 Scenario 2

In the second scenario (Table 4.2) the assistance system should support the worker during her daily assembly work. The worker has a certain degree of experience and she does not need instructions for every detailed step. The goal supported through adaptivity is performance support (G2). Adaptive functionalities of the assistance system could decrease the number of tasks which have to be confirmed through an explicit user interaction to a minimum.

We assume that the instruction has already been summarized as much as possible during previous assembly runs. Now the main targeted domain of adaptivity is the presentation (T3). If the system recognizes that Ute Graner executes all assembly steps within the targeted times (I2), the assistance system could offer to minimize the presented assembly instructions and only highlight individual steps which are different between the various variants of a product. By monitoring the assembly process through sensors and smart observation features, the company ensures that quality levels are kept high and the assembly is still documented through the system. Since the system can deliver concrete

Table 4.2 Persona for scenario 2

	Character:
	Ute Graner is employed in a company which manufactures household appliances. She works for many years in the factory, is very experienced and is paid by piecework. She is very fast but sometimes she is not concentrated and relies too much on her experience. Sometimes, when assembling new product variants with individual features she makes a few mistakes
First name: Ute **Last name:** Grauer **Age:** 44 **Mother tongue:** German	

information in case of individualized products and stays in the background during routine procedures, the worker feels safe to perform without errors. The goal is to reach a high acceptance of the system through offering good usability and only support the user when needed, not in every detail. We argue that the control about the adaptation should still be on the side of the user, in this case an objection-based approach could be chosen (C3). Still, the automation of the adaptations should be high, so the initiative to adapt will come from the system, based on intelligent analytic functionalities. To ensure that the user feels informed about the system's behavior, the visibility level of the performed adaptations should be low to high. The frequency is quite low in this case and adaptations might only triggered if the assembly times drop below a certain threshold and the adaptation may only be reversed, if the assembly times lay above the threshold for more than two or three assemblies (M3). Since the adaptations should explicitly increase the user acceptance of the system, we would choose a feedback based approach (E1), where the user can rate whether the adaptations were helpful and enhanced the system interaction.

4.3.3 Scenario 3

In this scenario (Table 4.3), the assistance system should support a worker with certain individual physical or cognitive conditions while carrying out an assembly process. This concerns the ergonomics goal of providing support that is adapted to the user's physical characteristics (G4).

Besides the experience of the worker, the system also needs to take into account the handedness and the visual limitation to adapt the assistance in order to support the worker in the best possible way. This should be done on the presentation and interaction level (T2&T3). With regard to the handedness, the system could adapt the arrangement of instructions and interaction elements to a layout optimized for left-handers (T3). The visual limitation could furthermore be addressed by selecting suitable visual representations or alternative output modalities (T2). Apart from physical or cognitive conditions, the

Table 4.3 Persona for scenario 3

	Character:
	Marius Waltz works as an assembly worker in a large industrial company. Due to a visual limitation, he is dependent on presentations with a high contrast ratio. As a left-handed person, he is also affected by the layout of the assembly stations, which is often designed for right-handed people.

First name: Marius
Last name: Waltz
Age: 28
Mother tongue: German

assistance could also be adapted based on personal preferences. Since most of the afore-mentioned features are more or less static, the system could simply provide a parametrization (C1) for the graphical user interface and might allow to switch input and output modalities, based on the devices that the production unit provides. Thus, the user initiates the adaptation and it is manually configured as a one-time adaption (M1). The personal preferences can be stored in the individual user model and if the company provides different production units with similar capabilities, these preferences are loaded as soon as the user logs into a system and experiences a personalized assistance for his work. These kind of adaptations are also best evaluated by manually asking the user to provide feedback (E1), whether the user experience has been enhanced through the adaptation.

4.4   Building Blocks for Adaptive Assistance

In this section, we present a reference architecture for assistance systems designed to provide the human-centered adaptivity described in Sect. 4.2. For this purpose, we first discuss how adaptivity is implemented in existing adaptive assistance systems and address the fundamental cyclical approach for the implementation of adaptivity in interactive systems. Afterwards, we describe the components of our proposed reference architecture and discuss the methods for its implementation.

4.4.1   Analysis of Existing Concepts and Implementations

Although we could find several approaches and some conceptual work about adaptive assistance in the literature, only very few publications provide details about their software architecture and even less describe the adaptation loop which is crucial for an ongoing personalization of an assistance system.

Related Work. The assembly assistance systems presented by Funk et al. [4] and Bächler et al. [40] provide instruction sequences with different levels of detail for distinct experience levels of the users. The classification of the users is done by analyzing the processing times and error rates from previous process executions. Through this, the level of detail of the instructions gradually decreases during the user's improvement. The system architecture presented by Aehnelt and Bader [41], uses a cognitive architecture to adapt the assistance. The system is able to adapt the instruction sequences in a self-organizing way based on contextual background knowledge and the detected situation. The architecture also describes a machine learning component that continuously adapts the contextual background knowledge to the recognized situations to optimize itself for future applications. The system architecture for an adaptive assistance system presented by Bannat et al. [42] features an adaptive process modeling to generate different instruction sequences. It also includes modeling of the user's cognitive state, work history, and preferences as well as a database with scientifically collected information on cognitive processes and behaviors during the execution of assembly tasks. The system selects the appropriate instructions based on the continuous recording of the cognitive state and the known user information. An adaptive assistance system presented by Rönick et al. [34] also uses information about the user's height and reaching distance to adjust the working height of the mounting surface and thus improve the workstation ergonomics. The system also allows the assistance content shown via a tablet PC to be adapted according to personal preferences. Galaske and Anderl [43] presented an approach for an adaptive work assistance system based on a model for individual worker profiles. The profile includes authentication information, competence information, and the user's physical characteristics usable to adapt the level of assistance. Besides the information stored in the profile, the worker is further able to manually adapt the level of assistance to a certain degree. Burggräf et al. [44] presented a virtual asset representation to enable adaptive assembly assistance. The representation includes a skill matrix to describe the level of expertise of a worker at a specific process. Furthermore, the system allows workers to adapt the level of assistance independent from their skill level. Gellert and Zamfirescu [45] presented an adaptive assistance system that is able to adapt assembly assistance via a context-based predictor. The predictor uses information on the workers' condition, characteristics, preferences, and behaviors during the assembly process.

While the systems described above already show approaches for the implementation of individual aspects and dimensions of the human-centered adaptivity described in Sect. 4.2, none of the systems offers a holistic solution and provides enough details about the implementation for reproducibility.

Adaptation Loop Adaptivity describes the ability of a system to continuously adapt itself to changing situations in order to increase its own efficiency with regard to achieving a certain goal. In the development of interactive systems, adaptivity can generally be seen as a cyclical process through which a system continuously adapts to changing conditions. Therefore, several approaches exist, a popular one is called MAPE-K [46]. While that approach divides the process into four phases called monitor, analyze, plan and execute, a similar model from Streicher and Smeddinck [12] labels the four successive phases slightly different, but describes the same cyclical process:

1. *Capture:* In the first phase, the system collects information about the current state of itself, the user, and the environment. This information can either be collected via external devices and sensors or is available to the system in the form of performance data and status information from previous cycles.
2. *Analyze:* In the second phase, the system analyses the acquired information from phase one to determine the context of the system. To do this, the system analyses the acquired and the stored information on the reachable states of the system to identify the situation.
3. *Select:* In the third phase, the system selects a suitable adaptation based on the situation identified in phase two. For this purpose, the system has one or more adaptation strategies, which define rules for the selection of the appropriate adaptation. At the same time, the system needs one or more adaptive features that can be adapted based on the strategy.
4. *Apply:* In the fourth phase, the system adapts itself according to the adaptation selected in phase three. This involves changing the settings of the dimensions affected in the selected adaptation.

These four phases form a continuous loop that provides the adaptive functionalities for an assistance system. They are realized through the implementation of several software components, the building blocks, which constitute our reference architecture presented in the following section.

4.4.2 The Reference Architecture

Based on the design of existing adaptive assistance systems addressed in Sect. 4.4.1 and taking into account the cyclical approach defined in the paragraph, we want to introduce the reference architecture for the implementation of human-centered adaptive assistance systems (see Fig. 4.4).

The system architecture consists of seven components: Input Modalities, Situation Detection, Decision Making, Adaptation, Execution, Output Modalities, and Knowledge Base. Besides the components of the assistance system, Fig. 4.4 also features a production unit that provides the user with the relevant parts and tools for the supported assembly processes. The knowledge base of the assistance architecture is also connected to a central enterprise data management system (right side) which provides the assistance system with resources on the users, the processes, and the production environment via various subsystems.

The cyclical adaptation process described in Sect. 4.4.1 is distributed over various components of the assistance architecture. Phase 1 and phase 2 for capturing and analyzing

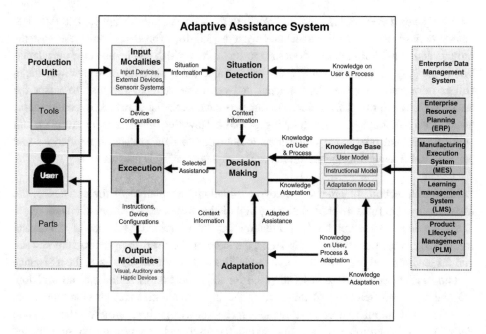

Fig. 4.4 Overview of the reference architecture for human-centered adaptive assistance systems. The boxes visualize the key components and the information flow is described by the arrows in between

the situation are carried out by the input modalities and the situation detection component. The selection of the adaptations in phase 3 takes place in the adaptation component. The application of the adaptations in phase 4 is finally implemented by the execution component and the output modalities.

In the following sections, we follow the adaptation cycle, describe the roles of the individual components of the reference architecture, and provide options for their implementation.

Input Modalities In order to correctly evaluate the current situation and provide adaptive assistance, process information needs to be gathered. Input modalities are systems that collect information during the assembly process. This primarily includes input devices that allow the user to navigate through the assembly process. In this context, assistance systems widely use camera systems or specific sensor devices in combination with motion detection and hand-tracking to collect data on the user's behavior with a high level of detail [4, 9, 47]. In this matter, however, the quality of the underlying data is very important. The correctness of the detected behavior can depend on the data quality of the device, the surrounding area, and the applied method for the motion detection. Another approach lies in the use of less complex sensors, like haptic buttons or pedals. These sensor systems provide less complex information but are more robust to the individual surrounding. In addition to the primary input devices, it is also possible to connect digital tools like electrical screwdrivers or scanners and test systems such as scales or optical and technical quality control devices to the assistance system [35, 48]. The data from these systems provide additional information about the work progress which can be used for the adaptation of the system. Previous studies also investigated the application of sensor systems to collect information on the physiological and cognitive state of the user to adapt the assistance accordingly. This includes parameters such as the heart rate, skin conductance, electroencephalography, or eye-tracking [49–51]. However, these sensors are often not applicable in real production environments both due to data protection regulations and personal rights as well as the sensitivity and complexity of their hardware.

Situation Detection In order to achieve a situational adaptation of the assistance, the assistance system must be able to identify and distinguish between different situations.

At the implementation level, a situation can be described by a set of contextual information, which contains details about the state of the user, the process, the system, and the environment [5]. With the input modalities and the knowledge base, the assistance system has two data sources at its disposal for the collection of situational information. Depending on the selection of the devices the input modalities can provide situational information about all four context elements. The knowledge base provides the situation detection with additional information about the user and the process to support the identification of the situation. This includes the user's process history, which provides run-time information such as processing times and error rates, as well as information about the current process step and the process sequence.

The situation detection component provides a logic that processes and analyzes the information from the two data sources to extract suitable context information which is then transferred into a uniform structure that can be used for further processing and reasoning. Depending on the amount of information included in the analysis and the applied method, the component is capable of extracting complex representations of the situation.

In a first step, the situation detection logic needs to translate the raw data provided by the input modalities into information that can be used by the other components of the assistance system. This includes processes like the translation of captured hand positions from a hand tracking algorithm into events for accessing component boxes or virtual buttons or the interpretation of data provided by external devices and environmental sensors. In this context, some existing assistance systems also use sensor fusion approaches to combine the raw data from different sensors to increase the accuracy of the captured data [42, 52, 53].

Subsequently, the logic needs to analyze which effects the extracted information has on the current process state, taking into account the information about the required activities and goals of the current step. Activation of a haptic or virtual button, for example, can have different meanings in distinct process steps such as confirming a performed assembly or a solved error. The results of the analysis are then stored in a data structure usable for the other components of the architecture.

Regarding the design of the data structure, Wolf et al. [54] introduce a context model based on object-oriented entities for real-world objects.

The situation detection of the assistance system by Aehnelt and Bader [55] uses a cognitive architecture in conjunction with a stochastic approach based on Hidden Markov Models to determine the current state of the process. The input modalities are defined as operators that can be used directly by the cognitive architecture. Finally, the data structure holding the context information is forwarded to the decision-making component to select the assistance information to be presented in the next step.

Decision Making Based on the identified situation provided by the situation detection, the assistance system needs to determine a suitable selection of assistance to be presented to the user in the next step. The decision-making component needs to provide a logic that selects the appropriate assistance information for the given situation based on the contextual information and the information from the knowledge base. In a first step, the logic needs to analyze the context information to decide whether the situation at hand represents a legitimate state concerning the defined process sequence. If this is not the case, the logic has to initiate a suitable error handling and provide the user with the necessary assistance to resolve the error. In the case of a legitimate state, on the other hand, the logic needs to initiate a decision process that determines an appropriate selection of the assistance for the next step in the process sequence. This decision process incorporates information about the context as well as information about the user and the process from the knowledge base.

The human-centred adaptation of the assistance is achieved through an exchange between the decision-making component and the adaptation component. During this exchange, the decision-making component forwards the contextual information to the adaptation component, which performs the corresponding adaptation. The result is then passed back to the decision-making component to be included in the decision-making process. After the appropriate assistance has been selected by the decision-making process, it is transferred to the execution component to be presented to the user.

There are various approaches for implementing the decision-making logic including rule-based algorithms, cognitive architectures [56, 57] and solutions based on artificial intelligence methods [58]. Rule-based approaches thereby compare the existing context information based on predefined rules or threshold values. They are thus primarily suitable for context information with limited complexity. For more complex context information, on the other hand, cognitive or AI-based approaches are more suitable. Cognitive architectures are also suitable for more demanding scenarios which go beyond rather simple process-based assistance.

Adaptation The adaptation component represents the part of the assistance system that is responsible for the implementation of human-centered adaptivity. As described in Sect. 4.2, four goals can be identified for human-centered adaptivity: Learning Support, Performance Support, Motivation, and Ergonomic Support. To support these goals, the adaptation component must provide appropriate procedures for adapting the assistance. In

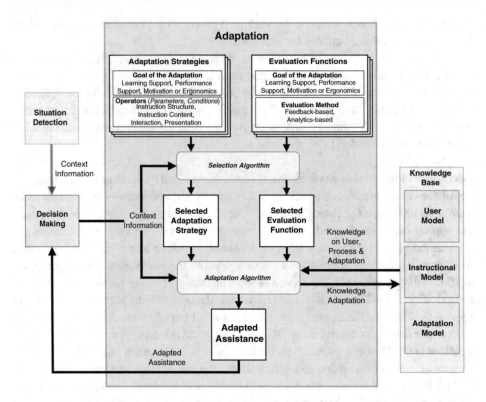

Fig. 4.5 Overview of the adaptation component. The boxes at the top show collections of applicable methods. The boxes with rounded corners use these as inputs to generate an output for further processing. The arrows visualize the information flow

addition, the component must provide methods for evaluating the success of the adaptations concerning the pursued goal to achieve a goal-oriented adaptation of the assistance. Figure 4.5 shows an overview of the adaptation component and its connections to the other components of the reference architecture. Besides the context information it receives from the decision-making component, it further has access to the information of the user model, the instructional model, and the adaptation model stored in the knowledge base.

The procedures for adapting the assistance are represented by a set of adaptation strategies. In the same way, the methods for the evaluation are expressed by a set of evaluation functions. The adaptation of the assistance is carried out in two steps: First, the adaptation component uses a selection algorithm to determine a suitable adaptation strategy and the associated evaluation function. In a second step, the selected strategy and evaluation function are used by an adaptation algorithm to adapt the assistance based on the available context information and the stored knowledge.

Adaptation Strategies Adaptation strategies define the procedure for adapting the assistance to pursue one of the specified goals of human-centered adaptivity. They determine how and in what time frame the adaptation of the assistance will be carried out and which information will influence this determination. They also specify which adaptation targets are influenced through the adaptation. Therefore, they contain a set of operators that are responsible for the adaptation of specific aspects of the different adaptation targets (Instruction Structure, Instruction Content, Interaction and Presentation). For example, to adapt the structure of an instruction to be more conducive to learning, an adaptive assistance system needs operators which modify the structure of the instructions. Oestreich et al. [59] present a list of operators to modify the structural design of the instruction. Multiple tasks can be composed together, complex tasks can be split up into smaller tasks with less information to be processed at once and sequences of tasks might be repeated to improve memorizing of a certain section of the overall assembly process.

Other operators for instance would modify the current user interface configuration or select a suitable instruction content, depending on the strategy and current evaluation results. The operators depend on certain parameters and conditions that affect the form in which the adaptation is carried out. The parameters of the operators are provided by the adaptation model in the knowledge base. The conditions, on the other hand, are composed of the contextual information and the information about the user and the process from the knowledge base. They control the application of the adaptation to not violate surrounding constraints.

There are numerous possibilities for the implementation of adaptation strategies. The following examples are therefore intended to provide a brief overview of the possibilities of what potential strategies for the adaptation of the assistance might look like. For the performance support, for example, a comparatively simple strategy as presented by Funk et al. [4] and Bächler et al. [40] could be used to gradually fade out elements of the displayed instructions with increasing experience measured via processing times and error rates. In this way, the level of assistance reflects the experience level of the user, who is not distracted by information that has already been memorized. A more comprehensive strategy for a learning support scenario, on the other hand, could extend the process sequence by additional knowledge, quizzes, or tests oriented to the user's learning progress [60]. A strategy for improving the ergonomic support of the assistance system, as described by Rönick et al. [34], could be to adjust elements in the work environment such as the height of the work table based on the user's body measures stored in the knowledge base or determined using a sensor system. Strategies to increase the motivation of the user could further use motivational theories and approaches from gamification to make the process more enjoyable as proposed by Dostert and Müller [31].

Evaluation Functions The evaluation functions serve to evaluate the success of an applied adaptation of the assistance concerning the purpose of a certain goal. This allows the assistance system to determine whether the applied assistance can be maintained or whether it needs to be re-adapted. Each goal thereby requires its own evaluation function, since the respective success is represented by different target parameters and different conditions are included in the evaluations. As described in Sect. 4.2, the evaluation of the adaptation can either be performed based on explicit user feedback or analytics. In a feedback-based evaluation, for example, the user could be asked to rate the last adaptation on a Likert scale after completing a process. An analysis-based method, on the other hand, could examine performance-oriented parameters such as processing times and error rates or parameters about the user's physical and mental workload to assess the success of the adaptation.

Selection Algorithm To apply the human-centered adaptivity, the adaptation component must first select a suitable adaptation strategy and an associated evaluation function. The selection is carried out by a selection algorithm that analyzes the available contextual information to identify the underlying goal of the adaptation. For example, when the user starts a process being in a learning mode, the selection algorithm will choose a strategy and an evaluation function for the learning support. After the selection, the adaptation strategy and evaluation function are forwarded to the adaptation algorithm.

Adaptation Algorithm Once an adaptation strategy and an evaluation function have been chosen via the selection algorithm, they are used by an adaptation algorithm to carry out the adaptation of the assistance. The algorithm implements a cyclical adaptation process that adapts the assistance according to the situation at hand.

In the first cycle, the selected adaptation strategy is applied to determine an initial adaptation of the assistance which is then provided to the decision-making component for its execution. The strategy thereby uses either available context information or information about the user and the process stored in the knowledge base and the corresponding parameters from the adaptation model. At each following cycle, the adaptation algorithm first applies the selected evaluation function to evaluate whether the last adaptation was successful regarding the goal of the adaptation. If a further adaptation of the assistance is required, the adaptation process applies the selected adaptation strategy.

Afterwards, the adapted assistance is communicated to the decision-making component for its execution and the models inside the knowledge base are updated to hold the information extracted from the evaluation and to persist personalized adaptations of the user interface or the instruction.

Execution After a suitable assistance has been selected by the decision making component (see Fig. 4.4), it has to be executed to present new instructions to the user and to adjust the interactions and other adaptive elements of the assistance system. For this purpose, the assistance information has to be forwarded to the corresponding output modalities and input modalities may have to be reconfigured.

In this context, the execution component serves as a central interface for the modalities.

It adapts the assistance according to the characteristics and features of the existing modalities and forwards the information to the appropriate devices. For example, a stationary display usually has a smaller area for displaying instructions compared to in-situ projections. Consequently, the arrangements of the displays need to be adapted differently. Furthermore, compared to in-situ projections, stationary displays do not offer the possibility of spatially overlaying the working environment. Corresponding information must therefore be presented differently on the display.

Output Modalities To communicate the assistance to the user, the assistance system needs suitable output channels to match the individual preferences and needs of the user. In this context, output modalities are devices through which assistance information and feedback information can be communicated to the user. Assistance systems primarily rely on visual presentations via stationary displays, tablet PCs, smartwatches, head-mounted displays or in-situ projections [9, 61, 62]. In recent years, assistance systems thereby have increasingly relied on augmented reality solutions to spatially overlay the working environment [63, 64]. However, apart from visual devices, assistance systems may also use auditory devices in the form of speakers or haptic-tactile devices such as vibrating wristbands or gloves to present instructions and feedback information to the user [65, 66].

4.4.3   Knowledge Base

The knowledge base comprises three data structures which contain the required information about the worker (user model), the instructions (instruction model), and the adaptation strategies (adaptation model). The e-learning domain uses similar models, which are named slightly different. They have a learner model (user model), a domain model (instruction model) and a tutor/pedagogic model (adaptation model) [67]. The following sections will describe which parts have to be considered for the implementation of the models.

User Model A foundation crucial for human-centered adaptivity is a dynamic user model. Adaptations might either be initiated by the model itself or by external analytic components which observe the model and all applied changes.

Recently, the concept of a digital twin is being adopted for this, and we see rising use of the term digital twin of the human [33, 68–71]. This term emphasizes the bidirectional information flow between the user and the model: On the one hand, the model is updated through monitoring the user of a system and interpreting his/her input, while on the other hand, the digital twin adapts the system (behavior) proactively and thus has an influence on the user.

Still, the central question remains: What has to be part of the user model? Unfortunately, the unsatisfying answer seems to be: It depends. Based on the adaptations that should be executed, triggers that are suitable to support these must be chosen and this is in most times context-specific. The literature provides a wide range of possible parameters for a user model. Although no holistic definition of the digital twin of the human can be found yet, the literature provides the following examples: Gräßler and Pöhler [68] include skills, motivation, character, and other factors in their model. Josifovska et al. [33] add demographic, social, psychological, and cognitive characteristics to the digital twin of the user, as well as physical interaction and health data. Another very sophisticated user model can be found in Schwarz et al. [72]. They distinguish between individual factors (short and long term), physiological, and behavioral reactions in the user state and thus emphasize the need for real-time capabilities of the user model. Galaske and Anderl [43] furthermore propose a concept which combines weighted parameters to an overall privilege value that determines the access rights for the user in the instructions. Based on the privilege level a user might be allowed to skip the instruction steps, decrease the level of detail in an instruction, or might even freely decide on the usage of the system.

While some of these features of a user model are necessary for adapting the user interface and used modalities, other features become important when supporting learning with assistance systems. We need to know the characteristics a learner provides when he/she starts working with the system. This may include available skills, learning preferences, cognitive abilities, and furthermore. In the interaction with the system further data gets important: The historical data from previously executed processes and assemblies, as well as interaction data with the system. Extended with intelligent monitoring and analytic functionalities, this builds the foundation for adaptive assistance.

When implementing and using a user model, the evolution of the user model has to be regarded as well. If an adaptive system requires a complete user model, with an extensive interaction history, the system will fail if a new user starts working with the system. Therefore, the developers have to take care of the initialization of a user model. This can be realized through specific questionnaires which have to be answered before first use of a system or the adaptive features explicitly take into account that a user model will not always provide all desirable features.

A user model should never be regarded as a static component where the underlying data model is fixed. Additional adaptive functionalities might be integrated into future versions and thus the user model has to be extendable and scalable.

In practical use further questions have to be regarded: Where is the user model stored? Who has access to which parts of the user model? Can the user manipulate his/her own user model? Is the user model transferable between different workstations with different requirements?

Instructional Model The instructional model can be seen as a road map, which is used to guide the user through the assembly process. In regard to this metaphor, adaptivity means alternate routes on the road map, which implies a more comprehensive instructional model. However, the structure of the instructional model is often influenced and limited by the structure of the model of the assembly sequence.

The creation process of the instructions varies from system to system and from company to company. One approach, which is also often used for not yet digitized instructions, is the manual one: The instructions are created manually with or without dedicated tool support and the information is gathered by the designer from different sources. If an assistance system is used, the system normally also provides tools for the creation of the instructional material. Vertical integration into the IT infrastructure of a company allows simplified access to the information needed for an instruction. Some 3D construction tools offer export functionalities to produce instructional material. Other examples of vertical integration are access to material lists for a product or to assembly sequences, modeled and managed by MES or ERP systems. Another approach for the creation of assembly instructions, explored by research, is the (semi-) automatic detection of assembly steps by demonstration. A worker or a foreman uses the assistance system itself to produce the instructional material at the workplace, by interactively creating the instructions for each step, or more advanced systems try to create the instruction based on video processing from observing a worker during an assembly [73].

No matter how the instruction was created, its underlying model is crucial for adaptive features. Figure 4.6 illustrates different dimensions of the instructional model and lists corresponding possible implementations for adaptive assembly assistance. While the fulfillment of the first two dimensions (1, 2) already allows to instruct and to support a worker at a basic level, the remaining four dimensions (3, 4, 5, 6) enhance the user experience and allow real learning support and even motivational aspects in future adaptive assistance systems. The following paragraphs will elaborate on the individual dimensions in detail.

1. *Assembly Sequence.* Fundamentally, an assembly process can be seen as a series of operations, in which components are handled, manipulated, and mated to form a finished product. In the simplest case, an assembly process is a linear sequence, where the order of the operations is implied by the physical structure of the finished product. More complex assembly processes however, often allow multiple ways to sequence the individual operations to finish the product, which cannot be modelled using a simple sequence. In these cases, the relationships between parts are often represented in a graph-structures [74, 75] or with formal task models [76]. The nodes of the graph hold a unique representation of the state of the assembly, while the edges of the graph represent the transitions between the assembly steps. While these models can capture the fundamental sequencing and conditional logic, their richness in regards of semantics, hierarchical organization, and most important executability is limited. One solution for this is the use of a graphical modeling language, which is normally used to describe business processes, the "Business Process Model and Notation.[1] Its semantics cover most requirements which a designer of an instruction is faced with, like conditional sequence flows, parallel execution, calling other process models and further more. As the underlying foundation for the whole instruction, a manipulation of the modeled assembly

---

[1] Business process modeling and notation in version 2, is a graphical modeling language which creates executable models, standardized by the Object Management Group (OMG) https://www.omg.org/spec/BPMN/2.0/About-BPMN

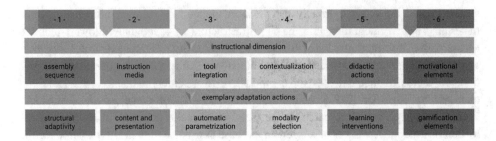

Fig. 4.6 Dimensions of an instructional model for assembly and the adaptive counterparts

sequence is a first possibility for implementing adaptivity. An example of how the structural adaptivity can be realized is explained in the context of the adaptation strategies in Sect. 4.4.2.

2. *Instruction Media.* Instruction media illustrates the actions, which need to be executed in an assembly step. Well-designed instructions allow an intuitive understanding of what needs to be done and how [77], while badly designed ones promote confusion and assembly errors. However, different types of instruction require different amount of information processing by the worker to convert the instructions into actions. For workers with lower experience levels, instructions with a lower information density are often more suitable, like for example video sequences. Here workers, can mimic the actions taken in the video. Images, which show the result of the assembly step take less effort for information processing but have a higher information density and thus are more suitable for workers with more experience. A model-driven implementation ensures that the instruction media has a clear structure which describes which content is referred to for each task and may even include information about adaptation options. The latter is part of what we call adaptivity by design. During design time, possible adaptive functionalities are modeled. With their adaptive learning objects Apoki et al. [78] give an example for that. Since it increases the modeling effort for the instruction designer, this can be seen critical and for this chapter, we focus on adaptivity during run-time. However, the right choice of instruction media for the individual user is another step for adaptive and personalized instructions.

3. *Tool Integration.* In many cases, various tools are needed to complete an assembly task. This ranges from simple hand tools to more complicated external devices with digital interfaces. These interfaces can be used to either acquire information about the assembly process and use this information as a trigger for further actions, or to configure devices based on the individual needs of the worker. The integration of these tools into the assistance system can enhance the support of the worker if the parametrization is automated and the user can focus on the task at hand and furthermore it is ensured, that the right parameters were used for the task at hand.

4. *Contextualization.* A further enhancement of the instructions can be that they are adapted to the current context and situation that the user is in. This could mean that

an in-situ projection is used where the information is not only displayed in the field of view of the current user but where it is projected directly onto the work piece. AR glasses are another example of localized instructional content if the instruction is connected to objects in the real world. For both use cases, the location to which a piece of information belongs must be persisted in the instructional model.

5. *Didactic Actions.* Learning support is one of the major application areas of assembly assistance systems. This means the didactic design of assembly instructions is of great importance for this application [79]. While assembly instructions for consumers are often rather intended for a one time use without the goal to learn the process, the requirements for an assistance system are different. In most cases the goal should be to train the worker to be able to execute the assembly without the assistance after a certain time. Still, other requirements justify the use of the assistance system beyond the initial learning phase. As part of the didactic concept, it might be suitable to add further work steps and integrate them into the workflow. These steps, sometimes also called learning interventions, can then be used to include questionnaires and quizzes into the workflow. To be able to integrate these interventions dynamically into the execution, they must be part of the instructional model.

6. *Motivational Elements.* If the assistance system should also implement motivational enhancements, these might also be grounded in the instructional model. For example, the annotation of experience points for each task are conceivable. If the user interface implements gamification elements, these experience points can then be integrated into the calculation of the overall experience and success factors.

Whenever a product is changing, it is necessary to also change the corresponding instructional model. Given the number of products in a frequently changing production environment, this can result in a high effort for maintenance. Therefore, the instructional model should also support versioning and the management of older versions of the instruction.

Adaptation Model The adaptation model represents a data structure that provides predefined parameters for the configuration of the adaptation strategies implemented by the adaptation component. The parameters are used to configure the operators that are addressed by the adaptation strategies. Since the strategies have different goals and combinations of operators, the adaptation model must provide at least one corresponding set of parameters for each strategy.

Parameter values can be found by conducting an adaptation design planning supported by the morphological box (see Fig. 4.2). Parameters describe for instance, whether the adaptation should be controlled by the system or by the user, whether it should be made visible explicitly and when/how often it should be applied. Depending on the concrete adaptation operators, further parameters can be suitable, for example a parameter controlling the degree of adaptation which should be executed in one adaptation cycle.

By providing multiple sets of parameters with distinct parameter settings for each strategy, the adaptation model could cover different contexts. This would allow the adaptation component to choose between different parameter sets to find the best settings for a situation. For this purpose, the adaptation model could store additional information on the success of the parameter sets at specific contexts, provided by the evaluation function of the adaptation component. This information could then be used to select the best fitting parameter set.

4.5   Algorithms for Adaptive Behavior

The adaptive behavior of an assistance system is realized by selecting and combining suitable methods for implementing the adaptation logic inside the components of the architecture. The cyclic process of adaptation, mentioned in Sect. 4.4.1, especially steps two and three, are facilitated using different methods of modeling and data processing. This section will present an overview of the methods which can be used to implement adaptive assembly assistance and discuss their advantages and disadvantages in terms of the effort required for implementation and the flexibility of their application.

When analyzing data from an assembly process, the first issue is finding situations that require an adaptation of the system, i.e. identifying adaptation triggers. The first possible trigger for an adaptation occurs when the user logs into the assistance system. In this case, the system is restricted to the information on the individual experiences, skills, and needs of the user specified in the user model. During the assembly, process data can be collected and analyzed to identify more adaptation triggers. In many cases, these are accompanied by changes in the process data, e.g. competition times or error rates. A change in the data can indicate a change in the performance of the user, which the system can then adapt to. However, also non-changing data can be a trigger for adaptation. If the performance level of the user remains stable over a longer period of time or over a set amount of repetitions, the assistance system could adapt by removing additional information that might no longer be needed. In both cases, changing and non-changing data, methods are needed to analyze the process data. There are several methods in all of these categories, which can be used as tools to implement adaptivity. However, there is most likely not one single method, which can be used to cover the whole spectrum of adaptive assembly assistance. A combination of multiple of these methods can be used to reach a more complex level of data-processing and allows the creation of more abstract measures like performance- or experience levels.

4.5.1   Rule-Based Approaches

Rule-based approaches form the most basic method to implement adaptivity. The syntax of a single rule can incorporate an If-Then-statement to define the conditional behavior of a system in a given situation [80]. However, a single rule can only define one conditional aspect of the system's behavior regarding one parameter. For a more precise assessment of a given situation and a more advanced adaptive behavior, multiple rules can be put in a sequence. Considering the cyclic process of adaptation, a rule-based approach can be used to implement both, the analysis and the selection. For example, a rule-set can be used to define experience levels in regard to completion time or error rates of the assembly [43]. Another rule-set can then link the experience level to a corresponding type of assistance, which is provided in this situation. The implementation of an adaptive assembly assistance system requires expert knowledge about the process, which needs to be incorporated into the rule-set. To formalize the rule sets, their modeling should follow a common specification. In the field of business enterprise systems, various rule engines are available and notations like the Decision-Model-Notation (DMN) are standardized by the Object-Management-Group (OMG). The existing tool support for modeling the rules allows their specification outside the assistance software architecture itself and thus provides extendibility and maintainability.

For many industrial applications, this can still be a significant drawback, because of the required modeling effort, which also needs expert knowledge about the whole system architecture. Additionally, rule-based approaches are not very suitable for situations, which require a high amount of flexibility. Because the behavior of the system is given by the different rules, more flexibility results in a more complex set of rules which adds to the already significant maintenance effort.

4.5.2   Methods of Machine Learning

Models allow a formal understanding of a part of the process or even the whole process.

They can either be generated prior to the operation of the assistance system, or they can be generated during operation. Depending on the model, they can be extended and updated with continuously gathered data, which makes them more flexible than rule-based approaches. There are different types of models that can be used for different applications.

Regression Models Regression analysis is a statistical tool to model the relationship between one or more independent and one dependent variable. Typically, regression analysis is used to predict the outcome of the dependent variable or analyze the relationship between the variables. The regression model is created by fitting a function model to a distributed data set. To estimate the quality of the fit, different measures are available, like the least-squares or the maximum-likelihood. This qualitative estimate of the fit is then used as a cost function to optimize the fit of the regression model using a curve-fitting algorithm like the Gauss-Newton or the Levenberg-Marquard. In the domain of adaptive assembly assistance, regression models can be used to analyze and predict the time-dependent progression of certain aspects of the process, like for example the learning process of the worker.

When a worker starts a task which he has no prior experience in, his required completion time decreases, as he gathers more experience. A regression model built from the process data can be used to analyze the learning process of the worker [81]. This also allows an estimate of the required completion time for upcoming units, as Fig. 4.7a illustrates. Alternatively, it would also be possible to analyze the learning process in regard to a predefined target performance. If, for example, the learning process is not as fast as estimated, the assistance system could provide different learning interventions to advance the learning process. Because regression models can be built during the process using the collected data, they are highly flexible and represent the specific aspect they are modeled for, e.g. a specific user. The only expert knowledge required is the general model which is used to represent the correlation. However, since regression is a statistical tool, it is only as reliable as the data which is used for building it, which makes the model prone to errors. This error needs to be taken into account when using these models.

Classification The second group of models that can be used are classification models. Classification is the task of organizing data points into categories based on their attributes. A requirement for the classification is available training data which is organized into categories by a corresponding classification model. A classifier can then predict the category of new data points. A popular example of this task is organizing e-mails into "spam" and "no-spam" categories. In this case, the training data includes e-mails that are labeled as "spam" or "no-spam". The classification model is usually generated using machine learning algorithms, which process the training data and build the classification model. When the model is completely finished, the classifier can use the model to predict the label of a new data point, i.e. classify a new mail as "spam" or "no-spam".

In some implementations for adaptive assembly assistance, the adaptive behavior is based on a defined performance state of the user. This performance state can be defined using assembly completion times or error rates [4]. This can alternatively be achieved using classification methods based on machine learning. In this case, the users working with the assistance system can be classified into different performance levels using the same measures. However, classification methods can also be used for a more elaborate approach of the estimation of the user-state, which takes more aspects into account. Research has shown that physiological data, like heart rate or skin conductivity, can be used for stress detection and classification of the human's emotional state [82, 83], like shown in Fig. 4.7b.

This data can also be used to implement adaptive behavior in an assembly assistance system [49]. Another aspect that can be used for classification is the eye-movement of the user. This enables the detection of the user's attention towards certain areas of the workplace and can also be used to detect the mental state of the user or physical effects like drowsiness [84]. Given the possibilities offered by classification algorithms, they could potentially be used to implement more elaborate approaches of adaptive assembly assistance by improving the analysis of the situation and the selection. The major drawback of this method is the requirement for training data. As long as the training data is gathered adaptive system behavior cannot be implemented. Additionally, the quality of the classification also depends on the quality of the training data. Another drawback is the required

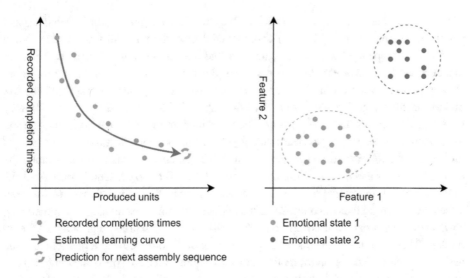

(a) Regression model of a learning process with prediction

(b) Classification model for emotional state of the user based on physiological data [82]

Fig. 4.7 Example applications of methods of machine learning for adaptive assembly assistance. (**a**) Regression model of a learning process with prediction. (**b**) Classification model for emotional state of the user based on physiological data [82]

amount of expert knowledge. To implement a classification, there are several methods available for supervised (KNN) and unsupervised (Artificial Neural networks, K-Means) learning. The method selected for the implementation needs to be chosen and adjusted to the structure of the data and the given situation. Otherwise, the quality of the classification decreases.

Behavior Models The different input modalities of an adaptive assistance system provide data, which can be used to assess the current situation and derive suitable adaptive assistance. However, the individual behavior of the user during the assembly process should also be taken into account. For example, two different users with the same capabilities and experience might still have different work paces. Neglecting the individual behavior of the user can lead to a false assessment of the situation. An assembly process can be interpreted as a series of distinct tasks, like picking parts or different assembly operations. When a worker conducts an assembly process, the series of events is put in a unique timed sequence, which represents the behavior of the worker in the given process. If the mental state of the worker changes due to effects like fatigue, this can also result in a different behavior, which in turn can indicate the necessity for adaptation. In the domain of cyber-physical production systems behavior models are often used for system diagnosis.

With this approach, a behavior model is used to simulate the system's behavior and compare the simulation results with data gathered from the physical system [85]. When there is a discrepancy between the data from the real production unit and the behavior model, an error in the production process can be detected. The concept of this approach has already been converted to the domain of manual assembly by Sehr et al. [81]. They use online machine learning algorithms, to build an individual behavior model for every worker and assembly task respectively. They also take into account the difficulties during the modeling phase, such as the learning behavior of performance fluctuations. The learned model is a deterministic timed automaton, which represents the different states along the assembly process and the corresponding time constraints for every assembly step. After the modeling phase, the model can be used to detect anomalies in the behavior of the user, which are then used to deduce measures to adapt the system to the situation. Alternatively, stochastic models can be used to model the behavior of the worker, e.g. hidden Markov models. This type of model was used by Gellert et al. [86] to develop a prediction-based assistance system. The behavior models in the example above are specialized towards one user and one process, which means they can be used for a very individual adaptation of the assistance system.

However, this also means that many different models are required to cover every combination of worker and assembly process. Similar to the classification models, behavior models also require a learning phase in which data from the process is gathered to train the model. However, the amount of data needed is influenced by many factors like the assembly process, the worker, the learning algorithm, and the configuration. Consequently, the time needed to finish the model can hardly be estimated in advance.

Cognitive Architectures When thinking about adaptive assembly assistance, certain aspects of different assembly tasks have similarities that are inherited from the fundamental elements of assembly work, like different types of part mating operations. This leads to the thought, that a more abstract representation of an assembly work can provide reusable knowledge and lead to a model for adaptive assembly assistance which can be generalized for multiple tasks. Cognitive architectures resemble the structure of human cognition. In artificial intelligence, they provide a framework to model and process knowledge in order to enable intelligent system behavior [87]. Popular examples for cognitive architectures which also support a learning functionality are ACT-R, CHREST, CLARION, HTM, and Soar [88]. In the domain of adaptive assembly assistance, cognitive architectures can be used to hold and process knowledge about the assembly process as well as contextual background knowledge about assembly work. In this case, cognitive architectures can be used for situation detection and the selection of adaptive measures. Cognitive architectures have already been applied in the domain of adaptive assembly assistance [55]. In this work, the Soar architecture was used to combine different assistance goals and knowledge types to provide situation-aware assembly assistance. Similarly, Bannat et al. [42] propose a cognitive assistant, which operates similarly to a cognitive architecture, but has a different structure. Cognitive architectures represent a more holistic approach towards adaptive assembly assistance because they are able to incorporate more aspects into the decision-making progress than simpler models, by adding more knowledge to them. Since the knowledge representation is more abstract, it can be transferred to multiple assembly processes, which increases the flexibility and therefore the usability in an industrial application. However, the knowledge needed for the functionality of the system must be formulated using the architectures' unique syntax. This involves a great effort for the implementation.

4.5.3   Suitable Adaptation Algorithms for the Scenarios

The uniqueness of each user and the variety of different assembly tasks make adaptive assembly assistance a very challenging encounter. The methods used to implement the adaptive behavior of the system need to be selected according to the available process data and the scenario the adaptive assistance system in used for. In Sect. 4.3, three exemplary applications for adaptive assembly assistance were presented. Hereafter, each of the three scenarios will be analyzed to select methods, which can possibly be used to implement adaptive assembly assistance.

Learning Support Multiple repetitions of the same task help the worker to get more familiar with the task. Consequently, the user memorizes more assembly steps and thus requires less time to complete the task. As a measure for adaptive assembly assistance, the provided information can be reduced as the learning process continues. Additionally, the learning process can be interrupted by learning-interventions, which help to ensure important milestones in the learning process. In both cases, it is necessary to identify the progress of the learning process. Kölz et al. [89] used a rule-based approach to define three states along the learning process. In this case, predefined thresholds for completion times determine the user's performance. Another approach is presented by Sehr et al. [81] where a regression model is used to analyze the learning process of the user. The result of such an analysis could possibly also be used with a cognitive architecture, to achieve a more wholesome assessment of the situation.

Performance Support In this scenario the user is already well experienced and familiar with the task. Still the performance of the worker can change due to a lack of concentration or external distractions. In this case, the assistance system should detect changes in the user's behavior and adapt accordingly. However, a precise adaptation depends on the correct identification of the underlying effect, that caused the change. A rule based approach targeting e.g. the rate of errors during the assembly process [4] is relatively easy to implement but has limited capabilities in analyzing the causative effects. In some cases, the user might only have issues with a particular step rather than the whole assembly process. To identify the situation more precisely, behavior models can be used. This allows targeting the adaptive behavior towards the particular assembly step, which the user has problems with, e.g. by giving additional information only during this step. A change of the user's performance can also be induced by stress. An individual classification model can identify a rising stress level. This information could then be used to trigger adaptive measures to counteract this effect.

Ergonomic Support This scenario concerns users with individual physical or cognitive conditions, which limit their performance, like a reduced eyesight. Often, the most efficient solution is allowing the user to reconfigure interactions with the assistance system, e.g. changing the font size of instructions or changing provided error-feedback from visual to auditory [65]. The preferred settings for each user can then be implemented when a user logs in. This functionality can be modeled using a rule-set, in which the user can define his preferences.

4.6   Summary and Conclusion

This chapter gives an overview of adaptive assistance system architecture and starts by presenting a guideline for designing adaptive assistance systems. Afterwards, three scenarios are described to open up a design space of adaptive solutions and illustrate it with real-world scenarios. The main contribution is located in Sect. 4.4, where we present a reference architecture for adaptive assistance systems. We identify the building blocks such a system needs and describe how these components can be implemented. By abstracting from a concrete assistance system implementation, we describe which alternative solutions can be chosen for the individual parts and present several tools and methods for their realization.

Referring back to the requirements formulated in the introduction, the presented architecture helps to guide workers through complex work processes (1). If the adaptation is designed with the help of the morphological box, the goal of the adaptive system is to reduce cognitive load (2) by individualizing the instructions based on the current experience and skill level of the worker (3). Further, the adaptive functionalities allow the support of different groups of workers (4) through giving personalized assistance for a number of different situations (5), which are automatically detected by the situation detection component. Another highlight of the reference architecture is the abstraction of specific interfaces for in and output which supports the implementation of any desired modalities (6) and does not limit the assistance system to certain in- and output devices.

While the research about adaptive assistance systems for the shop floor is still rather sparse, they will probably play an important role in the future of industrial user interfaces. A key question that remains is whether adaptivity pays off in regard to business goals and economics. Focused learning support might slow down a user in the first place, but we argue that in the long run, a qualified, sustainable training will also be beneficial for the companies. Still, the current goals of adaptivity in a system might be conflicting with current business goals and success will only be visible if the right key performance indicators are observed.

The use of adaptive functionalities will have different consequences for different stakeholders: For the developers of the system, adaptivity requires more complex architectures and it increases the effort to build a stable system that offers a good user experience. Still, the requirements help to generalize functionalities of the system and improve the versatility of the finished software product. The users profit from the possibilities of personalized performance support, an enhanced human-machine-interaction, and better learning support. The assistance should motivate the workers to qualify themselves for a broader spectrum of tasks and help to reduce the cognitive load in complex assembly situations. From a manager's perspective, the systems can be beneficial if they are integrated into the workstations in the assembly lines because they can reduce error rates, improve quality and enhance traceability. Furthermore, they allow flexible scheduling of staff to assembly stations, since even untrained workers can be integrated into the production. Thus, from a business perspective, the introduction of adaptive assistance can be beneficial in the long run.

Considering the user acceptance of the systems, it is crucial to follow a human-centered approach. Our guideline for designing the adaptivity in the form of a morphological box helps to keep the user in focus. While adaptivity can increase the autonomy of a system itself, it largely depends on its concrete design, whether the autonomy of the user is enhanced or limited by the adaptivity. This is why we included the aspects of automation, control, and visibility of adaptivity in our design guideline. Interdisciplinary research with disciplines like psychology and sociology can enhance the development, introduction, and acceptance of these systems. This ensures that the workers' autonomy is preserved and helps to design systems where users feel treated fairly and understand why an adaptation has been made or at least been offered.

Our contribution in this chapter presents a rather conceptual approach and lacks evidence of a concrete implementation and evaluation. While we tried to give implementation options for some of the components, an overall implementation concept would help to understand how to realize an adaptive assistance system as a whole. The concepts of the reference architecture arise from the experience from various research projects and assistance system implementations which realize the presented approach at least to some extent. An ideal assistance system would be capable of combining real-time interventions, personalized parametrization and long-term adaptations.

We also did not elaborate on the topic of deployment and operation of the assistance system instances. In larger companies, the assistance systems might be deployed for various workstations and eventually with requirements for interconnecting the individual instances. Furthermore, the realization of an assistance platform with physical or cloud provisioning holds many interesting research questions from a software engineering perspective. In this context, the topics of data protection and privacy in those systems are also very relevant (see [90] for details). Adaptive functionalities can only work if data about the user is gathered and analyzed. It is important to control the access to this data and to ensure that the goal of these systems is not the observation of the users in the first place. Still, whenever possible, data should be anonymized and it should be possible for the user to access his/her personal data at any time and to have the right to delete it when requested. Since European and especially German data protection and privacy laws are very strict, they have to be respected in the development process. But close cooperation with employees, unions, and all stakeholders during the development and introduction phases in a company can help to reduce fear and rejection.

Summarizing the contribution, the design guideline, the reference architecture, and the presented selection of algorithms open up the space for discussions regarding the overall approach and implementation options for the individual components of the architecture. The emphasis of the integrated, automated evaluation of the adaptations ensures that the technical implementation always keeps a human-centered focus.

Acknowledgments This work was supported through the research program "Design of Flexible Work Environments - Human-Centric Use of Cyber-Physical Systems in Industry 4.0", by the North Rhine-Westphalian funding scheme "Forschungskolleg" and the Ministry of Economic Affairs, Innovation, Digitalisation and Energy North-Rhine Westphalia (MWIDE) within the funding initiative it's OWL managed by the project management agency Jülich (PTJ) under grant agreement number 005-2011-0241. The work was further supported through the research project „Technology-enabled inclusion through human-centred systems analysis and assistance in industry", by the German Federal Ministry of Education and Research (BMBF) within the funding scheme „FHprofUnt" managed by the project management agency VDI Technologiezentrum under grand number 13FH110PX6. The avatar graphics used for the personas in Sect. 4.3 were designed by *macrovector_official/Freepik*.

References

1. Bläsing, D. & Bornewasser, M. (2021). Influence of increasing task complexity and use of informational assistance systems on mental workload. *Brain Sciences, 11*.
2. Büttner, S., Mucha, H., Funk, M., Kosch, T., Aehnelt, M., Robert, S., & Röcker, C. (2017). The design space of augmented and virtual reality applications for assistive environments in manufacturing: A visual approach. In *Proceedings of the 10th International Conference on PErvasive technologies related to assistive environments*.
3. Mark, B. G., Gualtieri, L., Rauch, E., Rojas, R., Buakum, D., & Matt, D. T. (2019). Analysis of user groups for assistance systems in Production 4.0. In *2019 IEEE International Conference on Industrial Engineering and Engineering Management (IEEM)*.
4. Funk, M., Dingler, T., Cooper, J., & Schmidt, A. (2015). Stop helping me – I'm bored! Why assembly assistance needs to be adaptive. In *Proceedings of the 2015 ACM International Joint Conference on Pervasive and Ubiquitous Computing and Proceedings of the Symposium on Wearable Computers - UbiComp'15*. New York, USA.
5. Schmidt, A., Beigl, M., & Gellersen, H.-W. (1999). There is more to context than location. *Computers & Graphics, 23*, 893–901.
6. Kosch, T., Abdelrahman, Y., Funk, M., & Schmidt, A. (2017). One size does not fit all: Challenges of providing interactive worker assistance in industrial settings. In *Proceedings of*

the *2017 ACM International Joint Conference on Pervasive and Ubiquitous Computing and Proceedings of the 2017 ACM International Symposium on Wearable Computers*. ACM.

7. Apt, W., Schubert, M., & Wischmann, S. (2018). *Digitale Assistenzsysteme: Perspektiven und Herausforderungen für den Einsatz in Industrie und Dienstleistungen*. Institut für Innovation und Technik.

8. Kleineberg, T., Hinrichsen, S., Eichelberg, M., Busch, F., Brockmann, D., & Vierfuß, R. (2017). *LEITFADEN: Einführung von Assistenzsystemen in der Montage*. Springer.

9. Büttner, S., Sand, O., & Röcker, C. (2017). *Exploring design opportunities for intelligent worker assistance: A new approach using projection-based AR and a novel hand-tracking algorithm* (Vol. 10217). Springer.

10. Sochor, R., Kraus, L., Merkel, L., Braunreuther, S., & Reinhart, G. (2019). Approach to increase worker acceptance of cognitive assistance systems in manual assembly. In *52nd CIRP conference on manufacturing systems* (Vol. 81, pp. 926–931). Elsevier.

11. Fischer, H., Senft, B., & Stahl, K. (2017). Akzeptierte Assistenzsysteme in der Arbeitswelt 4.0 durch systematisches Human-Centered Software Engineering. In *Wissenschaftsforum Intelligente Technische Systeme (WInTeSys)*. Paderborn.

12. Streicher, A., & Smeddinck, J. D. (2016). Personalized and adaptive serious games. In R. Dörner, S. Göbel, M. Kickmeier-Rust, M. Masuch, & K. Zweig (Eds.), *Entertainment computing and serious games* (Vol. 9970, pp. 332–377). Springer.

13. Burgos, D., Tattersall, C., & Koper, R. (2007). How to represent adaptation in e-learning with IMS learning design. *Interactive Learning Environments, 15*, 161–170.

14. Oestreich, H., Wrede, S., & Wrede, B. (2020). Learning and performing assembly processes. In *Proceedings of the 13th ACM International Conference on PErvasive technologies related to assistive environments*. ACM.

15. Browne, D. P., Totterdell, P., & Norman, M. (Eds.). (1990). *Adaptive user interfaces*. Academic Press.

16. Holmes, W., Anastopoulou, S., Schaumburg, H., & Mavrikis, M. (2018). *Technology-enhanced personalised learning: Untangling the evidence*. Robert Bosch Stiftung.

17. Ras, E., Wild, F., Stahl, C., & Baudet, A. (2017). Bridging the skills gap of workers in Industry 4.0 by human performance augmentation tools. In *Proceedings of the 10th International Conference on PErvasive technologies related to assistive environments*. ACM.

18. Hodaie, Z., Haladjian, J., & Bruegge, B. (2018). TUMA: Towards an intelligent tutoring system for manual-procedural activities. In R. Nkambou, R. Azevedo, & J. Vassileva (Eds.), *Intelligent tutoring systems* (Vol. 10858, pp. 326–331). Springer.

19. GCSM. (2013). *Learnstruments in value creation and learning centered work place design*. Universitätsverlag der TU Berlin.

20. Heinz, M., Büttner, S., & Röcker, C. (2020). Exploring users' eye movements when using projection-based assembly assistive systems. In *International conference on human-computer interaction*. Springer.

21. Prinz, C., Kreimeier, D., & Kuhlenkötter, B. (2017). Implementation of a learning environment for an Industrie 4.0 assistance system to improve the overall equipment effectiveness. *Procedia Manufacturing, 9*, 159–166.

22. Oestreich, H., Töniges, T., Wojtynek, M., & Wrede, S. (2019). Interactive learning of assembly processes using digital assistance. *Procedia Manufacturing, 31*, 14–19.

23. Letmathe, P., Schweitzer, M., & Zielinski, M. (2012). How to learn new tasks: Shop floor performance effects of knowledge transfer and performance feedback. *Journal of Operations Management, 30*, 221–236.

24. Wilschut, E. S., Könemann, R., Murphy, M. S., van Rhijn, G. J. W., & Bosch, T. (2019). Evaluating learning approaches for product assembly. In *Proceedings of the 12th ACM*

International Conference on PErvasive technologies related to assistive environments – PETRA'19. ACM.

25. Baker, R. S. J., D'Mello, S. K., Rodrigo, M. M. T., & Graesser, A. C. (2010). Better to be frustrated than bored: The incidence, persistence, and impact of learners' cognitive–affective states during interactions with three different computer-based learning environments. *International Journal of Human-Computer Studies, 68*, 223–241.

26. Heinz, M., Büttner, S., & Röcker, C. (2019). Exploring training modes for industrial augmented reality learning. In *Proceedings of the 12th ACM International Conference on PErvasive technologies related to assistive environments – PETRA'19*. ACM.

27. Müller, B. C., Nguyen, T. D., Dang, Q.-V., Duc, B. M., Seliger, G., Krüger, J., & Kohl, H. (2016). Motion tracking applied in assembly for worker training in different locations. *Procedia CIRP, 48*, 460–465.

28. Büttner, S., Prilla, M., & Röcker, C. (2020). Augmented reality training for industrial assembly work-are projection-based AR assistive systems an appropriate tool for assembly training? In *Proceedings of the 2020 CHI conference on human factors in computing systems*. ACM.

29. Hinrichsen, S., & Bornewasser, M. (2019). How to design assembly assistance systems. In W. Karwowski & T. Ahram (Eds.), *Intelligent human systems integration 2019* (Vol. 903, pp. 286–292). Springer.

30. Hold, P., Erol, S., Reisinger, G., & Sihn, W. (2017). Planning and evaluation of digital assistance systems. *Procedia Manufacturing, 9*, 143–150.

31. Dostert, J. & Müller, R. (2020). Motivational assistance system design for industrial production: From motivation theories to design strategies. *Cognition, Technology & Work*.

32. Schulz, A. S., Schulz, F., Gouveia, R., & Korn, O. (2018). Branded gamification in technical education. In *2018 10th international conference on virtual worlds and games for serious applications (VS-Games)*. IEEE.

33. Josifovska, K., Yigitbas, E., & Engels, G. (2019). A digital twin-based multi-modal UI adaptation framework for assistance systems in Industry 4.0. In *Human-computer interaction. Design practice in contemporary societies. HCII*. Springer.

34. Rönick, K., Kremer, T., & Wakula, J. (2018). Evaluation of an adaptive assistance system to optimize physical stress in the assembly. In *Congress of the International Ergonomics Association*. IEA.

35. Petzoldt, C., Keiser, D., Beinke, T., & Freitag, M. (2020). Functionalities and implementation of future informational assistance systems for manual assembly. In *Subject-oriented business process management*. The Digital Workplace - Nucleus of Transformation.

36. Taoum, J., Raison, A., Bevacqua, E., & Querrec, R. (2018). An adaptive tutor to promote learners' skills acquisition during procedural learning. In *WeASeL 2018 - 1st international workshop eliciting adaptive sequences for learning* (Vol. 1, pp. 12–17). HAL CCSD.

37. Korn, O. (2018). Autonomie beim Einsatz kontextbewusster Systeme: Der Weg zum Emotionsbewusstsein. In T. Breyer-Mayländer (Ed.), *Das Streben nach Autonomie* (pp. 203–214). Nomos Verlagsgesellschaft mbH & Co.

38. Sauer, M. (1990). *MTM-UAS, universelles Analysiersystem: eine programmierte Unterweisung*. Dt. MTM-Vereinigung.

39. Nelles, J., Kuz, S., Mertens, A., & Schlick, C. M. (2016). Human-centered design of assistance systems for production planning and control: The role of the human in Industry 4.0. In *2016 IEEE International Conference on Industrial Technology (ICIT)*. IEEE.

40. Bächler, L., Bächler, A., Kölz, M., Hörz, T., & Heidenreich, T. (2015, 2015). Über die Entwicklung eines prozedural-interaktiven Assistenzsystems für leistungsgeminderte und-gewandelte Mitarbeiter in der manuellen Montage. *Kognitive Systeme*, 2015.

41. Aehnelt, M., & Bader, S. (2016). Providing and adapting information assistance for smart assembly stations. In *Proceedings of SAI intelligent systems conference*. Springer.
42. Bannat, A., Wallhoff, F., Rigoll, G., Friesdorf, F., Bubb, H., Stork, S., Müller, H. J., Schubö, A., Wiesbeck, M., Zäh, M. F., et al. (2008). Towards optimal worker assistance: A framework for adaptive selection and presentation of assembly instructions. In *Proceedings of the 1st international workshop on cognition for technical systems, Cotesys*. Springer.
43. Galaske, N., & Anderl, R. (2016). Approach for the development of an adaptive worker assistance system based on an individualized profile data model. In C. Schlick & S. Trzcieliński (Eds.), *Advances in ergonomics of manufacturing: Managing the enterprise of the future* (Vol. 490, pp. 543–556). Springer.
44. Burggräf, P., Dannapfel, M., Adlon, T., Hahn, V., Riegauf, A., Casla, P., Marguglio, A., & Fernandez, I. (2020). *Virtual asset representation for enabling adaptive assembly at the example of electric vehicle production*. Academic Press.
45. Gellert, A., & Zamfirescu, C.-B. (2020). Using two-level context-based predictors for assembly assistance in smart factories. In *International conference on computers communications and control*. IEEE.
46. IBM. (2006). *An architectural blueprint for autonomic computing*. IBM White Paper.
47. Bertram, P., Birtel, M., Quint, F., & Ruskowski, M. (2018). Intelligent manual working station through assistive systems. *IFAC-PapersOnLine, 51*, 170–175.
48. Rupp, S., & Müller, R. (2020). Worker assistance systems and assembly process maturity in the prototype and pre-series production. *Procedia Manufacturing, 51*, 1431–1438.
49. ElKomy, M., Abdelrahman, Y., Funk, M., Dingler, T., Schmidt, A., & Abdennadher, S. (2017). ABBAS: An adaptive bio-sensors based assistive system. In *Proceedings of the 2017 CHI conference extended abstracts on human factors in computing systems*. ACM.
50. Stoessel, C., Wiesbeck, M., Stork, S., Zaeh, M. F., & Schuboe, A. (2008). Towards optimal worker assistance: Investigating cognitive processes in manual assembly. In *Manufacturing systems and technologies for the new frontier* (pp. 245–250). Springer.
51. Kosch, T., Karolus, J., Ha, H., & Schmidt, A. (2019). Your skin resists: Exploring electrodermal activity as workload indicator during manual assembly. In *Proceedings of the ACM SIGCHI symposium on engineering interactive computing systems*. ACM.
52. Gollan, B., Haslgruebler, M., Ferscha, A., & Heftberger, J. (2018). Making sense: Experiences with multi-sensor fusion in industrial assistance systems. In *PhyCS*. ACM.
53. Gorecky, D., Worgan, S. F., & Meixner, G. (2011). COGNITO: A cognitive assistance and training system for manual tasks in industry. In *Proceedings of the 29th annual European conference on cognitive ergonomics*. IEEE.
54. Wolf, H., Herrmann, K., & Rothermel, K. (2009). Modeling dynamic context awareness for situated workflows. In *OTM confederated international conferences: On the move to meaningful Internet systems*. ACM.
55. Aehnelt, M., & Urban, B. (2015). The knowledge gap: Providing situation-aware information assistance on the shop floor. In *International conference on HCI in business*. HCI International.
56. Hollowell, J. C., Kollar, B., Vrbka, J., & Kovalova, E. (2019). Cognitive decision-making algorithms for sustainable manufacturing processes in Industry 4.0: Networked, smart, and responsive devices. *Economics, Management and Financial Markets, 14*, 9–15.
57. Aehnelt, M., & Bader, S. (2015). From information assistance to cognitive automation: A smart assembly use case. In *International conference on agents and artificial intelligence*. Springer.
58. Duan, Y., Edwards, J. S., & Dwivedi, Y. K. (2019). Artificial intelligence for decision making in the era of big data–evolution, challenges and research agenda. *International Journal of Information Management, 48*, 63–71.

59. Oestreich, H., da Silva Bröker, Y., & Wrede, S. (2021). An adaptive workflow architecture for digital assistance systems. In *Proceedings of the 14th ACM international conference on PErvasive technologies related to assistive environments*. ACM.

60. Ullrich, C. (2016). Rules for adaptive learning and assistance on the shop floor. In *13th international conference on cognition and exploratory learning in digital age* (pp. 261–268). IADIS Press.

61. Funk, M., Kosch, T., & Schmidt, A. (2016). Interactive worker assistance: Comparing the effects of in-situ projection, head-mounted displays, tablet, and paper instructions. In *Proceedings of the 2016 ACM international joint conference on pervasive and ubiquitous computing*. ACM.

62. Blattgerste, J., Strenge, B., Renner, P., Pfeiffer, T., & Essig, K. (2017). Comparing conventional and augmented reality instructions for manual assembly tasks. In *Proceedings of the 10th international conference on pervasive technologies related to assistive environments*. Springer.

63. Fraga-Lamas, P., Fernandez-Carames, T. M., Blanco-Novoa, O., & Vilar-Montesinos, M. A. (2018). A review on industrial augmented reality systems for the industry 4.0 shipyard. *IEEE Access, 6*, 13358–13375.

64. Fite-Georgel, P. (2011). Is there a reality in industrial augmented reality? In *2011 10th IEEE international symposium on mixed and augmented reality*. IEEE.

65. Kosch, T., Kettner, R., Funk, M., & Schmidt, A. (2016). Comparing tactile, auditory, and visual assembly error-feedback for workers with cognitive impairments. In *Proceedings of the 18th international ACM SIGACCESS conference on computers and accessibility*. ACM.

66. Günther, S., Kratz, S., Avrahami, D., & Mühlhäuser, M. (2018). Exploring audio, visual, and tactile cues for synchronous remote assistance. In *Proceedings of the 11th PErvasive Technologies Related to Assistive Environments Conference*. ACM.

67. Wilson, C., & Scott, B. (2017). Adaptive systems in education: A review and conceptual unification. *The International Journal of Information and Learning Technology, 34*, 2–19.

68. Graessler, I., & Poehler, A. (2017). *Integration of a digital twin as human representation in a scheduling procedure of a cyber-physical production system*. IEEE.

69. Josifovska, K., Yigitbas, E., & Engels, G. (2019). Reference framework for digital twins within cyber-physical systems. In *IEEE/ACM 5th International Workshop on Software Engineering for Smart Cyber-Physical Systems (SEsCPS)* (Vol. 2019, pp. 25–31). IEEE.

70. Combemale, B., Kienzle, J. A., Mussbacher, G., Ali, H., Amyot, D., Bagherzadeh, M., Batot, E., Bencomo, N., Benni, B., Bruel, J.-M., Cabot, J., Cheng, B. H. C., Collet, P., Engels, G., Heinrich, R., Jezequel, J.-M., Koziolek, A., Mosser, S., Reussner, R., et al. (2020). A Hitchhiker's guide to model-driven engineering for data-centric systems. *IEEE Software, 2020*, 0.

71. Berisha-Gawlowski, A., Caruso, C., & Harteis, C. (2020). The concept of a digital twin and its potential for learning organizations. In D. Ifenthaler, S. Hofhues, M. Egloffstein, & C. Helbig (Eds.), *Digital transformation of learning organizations* (pp. 95–114). Springer Nature.

72. Schwarz, J., Fuchs, S., & Flemisch, F. (2014). Towards a more holistic view on user state assessment in adaptive human-computer interaction. In *2014 IEEE International Conference on Systems, Man, and Cybernetics (SMC)*. IEEE.

73. Funk, M., Lischke, L., Mayer, S., Shirazi, A. S., & Schmidt, A. (2018). Teach me how! Interactive assembly instructions using demonstration and in-situ projection. *Assistive Augmentation, 2018*, 49–73.

74. Homem de Mello, L. S., & Sanderson, A. C. (1990). AND/OR graph representation of assembly plans. *IEEE Transactions on Robotics and Automation, 6*, 188–199.

75. Moore, K. E., Güngör, A., & Gupta, S. M. (2001). Petri net approach to disassembly process planning for products with complex AND/OR precedence relationships. *European Journal of Operational Research, 135*, 428–449.

76. Bader, S., & Aehnelt, M. (2014). Tracking assembly processes and providing assistance in smart factories. In *ICAART* (pp. 161–168). SCITEPRESS - Science and Technology Publications.
77. Agrawala, M., Phan, D., Heiser, J., Haymaker, J., Klingner, J., Hanrahan, P., & Tversky, B. (2003). Designing effective step-by-step assembly instructions. *ACM Transactions on Graphics, 22*, 828–837.
78. Apoki, U. C., Al-Chalabi, H. K. M., & Crisan, G. C. (2020). From digital learning resources to adaptive learning objects: An overview. In D. Simian & L. F. Stoica (Eds.), *Modelling and development of intelligent systems* (Vol. 1126, pp. 18–32). Springer.
79. Haase, T., Weisenburger, N., Termath, W., Frosch, U., Bergmann, D., & Dick, M. (2014). The didactical design of virtual reality based learning environments for maintenance technicians. In *Virtual, augmented and mixed reality. Applications of virtual and augmented reality. VAMR 2014. Lecture Notes in Computer Science* (Vol. 8526). Springer.
80. Tsalgatidou, A., & Loucopoulos, P. (1991). Rule-based behaviour modelling: Specification and validation of information systems dynamics. *Information and Software Technology, 33*, 425–432.
81. Sehr, P., Moriz, N., Heinz, M., & Trsek, H. (2021). Model-based approach for adaptive assembly assistance. In *2021 26th IEEE International Conference on Emerging Technologies and Factory Automation (ETFA)*. IEEE.
82. Wagner, J., Kim, J., & André, E. (2005). From physiological signals to emotions: Implementing and comparing selected methods for feature extraction and classification. In *2005 IEEE international conference on multimedia and expo*. IEEE.
83. Healey, J. A., & Picard, R. W. (2005). Detecting stress during real-world driving tasks using physiological sensors. *IEEE Transactions on Intelligent Transportation Systems, 6*, 156–166.
84. Jo, J., Lee, S. J., Park, K. R., Kim, I.-J., & Kim, J. (2014). Detecting driver drowsiness using feature-level fusion and user-specific classification. *Expert Systems with Applications, 41*, 1139–1152.
85. Vodenčarević, A., Büning, H. K., Niggemann, O., & Maier, A. (2011). Using behavior models for anomaly detection in hybrid systems. In *2011 XXIII International Symposium on Information, Communication and Automation Technologies*. IEEE.
86. Gellert, A., Precup, S.-A., Pirvu, B.-C., & Zamfirescu, C.-B. (2020). Prediction-based assembly assistance system. In *2020 25th IEEE International Conference on Emerging Technologies and Factory Automation (ETFA)*. IEEE.
87. Lieto, A., Bhatt, M., Oltramari, A., & Vernon, D. (2018). *The role of cognitive architectures in general artificial intelligence*. Elsevier.
88. Ye, P., Wang, T., & Wang, F.-Y. (2018). A survey of cognitive architectures in the past 20 years. *IEEE Transactions on Cybernetics, 48*, 3280–3290.
89. Kölz, M., Jordon, D., Kurtz, P., & Hörz, T. (2015). Adaptive assistance to support and promote performance-impaired people in manual assembly processes. In *Proceedings of the 17th International ACM SIGACCESS Conference on Computers & Accessibility*. ACM.
90. Varadinek, B., Indenhuck, M., & Surowiecki, E. (2018). *Rechtliche Anforderungen an den Datenschutz bei adaptiven Arbeitsassistenzsystemen*. Bundesanstalt für Arbeitsschutz und Arbeitsmedizin.

Deep Learning-Based Action Detection for Continuous Quality Control in Interactive Assistance Systems

5

Andreas Besginow, Sebastian Büttner, Norimichi Ukita, and Carsten Röcker

Abstract

Interactive assistance systems have shown to be useful in various industrial settings, in particular those involving human labor like manual assembly of workpieces. Current systems support workers based on different technologies like projection-based augmented reality, hand or tool tracking or automated inspections using computer vision techniques. While these technologies help to increase product quality significantly, existing solutions are not able to monitor the entire process, which makes it difficult to detect process errors.

In this paper, we present a deep-learning based approach for continuous on-the-fly quality control within an interactive assistance system. By using labeled video data of an assembly process, a model can be trained that automatically recognizes and distinguishes single actions and thus control the sequence of subsequent work processes. By integrating the system into the interactive assistance systems, users are made aware on any process errors.

A. Besginow (✉)
Institute Industrial IT, OWL University of Applied Sciences and Arts, Lemgo, Germany
e-mail: andreas.besginow@th-owl.de

S. Büttner
Institute Industrial IT, OWL University of Applied Sciences and Arts, Lemgo, Germany

Clausthal University of Technology, Clausthal-Zellerfeld, Germany

N. Ukita
Toyota Technological Institute, Nagoya, Japan

C. Röcker
Institute Industrial IT, OWL University of Applied Sciences and Arts, Lemgo, Germany

Fraunhofer IOSB-INA, Lemgo, Germany

Besides presenting the concept and implementation of our deep-learning integration into the assistance system, we describe the created industrial assembly-oriented dataset and present the results from our technical evaluation that shows the potential of applying deep-learning methods into interactive assistance systems.

CCS Concepts: • Human-centered computing → Interactive systems and tools; Gestural input; • Computing methodologies → Computer vision

---

Keywords

Action detection · Assistance system · Quality control · Deep learning · Artificial Intelligence

---

5.1   Introduction

Companies have automated their production processes throughout the past industrial revolutions, thus increasing their efficiency [1, 2]. But this is mostly the case for "off-the-shelf" products, which contrasts the trend we can observe nowadays, where consumers value customization of products [3, 4]. This is often not easily achievable since factories are optimized to produce, at most, few variations of a product. Interactive assistance systems have shown to fill that gap in regard to manual assembly. Workers receive instructions from the assistance system and are able to assemble a wide variety of products [5–9]. They also allow untrained personnel and people with disabilities to perform assembly work by making the assembly task more feasible [10–12]. Assistance systems have been improved with various techniques, like the use of tool detection [13], workflow prediction [8] and optical [14, 15] or pick-by-light [16] picking detection.

Some assistance systems apply Computer Vision (CV) methods to verify a correct assembly result based on images [17, 18]. But in cases of irreversible operations, e.g. fixing a component using a press, a mistake may mean the disposal of the whole workpiece. To avoid such mistakes, one could detect the worker's actions to detect if steps were skipped in the assembly. A task that is not easily achievable using classic CV methods, but can be tackled by applying Machine Learning (ML) methods.

We suggest the use of action detection, based on Deep Learning (DL), to automatically detect the actions executed by a worker. The field of ML, and DL in particular, has received a lot of attention over the last decades and has shown to perform well in various tasks, among them action and gesture recognition [19–22].

Through this implicit interaction [23, 24] we provide automatic progression of instructions and quality assurance by notifying users of mistakes. The automatic progression of instructions is useful due to the common problem that users are bored or even annoyed by instructions after they have learned the task [25]. Nevertheless, mistakes sometimes occur and for this, the quality assurance is always running and checks for mistakes. Based on the literature, we selected an architecture to use and trained it on a new dataset, which shows a exemplary assembly task.

This work has two main contributions:

1. A concept and prototypical implementation of an interactive assistance system with DL-based action detection capabilities.
2. The evaluation of the interactive assistance systems with with DL-based action detection capa- bilities in terms of detection performance.

The detection of individual user actions brings two potential benefits for future assistance systems: First, the work state can automatically be recognized by the system, which allows to show state-dependent information without explicit user interaction. Second, Quality Control (QC) can be improved through the "on-the-fly" detection of mistakes during work.

This paper is organized as follows: we begin by introducing other related work where action detection was applied in interactive systems (Sect. 5.2). Afterwards we briefly discuss our concept in Sect. 5.3 before presenting our dataset (Sect. 5.4). Then we present the selected ML system and go into details of our implementation in Sect. 5.5. Finally, we talk about our evaluation approach and the results (Sect. 5.6) before con- cluding the work with the limitations (Sect. 5.7) and pointing out potential directions for future research (Sect. 5.8).

5.2   Related Work

In this section we introduce other works in the field of Human-Computer Interaction (HCI) that make use of action detection, for different tasks, and briefly present their applied methods if available. We understand actions as goal oriented movements which either cause an effect on the environment (e.g. tightening a screw) or are performed only for the sake of the movement itself (e.g. sports), according to Jensenius [26].

Wang et al. [27] also apply a Convolutional Neural Network (CNN)-based method to determine the action classes of an industrial assembly. The data they use is the skeletal information based on the depth data from a Kinect camera. Based on this skeletal informa- tion they calculate distance and angle matrices that they feed into a CNN. In contrast, we process the raw RGB data in a fully convolutional Neural Network (NN) to perform the detection, without any additional feature extraction or skeletal detection.

Ni et al. [28] presents a lightweight approach based on hand trajectories to detect packaging actions. They search for topological structures in the trajectories, followed by a clustering and finally a gradient descent to detect the main representations for each cluster. After training, they execute a fine-tuning step which improves the parameters and makes the detection more precise. They have tested their approach based on a separated test dataset and show good performance, given the five different actions for the packaging task.

The system of Jeanne et al. [29] recognizes a user's movement in a Virtual Reality environment using a motion capturing system. It calculates the error between the user's gesture execution and an expert's prerecorded execution using dynamic time warping, which calculates the difference between the executions. The calculation is based on different criteria (e.g. speed and orientation) and compares the measurements to the expert examples. After each iteration, the participants were shown their errors to allow improvement. In a study, they show positive effects when training participants using the system.

In the work of Choi et al. [30], they implement a real time multi person tracker for five full-body actions: walking, running, sitting, standing and falling. To track people, they apply a modified version of the Mixture of Gaussians (MoG) algorithm, using weighted combinations of Gaussian distributions to segment persons in an image and allowing real time performance due to its fast computation speed. The result of the MoG was used to create motion history images, which summarize the movement over several images into a single image. Using the motion history image, an Multi Layer Perceptron (MLP) was trained to recognize the five action classes. With the MoG multiple people could be recognized and the action tracking could be applied to each person individually.

In the context of pick and place tasks Bovo et al. [31] present an approach that uses a combination of hand and gaze position to detect assembly errors with an accuracy of 97%. They modeled assembly steps as a retrieve-assembly pair and defined a manufacturing procedure as an (ordered) list of these steps. They fed this information into an Long Short Term Memory (LSTM) network to predict the pick/place location and to compare this with the correct action for the respective step.

Another work in this context was published by Fullen et al. [32]. Their focus was to compare different ML algorithms in their recognition performance and speed. Fullen et al. [32] evaluated various classical ML algorithms, e.g. Support Vector Machine (SVM) or a random forest classifier, in their ability to recognize user actions. The perceptron was the only NN model in the evaluation, without specifying more details about it. The dataset consisted of 3 dimensional hand coordinates for 408 steps, which were used to train the classifier. In the dataset, 2 action classes (pick and place) were used, which were evenly distributed during the full assembly (4 picks and 4 places). Their data was pre-processed to speed up the algorithm and increase inference performance. Due to the use of classical ML algorithms, their approach had a fast inference time. As opposed to their work, we focus on DL-based methods, which are able to generalize better and solve more complex tasks [33, p. 155].

Finally, there is also the large number of industrial and commercial applications on movement based interactions. Probably the best known device nowadays, that makes use gestures, is the Microsoft HoloLens.[1] It has been the centerpiece of a number of publications [34, 35], including industrial contexts [36, 37]. Examples for commercially available systems which make use of interactions, but not DL for their processes, are ARKITE HIM[2] or Schlauer Klaus.[3]

---

[1] https://www.microsoft.com/en-us/hololens

[2] https://arkite.nl

[3] https://www.optimum-gmbh.de

5.3   Concept

The aim of this work is to develop an assistance system prototype with action detection capabilities and to perform a technical evaluation of this proof-of-concept implementation, to examine the potential of action recognition for implicit interactions, such as on-the-fly quality assurance applications.

The development includes the selection of an appropriate DL-based action detection system to recognize actions (which we refer to as Action detection system (Ac-Det-Sys)), selecting or creating an appropriate dataset and implementing these capabilities into an assistance system.

5.3.1   Overall Architecture

In the standard workflow, shown in Fig. 5.1, a camera captures the working area from above. These images are forwarded to the Ac-Det-Sys, which performs a pre-processing before the data is given to the NN for action detection. After actions are detected, they are post-processed before they are transmitted to the assistance system to generate instructions. The instructions are either the standard assembly instructions or an error message in case of mistakes during the assembly. Instructions will be displayed using either in-situ projections in the working area or an additional display.

5.3.2   Assistance System

The assistance system is based on an existing workbench, which resembles a standard manual assembly workplace (cf. e.g. [38, 39], see Fig. 5.2).

Based on the information received from the Ac-Det-Sys the instructions, shown to the user, automatically progress when the correct action is executed. When the Ac-Det-Sys detects any discrepancies between the expected and the detected action, it informs the user of this and asks the user to reverse the last action. We assume the assembly task to be a strictly linear sequence of actions allowing no variation in the execution order.

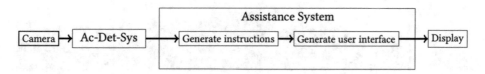

Fig. 5.1 The basic workflow of the complete system. The camera captures images and forwards them to the Ac-Det-Sys, where action predictions are created. The action predictions are then given to the assistance system, which uses them to generate instructions. Finally, a user interface that is displayed

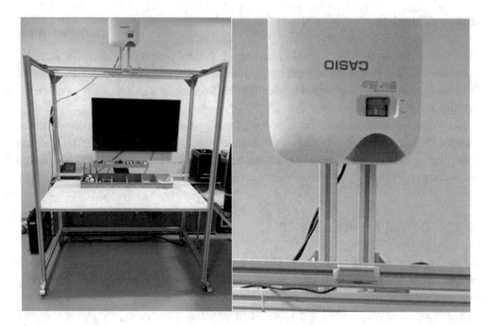

Fig. 5.2 The workbench mockup used as a base for the assistance system (left) and a close up of the camera at the bottom and projector at the top (right)

5.4   Dataset

To the best of the authors knowledge, no publicly available datasets for industrial assembly tasks exist (cf. [27, p. 2]). Therefore, a new assembly task was designed and a dataset was captured to train the NNs. The dataset is described below. It was captured and labeled using self-written software.

In the designed assembly task six components (see Fig. 5.3a–f) are combined to form and package a shower head (Fig. 5.3g). Some components have alternative versions to promote generalization beyond a specific component. The task consists of 13 steps with the six actions: pick, insert, fasten, connect, putIn and closeBag.

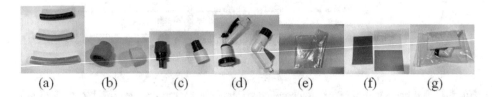

(a)   (b)   (c)   (d)   (e)   (f)   (g)

Fig. 5.3 (a–f) Show the components used in the assembly task and (g) shows the fully assembled workpiece. (a) Hose, (b) Nut, (c) Connector, (d) Shower head, (e) Bag, (f) Paper

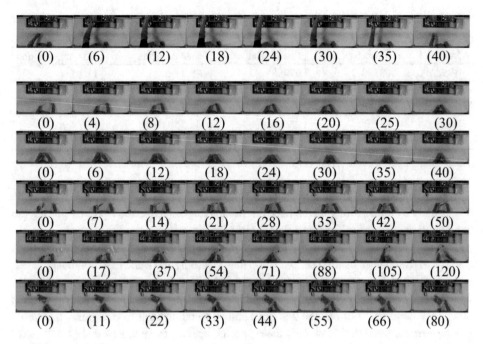

Fig. 5.4 Series of frames that show the pick, insert, fasten, connect, putIn and closeBag actions. The numbers below the frames depict the frame's position relative to the action start

One assembly consists of the following actions: 6 picks, 2 inserts, 1 fastening, 1 connecting, 2 putIn and 1 closeBag.

The dataset is recorded with the focus on being as authentic as possible, therefore the assemblies are executed as normally done, not considering the Ac-Det-Sys. Examples of the clips are shown in Fig. 5.4.

Pick. The most common step during the assembly is picking items from boxes at various coordinates. The picking action does not distinguish between the picked object, therefore picking from different boxes is always considered to be the same action.

Insert. Inserting actions always involves two objects, either a hose and a connector or a hose and a nut. The hose is either inserted into the nut, or it is attached to the connector. Since the motion is similar in both instances, the two executions are treated as the same action.

Fasten. After both the connector and nut are combined with the hose, the components are fastened, which is performed by repeatedly tightening the threads of the nut and the connector.

Connect. This action involves the shower head and the connector, which are moved towards each other and are connected to form the final shower head.

putIn. During a putIn action, an object is put into a transparent bag, to simulate the packaging process. It's either the assembled shower head or one of the three colored papers.

Table 5.1 Details about the recorded actions like the total number of instances, the mean duration of the clips and standard deviation in frames per action (fpa)

Action	Instances	Frame total	Mean duration	Std. dev. duration
Pick	3138	93055	29.7 fpa	11.1 fpa
Insert	1094	62513	57.1 fpa	40.0 fpa
Fasten	568	75852	133.5 fpa	39.0 fpa
Connect	528	22644	42.5 fpa	15.7 fpa
putIn	1042	67650	64.9 fpa	37.4 fpa
CloseBag	548	52680	96.1 fpa	66.7 fpa

CloseBag. closeBag is the final action and involves the use of both hands to close the bag, usually by either lifting the bag and sliding the fingers across the zip or by putting the bag flat on the workbench and pressing against the zip mechanism.

Dataset Details. The videos for the dataset were recorded at two different locations (Japan and Germany), using a similar setup and recording the same assembly task.

In total 12 volunteers were recorded executing 521 full assemblies of the workpiece, with 381 videos showing participant P_1 and the other 140 videos distributed between the other 11 volunteers.[4] For actions that have shown to be difficult to recognize, additional videos were captured that show only these specific actions, these videos make up 25 additional recordings performed by P_1. In 195 (of the in total 546) instances, a Kinect v2 with a resolution of 640 × 360 px was used to capture the videos. The other 351 videos were recorded with an Intel RealSense D415 and a resolution of 848 × 480 px. Both cameras recorded with 30 frames per second.

The total number of clips per action as well as detailed information about them can be found in Table 5.1. In total, 6918 action and 7459 no action instances were recorded. It can be seen that the majority of actions are picking actions, since these make up half of the assembly task.

The videos of the dataset are randomly split into a training, validation and test set with probabilities 0.7, 0.2 and 0.1. The distribution of the actions inside the datasets is approximately the same for all datasets and the videos of the different locations are also distributed fairly across the training, validation, and test sets.

5.5   Implementation

We begin with a description of the relevant hardware before continuing with the required software and libraries. Then we describe the Ac-Det-Sys we selected and conclude with the implementation details for the NN training and the assistance system.

---

[4] Due to organisational reasons the majority of videos captured participant P_1. This is non-optimal and the videos should be more equally distributed between all participants.

5.5.1 Hardware

The hardware has two main components: a computer and a workbench, which is equipped with a projector and a camera. The computer can be used to train the NN and run the Ac-Det-Sys, as well as the assistance system, in our case we trained on a separate computer. The workbench's periphery is connected to the computer to display the instructions provided by the assistance system.

The training was done on a tower PC equipped with an Nvidia GTX1080, with 8 GB VRAM and runs an Ubuntu 16.04.6 LTS. The final assistance system used an Nvidia GTX1060 with 6GB VRAM, also running an Ubuntu 16.04.6 LTS. The capture software could handle either a Kinect 2 camera or an Intel RealSense D415 camera as input.

The assistance system is based on an existing workbench mockup (see Fig. 5.7), which represents a common assembly workplace in the industry and is equipped with an Intel RealSense D415 RGB-D camera and a projector. Additionally, a monitor is connected as an alternative display which can be used, based on the user's preference.

5.5.2 Software

To realize the assistance system, various software needed to be written. This included tools to capture and label the dataset, as well as the actual implementation of the Ac-Det-Sys and the assistance system.

Since the capture and label tools can be developed as preferred, as long as the results are formatted according to Köpüklü et al. [21], we exclude these tools from discussion. Therefore we begin with the Ac-Det-Sys implementation, which is based on the code provided by Köpüklü et al. [21].[5] Aspects of the code, e.g. the training procedure, were modified and are also discussed below. The used libraries are: OpenCV (4.1.2), Pillow (6.2.0), PyTorch (1.3.0), h5py (2.10.0), PyRealSense (2.32.1.12.99), PyLibFreenect (presumably 0.1.2), Nvidia CUDA (10.1.243).

All software is written in, or has been modified to be compatible with, Python 3.6 except the PyLibFreenect library [40]. PyLibFreenect was used in an earlier version of the capturing software and is only supported until Python 3.5.

5.5.2.1 Machine Learning System
The architecture we uses was developed by Köpüklü et al. [21] and is based on two NNs to execute real-time gesture detection.

In contrast to an offline approach, where the NN evaluates a full dataset without any time constraints, this architecture specializes on real time data. There are certain difficulties associated with processing real-time data, like lacking the information that a gesture has

---

[5] https://github.com/ahmetgunduz/Real-time-GesRec

started or ended. In addition, it is unknown whether or not it was a unique execution, which potentially results in duplicate detections of a single gesture, or action. To tackle this problem, Köpüklü et al. [21] developed a detector-classifier architecture with a fast and lightweight detector NN and a deep and powerful classifier NN. The detector decides whether or not a gesture is occurring and acts as a switch to turn on the classifier, which then recognizes which class this gesture belongs to, differentiating between the trained gesture classes. The number of frames fed into the detector and classifier differs, with the detector receiving less frames than the classifier. The frames given to the two models overlap, as shown in Fig. 5.5.

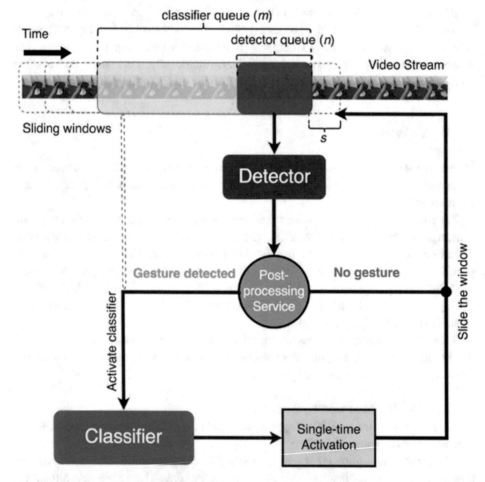

Fig. 5.5 The architecture used in [21]. The detector receives n frames (the detector queue) and decides whether or not a gesture occurred, turning on the classifier, which receives m frames (the classifier queue), to recognize which gesture it was, or continuing with the next frame. After a gesture was recognized by the classifier, a post-processing is applied to guarantee single-time activation, eliminating duplicate predictions

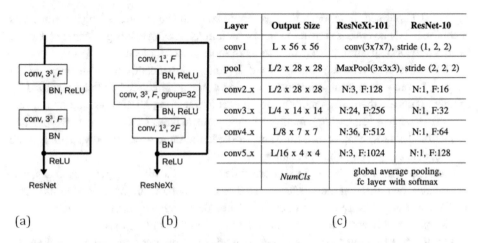

Layer	Output Size	ResNeXt-101	ResNet-10
conv1	L x 56 x 56	conv(3x7x7), stride (1, 2, 2)	
pool	L/2 x 28 x 28	MaxPool(3x3x3), stride (2, 2, 2)	
conv2_x	L/2 x 28 x 28	N:3, F:128	N:1, F:16
conv3_x	L/4 x 14 x 14	N:24, F:256	N:1, F:32
conv4_x	L/8 x 7 x 7	N:36, F:512	N:1, F:64
conv5_x	L/16 x 4 x 4	N:3, F:1024	N:1, F:128
NumCls		global average pooling, fc layer with softmax	

(a)   (b)   (c)

Fig. 5.6 The (**a**) ResNet and (**b**) ResNeXt blocks used in the detector and classifier architectures of [21]. In (**c**), number of feature channels is defined as F and the number of blocks in the corresponding layers are named N

Detector. The detector has to distinguish two classes: gesture and no gesture and based on this result, it turns on the classifier network. The overall accuracy of the system highly depends on the accuracy of the detector, since every False Positive (FP) or False Negative (FN) of the detector automatically results in an error in the classifier. Therefore the detection needs to be robust, lightweight and accurate. To make the detector lightweight, and therefore fast, it runs on fewer frames than the classifier and is based on an NN with fewer parameters, which, in turn, speeds up calculations. To reduce the number of FPs and FNs in the detector and make it more robust, it is trained using a weighted Cross-Entropy (CE) loss. The weighted CE loss applies weights on the loss outputs, which are multiplied on the regular CE output. A weight greater than 1 causes the loss for a class to be amplified. This forces the model to focus more on recognizing this class correctly, increasing the recall. Köpüklü et al. [21] weighted no gesture and gesture with 3 and 1, respectively. By doing so, they force the training process to put more emphasis on recognizing when no gesture occurs (precision), rather than recognizing gestures correctly (recall). The class weights for the CE have been chosen empirically, according to [21].

The detectors' architecture is based on ResNet blocks (see Fig. 5.6a), which have a very small feature size in each layer. This causes the network to have approx. 900k parameters, thus making it small and fast. For comparison, YOLOv3 [41], a state of the art object detection architecture for images (2D data), has approximately 62 M parameters [42].

ResNet [43] stands for Residual Networks and is based on the use of skip connections. Skip connections are identity mappings of the input which are added on the results of the nonlinear function F(x), e.g. the output of a convolution operation. This result is of the form

$$y = \mathcal{F}(x) + x$$

with $\mathcal{F}(x)$ a nonlinear function, x the layer input and y the layer output. Implementing this operation in NN layers has shown to improve training, allowing to design deeper networks. Deeper networks, in turn, have shown to learn more advanced features [43–45].

The ResNet architecture is usually used for image classification tasks and runs on 2D images. To run action recognition tasks, on videos, the architecture had to be modified, to be able to process them. This modification meant to transform all 2D convolutional layers to 3D layers, introducing a large number of parameters to the detector network. Despite this adaptation the underlying architecture still works in the same way, but now requires more training data compared to its 2D counterpart.

Classifier. The classifier, used to differentiate between the different gesture classes, was tested using two different architectures[21]. The C3D architecture by Tran et al. [46] and the ResNeXt architecture by Xie et al. [47]. They demonstrated better performance using the ResNeXt architecture, therefore it will be the main focus of discussion.

The ResNeXt architecture is adapted to work on 3 dimensional input data, like the ResNet architecture for the detector. In the case of ResNeXt, the number of trainable parameters has increased to approximately 47.5M, which demanded measures to ensure stable training. To tackle this issue, Köpüklü et al. [21] use the Jester dataset [48] to pretrain their model, before training on the actual dataset.

The ResNeXt architecture challenged the common belief that "deeper is better" [43–45] by introducing a new dimension, called cardinality.

Cardinality is the number of operations that are executed in parallel. In the case of [47]: applying a bottleneck architecture [43] (see Fig. 5.6b) on the data which scales down the images into a lower dimension and then scales them back up for the output. The operation can be chosen freely, but the same operation is applied in a single block. Where a block is defined as

$$\mathcal{F}(x) = \sum_{i=1}^{C} \tau_i(x)$$

with τ_i (x) an arbitrary function, in the case of [47] it is the bottleneck architecture from [43], x an input vector and C the cardinality, which defines the number of transformations. As shown in Fig. 5.6, the operations per block are defined by the size of the group, 32 in this case.

Xie et al. [47] show in their experiments that increasing the cardinality proved more effective in performance than making the network deeper or the layers wider. The ResNeXt architecture is based on ResNet, from which they adopted the idea of skip connections as well as the bottleneck architecture.

The two NNs Köpüklü et al. [21] used are called "detector", a fast and lightweight architecture, and "classifier", a deep and sophisticated architecture, and solve different recognition problems. The detector recognizes whether or not a gesture is currently occurring and, depending on this, turns on the classifier which then recognized exactly which gesture it was. The NNs that are used for the detector and classifier are freely selectable. In their evaluations, they tested the C3D architecture [46] and the ResNeXt architecture [47] for the classifier. For the detector, they used a lightweight ResNet network.

Their approach performed well on two hand gesture datasets, with accuracy ratings of 91.04% on the nvGesture dataset [49] and 77.39% on the EgoGesture dataset [50], for their overall architecture. Due to this good performance, as well as its real-time capable architecture, it was selected for this work.

Modifications. To better fit the task at hand, small modifications have been made to the training and pre-processing steps of the system, but the detector and classifier architecture have been left as is.

First, we implemented a custom dataloader, adapted from the dataloaders provided by Köpüklü et al. [21], for our dataset. Second, we modified the pre-processing by adding a temporal random skip and removing the leftmost and rightmost parts of the captured image. Cropping the image was done to reduce the amount of unnecessary information as the sides of a workbench are usually unused or, at most, rarely used.

The temporal transform (modification of the time-series information) has been extended with a frame skip of random size to simulate potential lag during the real-time operation.

5.5.2.2 Model Generation

The model generation is automated and different settings were tested throughout the procedure. Before training on the actual dataset, a pre-training was executed using the jester dataset [48]. After the actual training was completed, the final performance of the models was evaluated on the test dataset. The varied settings were the application of weighting on the action classes and also the variation of Nesterov momentum and dampening, which are mutually exclusive in PyTorch.[6] All other settings were left to standard as given in the code of [21].

This procedure is applied during both detector and classifier training and is repeated five times to compensate for randomness in the training. The best model, based on the test performance, was used in the assistance system.

5.5.2.3 Assistance System

The assistance system is based on the workbench mockup shown in Fig. 5.7 and the software running on the computer, including the Ac-Det-Sys.

The assistance system can run two modes: continuous display of instructions with automatic forwarding when actions are correctly executed; and a background task continuously performing the quality assurance based on the action detection.

---

[6] See https://github.com/pytorch/pytorch/blob/master/torch/optim/sgd.py#L67

Fig. 5.7 The user interface for the in-situ projection (left) and instructions displayed on an additional display (right)

Both modes can be used for quality assurance, since instructions are only forwarded if the correct action is executed. Otherwise an error message is displayed, asking to reverse the last action.

The Ac-Det-Sys runs the best performing NNs, based on the automated training, to predict actions and uses the same pre-processing as applied during validation in the training process. The predictions by the Ac-Det-Sys are compared against the expected step, which is stored in a step counter, to select the appropriate visualization for the user. If the user didn't execute the correct action, an error message is displayed. This process runs continuously until the assembly is completed.

The post-processing is based on a simplified version of the single time activation process described in [21]. The main difference between Köpüklü et al. [21] and our implementation is the recognition of early detections, which was not relevant for our purposes.

The instructions can either be displayed directly on the working area (in-situ) or on an external monitor, based on the user's preference (see Fig. 5.7).

Since the assembly is assumed to be strictly linear the steps must be performed in the predefined order stored in the assistance system. Assuming a wrong action has been executed, the user is shown an error message and asked to reverse the action and confirm the reversal. This is necessary, because the Ac-Det-Sys is not able to detect the reversal of actions.

5.6 Evaluation

To evaluate the assistance system we performed a technical evaluation, with the purpose of measuring the objective performance of the assistance system. First, we will outline the applied method before presenting and discussing the results.

5.6.1   Method

The NNs were trained using the previously described automatic training. We selected the best performing detector and classifier to run in the assistance system. The detector was trained using Nesterov momentum of 0.9 and a weighting of 1:3 for action and no action. The classifier was trained with a dampening of 0.9 and a weighting of 1:2:2:1:1:1 between the action classes.

To evaluate the assistance system, we ran the in-situ assistance mode and asked the participant to follow its instructions to assemble the workpiece. The person executing the assembly was participant P_1 (see section Dataset). We recorded 41 assemblies, including video, timestamps and confidence values for the predicted actions and later create Ground Truth (GT) labels used for evaluation.

The applied metric is the mean Average Precision (mAP), following the instructions for the THUMOS challenge [51, p. 20f], and is used to evaluate the system as a whole, rather than the individual NNs. We also calculated the Area-under-curve (AUC) for the interpolated precision-recall curve as an additional metric, since it considers the total number of FPs and FNs.

The assembly process and the evaluation follow these rules:

- The assembly is executed 41 times, without knowledge of prediction confidences or the recognized classes (no debug information) and is executed in the predefined order.
- Whenever an action is not recognized by the assistance system, the action is reversed after waiting for a brief moment.
- This is repeated until the assistance system correctly recognizes the action or detects ten mistakes.
- After ten mistakes, the assistance system automatically proceeds to the next assembly step (preventing infinite runs).
- If the assistance system proceeds to the next assembly step during a reversal, the previous action is shown manually and done once more to allow the subsequent assembly steps.
- The assistance system automatically records the assemblies and logs the predicted action. The logs show for each frame which action was predicted or if no action was predicted.
- The logfiles are then compared to the manually created GT to calculate their overlap.

5.6.2   Results

The Average Precision (AP) results, for the different actions as well as the AUC results are shown in Table 5.2. To calculate the mAP we therefore used

Table 5.2 Area-under-curves and average precisions for the different action classes. Calculated with an overlap threshold of 50% and 40%

Action class	Pick	Insert	Fasten	Connect	putIn	closeBag	Mean
AUC@50	0.23	0.14	0.68	0.53	0.6	0.16	0.39
AUC@40	0.56	0.20	0.74	0.66	0.71	0.47	0.56
AP@50	0.55	0.50	0.78	0.57	0.84	0.38	0.60
AP@40	0.79	0.61	0.78	0.64	0.87	0.74	0.74

$$AP\left(c\right) = \frac{\sum_{k=1}^{n} \left(P(k) \times rel(k)\right)}{\sum_{k=1}^{n} rel(k)}$$

$$mAP = \frac{1}{c} \sum_{c=1}^{C} AP(c)$$

where $c \in C$ are the classes, $P\left(k\right)$ is the precision cut-off for entry k and $rel\left(k\right)$ is 1 if k is a True

Positive (TP) and 0 otherwise. And to calculate the AUC the following is used:

$$p_{interp}\left(r\right) = \max_{r \geq r} p\left(\tilde{r}\right)$$

$$AP = \int_{0}^{1} p_{interp}\left(r\right) dr$$

with p_{interp} the interpolated precision at each recall, taking the maximum precision for recalls greater than r. The precision-recall curves used to calculate the AUC@0.5 are shown in Fig. 5.8. As can be seen in both, some actions are more difficult to detect than others, highlighting the fasten action as one of the easiest to detect and the insert action as one of the most difficult. Summarizing all the AP values for the actions, we have a total mAP of 0.60 for the evaluation. The values for the AUC calculation are lower due to their more strict consideration of FPs and FNs. Further we present the best performed run in Fig. 5.9. It shows the color-coded actions on a timeline, with the actual GT above them. The classification as TP, FP or FN is indicated above the actions by the circles and crosses in either the "Negatives" or "Positives" row. It is important to note that the instructions may have been forwarded in cases where the metric shows a FP and still be correct, due to the overlap being lower than 50%.

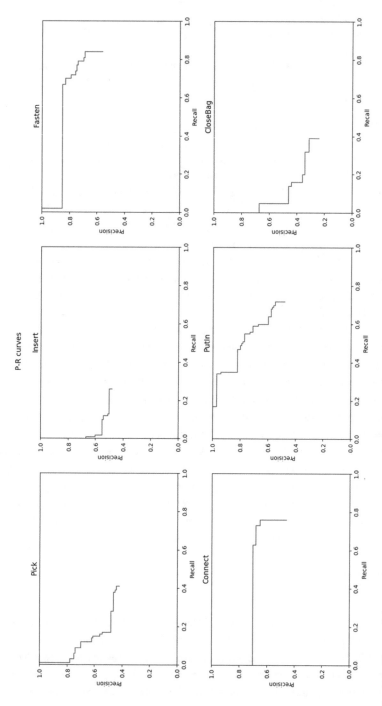

Fig. 5.8 The precision-recall curves for the action classes with 50% overlap. (Used for AUC calculation)

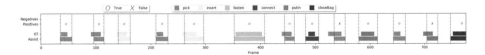

Fig. 5.9 The best performed run, visualized on a timeline. Red crosses indicate errors (either FP or FN), whereas green circles indicate correct detections (TP). (TP etc. calculated using 50% overlap)

5.6.3 Discussion

The results indicate that a reasonable starting level of detection is achieved using an out-of-the-box Ac-Det-Sys and applying it to our new dataset.

The performance of the insert action is lower than the other actions, which raises the concern that it was the wrong decision to summarize the two actions together in a single class. Based on the results we believe the recognition would have improved if the NN had been trained to differentiate between the classes.

On the other end, the fasten action was detected almost best, raising the question why that is the case, since it is the action which is, on average, the longest. This is potentially caused by the repeating motion which allowed the NN to learn very simple movement compared to other tasks.

Most surprising is the performance of the pick action which, despite roughly three times more training instances, performed below average in detection (AP@50 and AUC@50).

Despite the seemingly bad performance of the system, based on the evaluation results, the subjective performance for the assembly worker is better. Since classification in the evaluation is based on the rules of the THUMOS challenge (the common 50% intersection-over-union threshold), some actions are classified as FP even though they actually weren't. This can be seen in Fig. 5.9 where the user did not need to repeat a single action, but the evaluation considered 2 of the 13 actions as mistakes. We therefore calculated another mAP result with an overlap of 40%, which is the highest overlap before actions are mistakenly classified as FP in the best performed run. With this overlap we have an mAP of 0.74 (compared to 0.60 with 50% overlap) and the APs shown in Table 5.2. Therefore an evaluation with users, in cases of e.g. assistance systems, can be recommended to ensure that the objective performance translates correctly.

5.7 Limitations

As explained in section "Dataset", it is problematic that the distribution of videos is not equal among all participants, which is due to organisational restrictions at the time.

Due to this bias it is likely that the Ac-Det-Sys recognizes actions performed by P_1 slightly better. We believe that variations in the assembly (e.g., different movements of the arm, which happen naturally when assembling the same workpiece multiple times) may have corrected the bias slightly. It probably also had an impact on the evaluation performance and therefore the system shows a near-best case scenario in that sense, given a non-optimized Ac-Det-Sys.

Additionally, the applied Ac-Det-Sys is an out-of-the-box solution which wasn't tweaked or optimized to work with the dataset. This, in turn, introduces potential for improvement and could balance out the lower performance, that can be expected when using a more diverse and challenging dataset.

5.8 Conclusion and Future Work

In this work we extended an interactive assistance system prototype with action detection capa- bilities, trained the ML system and evaluated the assistance system in a technical evaluation. The results are promising, as this is one of the few works applying ML, and DL in particular, on the problem of optical quality assurance and user guidance in assembly situations. Without optimizing the post-processing and taking advantage of potential training modifications, a good performance is achieved. An analysis of pattern in the actions or underlying statistics and the exploitation of this information yields the potential to further improve the performance of the recognition, just like in [21]. Due to this, we expect a fully optimized system to perform better and make real time action detection a viable option for future assistance systems.

Still we do not recommend usage of this system in areas which are already performing well, in particular we highlight the picking detection. CV-based techniques already perform stable enough [14] and additionally allow DL-based techniques to focus on the other actions during training.

This work raises potential future research questions, for example in the field of ML the focus should lie on three tasks:

1. Creating larger, and more diverse, datasets for industrial assembly tasks
2. Optimizing the process to capture/create datasets
3. Designing better ML models to train action detection tasks

New datasets for industrial assembly tasks are the most important point to allow ML-engineers to develop models for these activities. In the optimal case, the datasets are big and show a large variety of actions usually performed during manual assemblies, with and without tools.

More important, but more difficult, is the optimization of processes to create datasets for training. Currently, it is not possible to reliably scale the usage of any DL-based action detection method, since the processes, cameras, lighting conditions, etc. differ in every company. Companies need to be able to use a pre-trained model for general action detection tasks, to quickly capture a dataset of their own processes, fine-tune the model by training it, and use it in their production. Right now, the most time consuming part is the dataset creation, which requires several workers to perform a task several times and then label the data to train a reliable model. Optimally, the videos are labeled as they are captured, which requires new developments, which are research topics themselves.

The development of new NNs for action detection is already a popular research field and it can be expected that this trend will continue. Therefore the existence of an industrial assembly focused dataset allows the researcher to test their NNs on a new kind of data compared to the standard datasets like THUMOS, Jester or NvGesture. Additionally, these efforts can be combined with research in the field of HCI, where the robustness and acceptance of action detection systems can be put to the test. For example, how does the Ac-Det-Sys react when the in-situ instructions display videos of the action, which are also common visualizations for instructions.

Finally, the impact of using such DL-supported user assistance in the field should be examined in user studies, as soon as robust Ac-Det-Sys are available.[7] Beyond a potential impact on work efficiency and quality, bringing intelligent systems into a work environment might affect the users' attitude towards work and the work itself. What is the user perception of the intelligent system? Will users perceive the system as a helper or rather as an intruder, that observes their work? Will they perceive it as a violation of their privacy? Applying and studying intelligent assistance systems in a work context will answer this questions regarding the acceptance of and trust in the new technology. Further research could relate to intercultural differences in the perception of intelligent agents for work assistance.

This discussion can be used to further facilitate the adoption of ML technologies in the industry and enable more research possibilities in the future.

References

1. Lasi, H., Fettke, P., Kemper, H.-G., Feld, T., & Hoffmann, M. (2014). Industry 4.0. *Business & Information Systems Engineering, 6*(4), 239–242.
2. Stock, T., & Seliger, G. (2016). Opportunities of sustainable manufacturing in industry 4.0. *Procedia Cirp, 40*(2016), 536–541.
3. Um, J., Lyons, A., KSL, H., Cheng, T. C. E., & Dominguez-Pery, C. (2017). Productvarietymanagement and supply chain performance: A capability perspective on their relationships and competitiveness implications. *International Journal of Production Economics, 187*(2017), 15–26.
4. Wan, X., & Sanders, N. R. (2017). The negative impact of product variety: Forecast bias, inventory levels, and the role of vertical integration. *International Journal of Production Economics, 186*(2017), 123–131.
5. Büttner, S., Funk, M., Sand, O., & Röcker, C. (2016). Using head-mounted displays and in-situ projection for assistive systems: A comparison. In *Proceedings of the 9th ACM international conference on pervasive technologies related to assistive environments* (pp. 1–8). Springer.

---

[7] Due to the current SARS-CoV-2 pandemic we were unable to execute a planned user study of the system.

6. Fellmann, M., Robert, S., Büttner, S., Mucha, H., & Röcker, C. (2017). Towards a framework for assistance systems to support work processes in smart factories. In *International cross-domain conference for machine learning and knowledge extraction* (pp. 59–68). Springer.

7. Funk, M., Bächler, A., Bächler, L., Kosch, T., Heidenreich, T., & Schmidt, A. (2017). Working with augmented reality? A long-term analysis of in-situ instructions at the assembly workplace. In *Proceedings of the 10th international conference on PErvasive technologies related to assistive environments (island of Rhodes, Greece) (PETRA'17)* (pp. 222–229). Association for Computing Machinery. https://doi.org/10.1145/3056540.3056548

8. Gorecky, D., Worgan, S. F., & Meixner, G. (2011). COGNITO: A cognitive assistance and training system for manual tasks in industry. In *Proceedings of the 29th Annual European Conference on Cognitive Ergonomics* (pp. 53–56). IEEE.

9. Sand, O., Büttner, S., Paelke, V., & Röcker, C. (2016). smARt. Assembly–projection-based augmented reality for supporting assembly workers. In *International conference on virtual, augmented and mixed reality* (pp. 643–652). Springer.

10. Baechler, L., Baechler, A., Funk, M., Autenrieth, S., Kruell, G., Hoerz, T., & Heidenreich, T. (2016). The use and impact of an assistance system for supporting participation in employment for individuals with cognitive disabilities. In *International conference on computers helping people with special needs* (pp. 329–332). Springer.

11. Funk, M., Kosch, T., & Schmidt, A. (2016). Interactive worker assistance: Comparing the effects of in-situ projection, head-mounted displays, tablet, and paper instructions. In *Proceedings of the 2016 ACM International Joint Conference on Pervasive and Ubiquitous Computing (Heidelberg, Germany) (UbiComp'16)* (pp. 934–939). Association for Computing Machinery. https://doi.org/10.1145/2971648.2971706

12. Heinz, M., Büttner, S., Jenderny, S., & Röcker, C. (2021). Dynamic task allocation based on individual abilities - Experiences from developing and operating an inclusive assembly line for workers with and without disabilities. *Proceedings of the ACM on Human-Computer Interaction, 5*, 19. https://doi.org/10.1145/3461728

13. Lai, Z.-H., Tao, W., Leu, M. C., & Yin, Z. (2020). Smart augmented reality instructional system for mechanical assembly towards worker-centered intelligent manufacturing. *Journal of Manufacturing Systems, 55*(2020), 69–81.

14. Büttner, S., Sand, O., & Röcker, C. (2017). Exploring design opportunities for intelligent worker assistance: A new approach using projection-based AR and a novel hand-tracking algorithm. In *European conference on ambient intelligence* (pp. 33–45). Springer.

15. Röcker, C. & Robert, S. (2016). *Projektionsbasierte Montageunterstützung mit visueller Fortschrittserken- nung. visIT Industrie 4.*

16. Baechler, A., Baechler, L., Autenrieth, S., Kurtz, P., Hoerz, T., Heidenreich, T., & Kruell, G. (2016). A comparative study of an assistance system for manual order picking–called pick-by-projection–with the guiding systems pick-by-paper, pick-by-light and pick-by-display. In *2016 49th Hawaii International Conference on System Sciences (HICSS)* (pp. 523–531). IEEE.

17. Büttner, S., Peda, A., Heinz, M., & Röcker, C. (2020). Teaching by demonstrating: How smart assistive systems can learn from users. In *International conference on human-computer interaction* (pp. 153–163). Springer.

18. Piero, N., & Schmitt, M. (2017). Virtual commissioning of camera-based quality assurance systems for mixed model assembly lines. *Procedia Manufacturing, 11*(2017), 914–921.

19. Benitez-Garcia, G., Haris, M., Tsuda, Y., & Ukita, N. (2020). Continuous finger gesture spotting and recognition based on similarities between start and end frames. *IEEE Transactions on Intelligent Transportation Systems, 2020*. https://doi.org/10.1109/TITS.2020.3010306

20. Benitez-Garcia, G., Haris, M., Tsuda, Y., & Ukita, N. (2020). Finger gesture spotting from long sequences based on multi-stream recurrent neural networks. *Sensors, 20*(2), 528. https://doi.org/10.3390/s20020528

21. Köpüklü, O., Gunduz, A., Kose, N., & Rigoll, G. (2019). Real-time hand gesture detection and classification using convolutional neural networks. In *2019 14th IEEE International Conference on Automatic Face & Gesture Recognition (FG 2019)* (pp. 1–8). IEEE.

22. Kopuklu, O., Kose, N., & Rigoll, G. (2018). Motion fused frames: Data level fusion strategy for hand gesture recognition. In *Proceedings of the IEEE conference on computer vision and pattern recognition workshops* (pp. 2103–2111). IEEE.

23. Atterer, R., Wnuk, M., & Schmidt, A. (2006). *Knowing the user's every move: User activity tracking for website usability evaluation and implicit interaction* (pp. 203–212). Springer. https://doi.org/10.1145/1135777.1135811

24. Schmidt, A. (2000). Implicit human computer interaction through context. *Personal Technologies, 2000*(4), 191–199. https://doi.org/10.1007/bf01324126

25. Funk, M., Dingler, T., Cooper, J., & Schmidt, A. (2015). Stop helping me - I'm bored! Why assembly assistance needs to be adaptive. In *Adjunct Proceedings of the 2015 ACM International Joint Conference on Pervasive and Ubiquitous Computing and Proceedings of the 2015 ACM International Symposium on Wearable Computers (Osaka, Japan) (UbiComp/ISWC'15 Adjunct)* (pp. 1269–1273). Association for Computing Machinery. https://doi.org/10.1145/2800835.2807942

26. Jensenius, A. R. (2007). *Action-sound: Developing methods and tools to study music-related body movement*. University of Oslo DUO research archive. Retrieved from http://urn.nb.no/URN:NBN:no-18922

27. Wang, Z., Qin, R., Yan, J., & Guo, C. (2019). Vision sensor based action recognition for improving efficiency and quality under the environment of Industry 4.0. *Procedia CIRP, 80*(2019), 711–716.

28. Ni, P., Lv, S., Zhu, X., Cao, Q., & Zhang, W. (2020). A light-weight on-line action detection with hand trajectories for industrial surveillance. *Digital Communications and Networks, 2020*. https://doi.org/10.1016/j.dcan.2020.05.004

29. Jeanne, F., Soullard, Y., & Thouvenin, I. (2016). What is wrong with your gesture? An error-based assistance for gesture training in virtual environments. In *2016 IEEE Symposium on 3D User Interfaces (3DUI)* (pp. 247–248). IEEE.

30. Choi, J., Cho, Y.-i., Han, T., & Yang, H. S. (2007). A view-based real-time human action recognition system as an interface for human computer interaction. In *International conference on virtual systems and multimedia* (pp. 112–120). Springer.

31. Bovo, R., Binetti, N., Brumby, D. P., & Julier, S. (2020). Detecting errors in pick and place procedures: Detecting errors in multi-stage and sequence-constrained manual retrieve-assembly procedures. In *Proceedings of the 25th International Conference on Intelligent User Interfaces* (pp. 536–545). Springer.

32. Fullen, M., Maier, A., Nazarenko, A., Aksu, V., Jenderny, S., & Röcker, C. (2019). Machine learning for assistance systems: Pattern-based approach to online step recognition. In *2019 IEEE 17th International Conference on Industrial Informatics (INDIN)* (Vol. 1, pp. 296–302). IEEE.

33. Goodfellow, I., Bengio, Y., & Courville, A. (2016). *Deep learning*. MIT Press. Retrieved from http://www.deeplearningbook.org

34. Hanna, M. G., Ahmed, I., Nine, J., Prajapati, S., & Pantanowitz, L. (2018). Augmented reality technology using Microsoft HoloLens in anatomic pathology. *Archives of Pathology & Laboratory Medicine, 142*(5), 638–644.

35. Kun, A. L., van der Meulen, H., & Janssen, C. P. (2017). *Calling while driving: An initial experiment with HoloLens*.

36. Evans, G., Miller, J., Pena, M. I., MacAllister, A., & Winer, E. (2017). Evaluating the Microsoft HoloLens through an augmented reality assembly application. In *Degraded environments: Sensing, processing, and display 2017* (Vol. 10197). International Society for Optics and Photonics.

37. Heinz, M., Dhiman, H., & Röcker, C. (2018). A multi-device assistive system for industrial maintenance operations. In *International cross-domain conference for machine learning and knowledge extraction* (pp. 239–247). Springer.
38. Bader, S., & Aehnelt, M. (2014). Tracking assembly processes and providing assistance in smart factories. In *ICAART* (Vol. 1, pp. 161–168). ACM.
39. Heinz, M., Büttner, S., & Röcker, C. (2020). Exploring users' eye movements when using projection-based assembly assistive systems. In *International conference on human-computer interaction* (pp. 259–272). Springer.
40. Yamamoto, R., Chinese, M., & Andersson, L. (2020). *r9y9/pylibfreenect2: v0.1.2 release*. https://doi.org/10.5281/zenodo.3835702
41. Redmon, J. & Farhadi, A. (2018). *Yolov3: An incremental improvement*. arXiv preprint arXiv:1804.02767.
42. Zhang, P., Zhong, Y., & Li, X. (2019). SlimYOLOv3: Narrower, faster and better for real-time UAV applications. In *Proceedings of the IEEE international conference on computer vision workshops*.
43. He, K., Zhang, X., Ren, S., & Sun, J. (2016). Deep residual learning for image recognition. In *Proceedings of the IEEE conference on computer vision and pattern recognition* (pp. 770–778). IEEE.
44. Simonyan, K. & Zisserman, A. (2014). *Very deep convolutional networks for large-scale image recognition*. arXiv preprint arXiv:1409.1556.
45. Szegedy, C., Liu, W., Jia, Y., Sermanet, P., Reed, S., Anguelov, D., Erhan, D., Vanhoucke, V., & Rabinovich, A. (2015). Going deeper with convolutions. In *Proceedings of the IEEE conference on computer vision and pattern recognition* (pp. 1–9). IEEE.
46. Tran, D., Bourdev, L., Fergus, R., Torresani, L., & Paluri, M. (2015). Learning spatiotemporal features with 3D convolutional networks. In *2015 IEEE International Conference on Computer Vision (ICCV) (2015)* (pp. 4489–4497). IEEE. https://doi.org/10.1109/iccv.2015.510
47. Xie, S., Girshick, R., Dollár, P., Zhuowen, T., & He, K. (2017). Aggregated residual transformations for deep neural networks. In *Proceedings of the IEEE conference on computer vision and pattern recognition* (pp. 1492–1500). IEEE.
48. Materzynska, J., Berger, G., Bax, I., & Memisevic, R. (2019). The jester dataset: A large-scale video dataset of human gestures. In *Proceedings of the IEEE international conference on computer vision workshops*.
49. Molchanov, P., Yang, X., Gupta, S., Kim, K., Tyree, S., & Kautz, J. (2016). Online detection and classification of dynamic hand gestures with recurrent 3D convolutional neural networks. In *2016 IEEE conference on Computer Vision and Pattern Recognition (CVPR)* (pp. 4207–4215). IEEE. https://doi.org/10.1109/cvpr.2016.456
50. Zhang, Y., Cao, C., Cheng, J., & Hanqing, L. (2018). Egogesture: A new dataset and benchmark for egocentric hand gesture recognition. *IEEE Transactions on Multimedia, 20*(5), 1038–1050.
51. Idrees, H., Zamir, A. R., Jiang, Y.-G., Gorban, A., Laptev, I., Sukthankar, R., & Shah, M. (2016). The THUMOS challenge on action recognition for videos in the wild. *arXiv*.

Advancements in Vocational Training Through Mobile Assistance Systems

6

Marc Brünninghaus and Sahar Deppe

Abstract

To ensure a sustainable education for trainees, vocational training has to keep up with the elevated digitalization of manufacturing and work. Complicated technologies used in production systems shift the requirements of job training towards a more tech-centered preparation for future employments. Paired with increased diversity among trainees regarding cultural and educational backgrounds, solutions to accommodate both slower and faster learners are sought after. After an initial exemplification of challenges, that vocational training faces, we discuss the general usage of assistance systems in training applications, in order to better the respective results. For a comparison, we examine state-of-the-art digital learning platforms, i.e. assistance systems, that were developed to enhance vocational training. Typically, these platforms offer more or less individually tailored training plans and implement various technologies and interfaces to improve learning speed or quality. Based on the results of this comparison, we propose a new, mobile and user-adaptive assistance system "XTEND for education", that involves all stakeholders of the training in the process and uses modern technologies to overcome the challenges of current vocational training. By building on top of node graph-based, dynamically generated assignments, AR-based inclusion of training material and equipment and a robust infrastructure to support it, the XTEND assistance system is shown to be capable of meeting requirements of vocational training in the realm of Industry 4.0.

M. Brünninghaus (✉) · S. Deppe
Fraunhofer IOSB-INA, Lemgo, Germany
e-mail: marc.bruenninghaus@iosb-ina.fraunhofer.de; sahar.deppe@iosb-ina.fraunhofer.de

Keywords

Vocational training · Assistance systems · Augmented reality · Mobile assistance

6.1 Introduction

The digitization of manufacturing and work is changing systems, processes, and technologies in the areas of production and logistics, as definitions of the term Industry 4.0 [1, 2]. The use of data glasses and augmented reality (AR) in production [3–8], concepts such as Big Data [9] for the design of powerful PLM (Product lifecycle management) systems [10], or intelligent algorithms [11–13] for decision support are prominent examples of the entry of digital technologies into the world of work. However, these changes do not only permanently alter the systems and their mechanisms, they also transform the skills required of individuals and the demands placed on employees in a modern working environment [14, 15]. Thus, the aspect of training is vital and must be considered as a central and decisive field of action in the implementation of Industry 4.0. Specifically, trainers and trainees must be prepared and educated for the use of new technologies [16, 17].

Operating with new technologies and complicated machines often requires a great amount of know-how, making training cost- and time-consuming. This is due to the lack of standardization by supervisors or colleagues who train or teach trainees. Additionally, lowered autonomy, inefficiencies in the human-technology interaction, and weak user experience are other drawbacks [18–20], which affect the training times. One solution to tackle these problems is using digital assistance systems [AS] in vocational training [21].

Digital assistance systems are intelligent technical systems, available in various forms [13]. Such systems are divided into stationary and mobile systems, that are designed to operate alongside the user by assisting him or her sensorily, cognitively, or physically whenever needed [22]. They provide benefits including automatically standardized training process, documented and certified training, and individuality and transparency for training organizations [16, 17]. Nevertheless, these systems still have some limitations. Focusing on one modality, either in-situ projection, tablet, or augmented reality, inflexibility, and incapability adapting to the trainee's level are some of these drawbacks.

To overcome the aforementioned problems, the XTEND assistance system [23] is proposed, which is employed in different applications and supports various training fields (electrical, mechanical, . . .).

This contribution investigates the limitations of assistance systems in vocational training and offers solutions for the current problems. It is structured as follows: In Sect. 6.2, the problems and challenges of vocational training along with their solutions provided by assistance systems are explained. Section 6.3 provides a review of the existing assistance systems and their benefits and limitations. In Sect. 6.4, the XTEND assistance system and its modules are introduced. The performance of XTEND is evaluated in two case studies, described in Sect. 6.5. Finally, a conclusion and outlook are indicated in Sect. 6.6.

6.2   Integration of Assistance Systems into Basic Training

Solving problems, that the basic vocational training faces, is a critical success factor of a good education and smooth transition of trainees into the productive operation of their respective companies. An assistance system can help reach the development goals that are focused on in modern training variations in order to overcome these challenges [4]. This section outlines the common problems in vocational training along with the solutions offered by assistance systems.

6.2.1   Embedding Complex Technical Systems

Operating with new technologies and complicated machines often requires a great amount of know-how, which results in a slower training speed in order to ensure safe handling of the machines involved [24]. Integrating an assistance system into basic training as early as possible exposes the trainee to new technology in their workplace from the earliest stages on [25]. This prolonged contact improves understanding the handling and utilization of modern technical systems. With these new technologies embedded in the training, the transition into real production is more straightforward because the required know-how is already acquired beforehand.

For their later job in a digitalized production, trainees need to learn how to operate increasingly complex machines. An AS helps them break down the complexity into simple guided steps and tasks. This is anticipated to lead to an increased training capacity in the same period of time, which results in a competence gain for the trainee.

In addition, without proper instructions, trainees are at risk of falsely operating machines that they are not fully trained for, yet. In order to avoid bodily injury and expensive or time-consuming repairs, AR-based AS reduce the risk of mishandling equipment [26]. In order to minimize improper use, AR content can aid trainees while working on real machines or with actual tools [7, 27]. This is done by giving visual, auditory, or haptic feedback through a device (e. g. tablet) and warn the user in case of possibly dangerous situations or actions. Moreover, AR can simulate equipment by placing virtual objects into a training scenario which the trainee can work on realistically.

6.2.2   Agility

In a fast-changing corporate setting, it is increasingly difficult for trainers to keep their programs up to date with the current state of the art. Technical, societal, or economical changes quickly find their way into production systems. Therefore, traditional "pen & paper"-training struggles to keep up due to slow adaption processes of training materials or disproportionate costs for minor improvements.

For this purpose, the proposed AS features a content generation tool for trainers (Sect. 6.4.2), that makes changes to existing exercises quickly and easy to model. With the use of node graphs [28] to generate single exercises, an iterative design concept to benefit incremental and adaptive changes is applied in this software tool. These changes can then be individually optimized and pushed to select trainees through the accompanying management tool (Sect. 6.4.3).

6.2.3   Inclusion of Different Levels of Education

In vocational training people with different backgrounds and skills are joined together in classes, where trainers have to go through the mandatory syllabus with everyone. Thus, there is only limited flexibility and time for them to address individual needs or potentials within the group. By equipping trainees with an AS, instructors can maximize trainees' potential by adapting their course content to individual learning capabilities [19, 29–31]. On one side, fast learners or trainees with higher education can go above and beyond the mandatory curriculum with more advanced exercises provided through the content management of the AS. On the other side, people without any extra prior knowledge or learning disabilities can be assisted on a fine-grained level. Because the trainer does not need to explain single exercises to the whole group in depth anymore they can use this time to go into detail and work on deficits with individual apprentices. For this, the integrated progress tracker (Sect. 6.4.6) can be used to dissect single exercises and discuss the trainee's work.

Moreover, in absence of the trainer, with the AS it is possible for the trainees to read up on topics they are interested in or that they might need to fulfill their due tasks. This is done by including optional, informative text chapters into exercises, that users can open on their devices at will. So, because trainees may read it on their own as opposed to the trainer teaching it as an obligatory lesson, the partial switch from information "push" to "pull" is anticipated to lead to increased learning motivation and speed [32, 33].

6.2.4   Place and Time-Independent Learning

Companies often operate on an international or even global scale. When dispersing on-site training locations across the globe, a consistent skill level after graduation is needed to ensure similar and stable product quality throughout the entire organization. Different challenges in the standardization of training arise from this. To accommodate heterogeneous trainees regarding place, language, skill level or the time zone they are in, a place and time-independent learning platform is desired. Thus, an AS is proposed here, which is divided into multiple distributed modules that can be used both on mobile and stationary devices. It is even possible for trainees to bring their own device (e. g. tablet) and use it to work on given exercises during their training. While it may come with other implications like company data protection, moving the responsibility for these devices away from the employer (i.e. the training facility) reduces procurement costs and device administration time. The distributed modules account for creating, storing, publishing, adapting, and executing exercises embedded into lessons and courses. In times of the global Covid-19 pandemic, remote (i.e. distanced) training, that is place and time-independent, becomes even more valuable.

6.2.5 General Appeal of Vocational Training

Over the past decades, the number of university students in Germany has increased significantly. Even adjusted for overall population growth and demographic change, it seems the traditional vocational training has become a more and more unpopular choice among school graduates [34]. Apart from the benefits of downstream employment opportunities, the general appeal of vocational training, as opposed to a university career, needs improvement to attract more prospective school graduates. The desire for an apprenticeship can be created or at least assisted by embedding modern technology into the training. Hence, vocational training that presents an opportunity for enhanced media competence has a greater appeal.

6.3 State of the Art

During the last decade, digital assistance systems have been increasingly employed in the workplace and for learning and vocational training. These systems can be compared in various ways. The ability to provide step-to-step guidance, creating (new) instructions, AR demonstration, data collection and processing are a few exemplary features for comparison. Table 6.1 summarizes the differences of currently available assistance systems for training that were developed due to the problems mentioned in Sect. 6.2.

The Elabo assistance system is designed for both basic and developed training in electronics and electrical engineering [35]. It is used equally in companies and technical colleges. Elabo is modular in design and can be flexibly adapted to changes in the training concepts. Nevertheless, this system cannot collect and analyze the data from machines and the learning process. Additionally, it only performs with the Windows operating system (OS).

The SurMe assistance system is intended to provide a haptic learning environment for simulating surgical procedures and to be used in the training and further education of doctors as part of an integrated training concept [36]. SurMe provides a location and time-independent platform for tutor-based learning where both trainees and teachers are involved.

Skyware Connected Knowledge [37] is an innovative solution for securing and retrieving expert knowledge in different work contexts. With a mobile application (app), employees can access knowledge documents such as manuals, plans and videos directly at the workplace. Moreover, they can digitally record their experiential knowledge during work and make it available to other colleagues. However, like Elabo, Skyware cannot gather and analyze data from assembled products and during the learning process.

REFLEKT ONE is a modular augmented reality platform for front-line users including the AR Viewer application [38]. It is employed in production, training, maintenance, and repair applications by providing visuals and guides on smartphones, tablets, and data glasses. Despite several advantages, REFLEKT ONE is unable to collect and process data.

Raumtänzer [39] is another system that assists employees' training by providing learning content and products' information digitally. It is accessible via the usual web browsers (desktop and mobile). However, this system is not flexible, meaning that new learning contents or work instructions cannot be created and added to the system. Additionally, this system lacks central storage.

The assistance system of Festo AG, APPsist [40], is mostly employed in maintenance and assembly processes [42]. However, this system is also tested for mechatronics training. The main advantage of APPsist is its adaptability, meaning that it automatically adjusts the application to the support needs of skilled users [40]. Moreover, APPsist is connected to individual machines via technical interfaces so that data from the production process can be transmitted on a mobile basis.

The assistant and learn systems from Christiani [41] proved their performance in the area of metal technology, electronic, and mechatronics training. Christiani offers several e-learning materials for trainees that are available on tablets or PC. The learning materials are presented as reading texts, animated graphics, pictures, and videos. One example of Christiani's assistant and learn system is the mMS Sorting System Compact 4.0, which is used in automation and mechatronics training. This system represents an industrial production process in miniature with the ability to show relevant production data on a computer's or tablet's screen. Regardless of their benefits, adding new products or instructions, and representing AR contents are challenging for Christiani's systems.

AWAre [19] is an adaptive assistance system, which is employed for industrial as well as educational aspects. It connects different user interfaces, and provides plugins to include sensors and actuators [19]. This system can be applied in stationary and mobile applications.

An analysis of different assistance systems (cf. Table 6.1) shows that all these systems are capable of providing a step-by-step guidance for users.

Nevertheless, editing or adding new instructions cannot be tackled by some of the systems. AR demonstrations offer possibilities to improve the quality of vocational education and training, however, not all the investigated assistance systems support this feature. Almost all tested systems are equipped with central storage and multi-platform support options. In order to achieve transparency and better integration into the company, documentation and maintenance processes are offered by most of the investigated systems. Options such as object tracking, data collection and processing are supported by a few of the mentioned systems.

Table 6.1 Comparison of assistance systems in the domain of learning and training

Features	Assistance system								
	Elabo [35]	SurMe [36]	Skyware [37]	REFKLEKT ONE [38]	Raumtänzer [39]	APPsist [40]	Christiani [41]	AWAre [19]	XTEND [23]
Step-by-step guidance	✓	✓	✓	✓	✓	✓	✓	✓	✓
Creating instructions	✓	✓	✓	✓	–	✓	–	✓	✓
Central storage	✓	✓	✓	✓	–	✓	✓	✓	✓
Multi-platform support	✓	✓	✓	✓	✓	✓	✓	✓	✓
Object tracking	–	–	✓	✓	–	✓	–	✓	✓
AR content playback	–	✓	✓	✓	✓	–	–	–	✓
Documentation of maintenance processes	✓	✓	✓	✓	–	✓	✓	✓	✓
Data collection and processing	–	–	–	–	–	✓	✓	✓	✓
Mobility	✓	–	✓	✓	✓	✓	✓	✓	✓

Limitations of the stated systems are tackled in XTEND [23]. XTEND is a modular assistance system, spanning over five individual modules. It proved its performance in a wide range of applications, such as industrial montage and assembly, maintenance, repair and installation, and training [43, 44]. The XTEND AS supports diverse end devices, e.g. different tablets, smartwatches or head-mounted displays. Additionally, it can be used in stationary as well as mobile applications. Mobile assistance systems, like XTEND, are not bound to a specific place and can be employed by trainers or trainees from different locations such as at home, on-site at work etc.

6.4   Design and Implementation Concept

To maximize the potential of assisted learning for trainers as well as trainees, the proposed assistance system is split up into modules, each serving its specific purpose to aid the vocational training in the best way possible. In this section, the design and implementation of these modules are described and their organizational and technical purposes are displayed.

6.4.1   The AS Modules

The XTEND AS for education is divided into five individual modules. They are working together closely to provide an effective and consistent training experience. Figure 6.1 shows the structure of the modules and their interaction with each other. The general work ow of the AS is also recognizable and is described in the following subsections.

6.4.2   Module 1: The Trainer Software

The trainer software is a standalone desktop app, that enables the trainer or other instructed persons to create a variety of content (i.e. assignments) for the trainees to access while using the assistance system. To make it as accessible as possible, the software is built to support multiple widespread operating systems, including Microsoft Windows and Apple's MacOS. After installing the app, the user can log into their centrally managed account and create new content to view in the trainee software. In order to support a vast variety of learning and training applications, the trainer software is built upon the principle of *Business Process Modeling and Notation* (BPMN). BPMN is perceived as a modeling language not only to visualize but to describe and define virtually any kind of business processes in a standardized way. According to the standard published by the Object Management Group [45], its "primary goal [... is to] provide a notation that is readily understandable by all business users, from the business analysts that create the initial drafts of the processes, to the technical developers responsible for implementing the technology

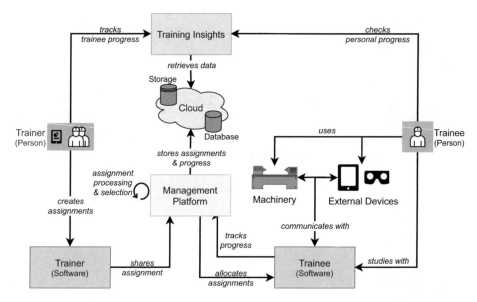

Fig. 6.1 Structure of the XTEND assistance system for education

that will perform those processes, and finally, to the business people who will manage and monitor those processes".

The trainer software is loosely based on this principle. It is using node graphs to define the look, content and ow of the assistance that the trainees receive. These nodes can be seen as placeholders (like empty shells) for user interface (UI) elements, that the trainer can then fill with meaningful content. This content can be visual elements like images, videos, PDF documents, texts but also - and that is the huge benefit of the XTEND assistance system - machine-specific and self-made database connections. These machine-specific connections can be used to communicate with equipment that the trainees work with during their training, e. g. a lathe or drill press. A generic *Device* node works as the connector between the trainer software and the plugin that runs on the machine itself. The per machine-implemented plugin translates abstracted, reusable commands coming from the AS to machine-understandable instructions. For example, if the trainer wants to make sure that his trainees have closed the guard before turning on the spindle of a lathe, they can use a device node to request the status from the connected machine whether or not the guard is closed and only then unblock the main switch to power it on. This way, a machine-independent interaction (apart from the machine model-dependent plugins) between trainee and used equipment can be realized via the AS. *Database* nodes enable the trainer to incorporate different kinds of questions into the exercises. If the trainee needs to type in the correct drill speed for a certain material, for example, their answer can be used for several tasks. First of all, it can be checked against predefined values for different materials in the connected database and feedback can be given based on the result. Secondly, the answer can later be reused to verify that the trainee has put in his own calculations into the machine

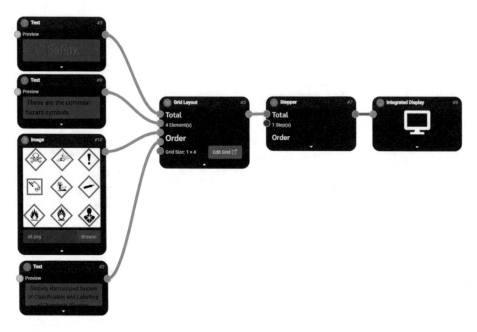

Fig. 6.2 A node graph of trainee instructions created in the trainer software

correctly and once more give feedback accordingly. This technique allows the AS to react to user input adaptively.

After filling the nodes with content, the trainer can connect them with edges, to define a flow or order in which they are shown or rather executed in the trainee software. Figure 6.2 shows a basic content view modeled in the trainer software. After completing the whole graph, the trainer can upload it to the management platform (cf. Sect. 6.4.3), which interprets and parses the graph into actual logical and UI elements for the trainee software.

6.4.3 Module 2: The Management Platform

The management platform is the central building block of the AS. It consists of a web frontend and a backend including several services which are the pillars, the complete AS is built upon. Next to technically necessary services like the communication infrastructure (REST API, mqtt broker), the web-frontend on the management platform lets selected users (e. g. trainers and administration personnel) manage the complete training operations. This includes tasks like

- new trainee registration,
- granting or managing permissions, classes and roles,

- assigning trainees and trainers to classes and courses as well as
- scheduling exercises.

Exercises can be allocated to courses, put in a specific order, and start as well as due dates can be specified. Additionally, different versions of exercises, e. g. for different trainee skill levels or handicaps like hearing impairments, can be assigned to single users and classes.

However, the most critical part of the platform is to process and select assignments created with the trainer software, in order to generate usable content for the trainee software (cf. Sect. 6.4.5). The trainee software runs on mobile devices and cannot interpret node graphs coming from the trainer software, this part is accomplished with the backend of the management platform.

Another service on the platform analyzes the node graph and deconstructs it back into its single nodes. These nodes, depending on their type, are parsed into UI elements, logic blocks, or raw data, which are all stored in the central database located on the local cloud server (cf. Sect. 6.4.4). The edges, which connect the nodes in the graph, are examined and relevant input and output data are connected via several tables in the database. This guarantees that every UI element gets the right data at the right time during runtime. After the initial analysis, all elements and blocks are parsed into a single human-interpretable view, which is later executed in the trainee software. This view is then persisted in the cloud storage for place and time-independent, network-wide access. Following the build process, the trainer is now able to give access to this assignment for their trainees through the management platform (cf. Fig. 6.3).

6.4.4   Module 3: Cloud Storage and Database

The cloud server is physically based on the same machine as the management platform. Thus, the name 'cloud' is easily misunderstood in this context. It means that it is a central server, always reachable from the local network, but configurable to be accessed from the internet. This server hosts a storage service for binary data, files like images and videos, and a service to store structured data (the database). These services are directly or indirectly used by all modules of the AS. It receives the assignments from the trainer software (processed by and through the management platform), the allocation of classes, courses, and assignments as well as user lists etc. from the management platform, individual progress of due tasks from the trainee software and status updates from connected machinery. It also provides data for all modules when needed, e. g. when a trainee opens an assignment on their mobile device and images are loaded from the cloud storage via Wi-Fi.

6.4.5   Module 4: The Trainee Software

The trainee software is an app available via the app stores for the two major OSes for mobile devices.

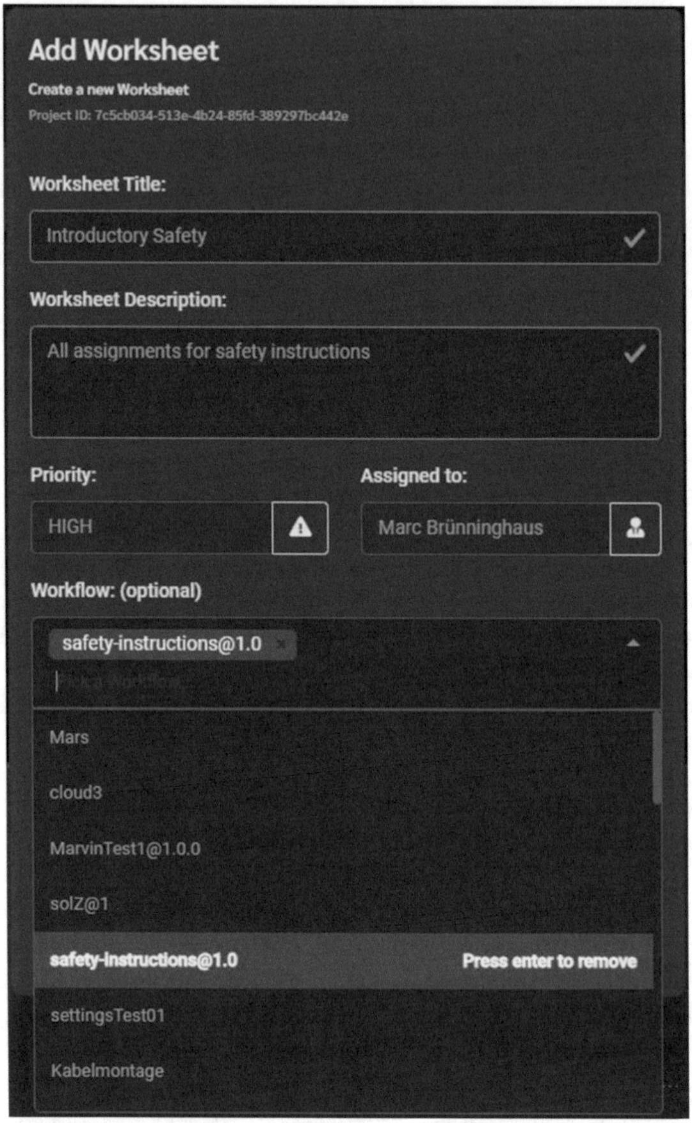

Fig. 6.3 Adding an assignment to a course on the management platform

It supports Apple's iOS and Google's Android OS, which have a combined market share of 99.18% by June 2021 [46]. This ensures, that the vast majority of users are able to download and install this app on their mobile devices. Especially if a training facility decides to use their trainees' own devices, this becomes a critical factor.

After logging into the account, the trainee sees their given assignments (which are allocated through the management platform (cf. Sect. 6.4.3)). It is also visible which assignments are still open and which are already completed. By opening an assignment, the software displays its contents on the device's screen. Now, the user is able to navigate through the exercises, read up on informative texts, answer the incorporated questions and interact with the assistance system in exactly the way the trainer intended and defined it. Theoretically, the possibilities here are endless because of the modular, reusable generation of content in the trainer software (cf. Sect. 6.4.2). What makes the XTEND trainee software so valuable for vocational training is its utilization of AR technology and its ability to communicate with external devices and machinery through its open communication standard. By utilizing Device nodes, the trainee can interact with machines directly through the app. The software wirelessly sends and requests data via its standardized interface to and from any device or machine, that implements this interface via a dedicated plugin. This way, commands like *Power On*, *Read Speed*, *Write Signal* or *Shutdown* can be send via the software at a specific point during the exercise. Hence, the theoretical experience during these exercises can be much more integrated into the practical work that is without AS strictly separated due to a lack of integration possibilities.

Another feature this software offers, are the AR-possibilities that can be used to transfer information in a better way than traditional media are capable of [7, 27, 47]. Information directly available at its source is known to be better understandable and memorable than information, e. g. in books, that is decoupled and dissociated from its context [48–50]. Figure 6.4 shows an application of AR in the XTEND software, where placement information and a description with detailed instructions for individual parts are placed directly at their target destinations in the assembly unit. Users can immediately see where the part has to go and what steps are necessary to secure it in place.

Similar to the communication with other machinery through Device nodes, the trainee software can interact with external in- or output devices like head-mounted displays or projectors with the integration of *External Display* nodes. These external displays offer additional AR functionality, e. g. in-situ-projection, and an even broader spectrum of possible training applications.

While working with the trainee software, the individual user progress can be automatically tracked and saved to the cloud, as well. For this purpose, the software sends updates with relevant data to the management platform, which in turn saves them to the database. From there, the trainee and their trainer can access this information using *Training Insights* (cf. Sect. 6.4.6).

6.4.6 Module 5: Training Insights

Training Insights is a web-based tool to see and review individual progress records of the participating trainees. The frontend uses an analytics engine in order to visualize training-relevant data like time spent in each exercise, wrong and correct answers, software usage frequency etc. [51].

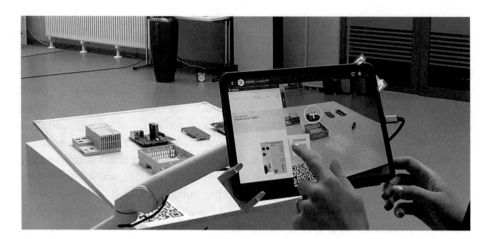

Fig. 6.4 An assembly use-case, with relevant information placed on top of their real-life counterparts via AR

This information can be used by trainers to make sure every exercise is completed in the given time by their trainees and gain an understanding of which exercises were difficult for certain trainees and which were easy [52]. Furthermore, based on this knowledge, which is not always willingly presented by the trainees, the trainer has the time and awareness to work on possible shortcomings that may become visible throughout the training.

Trainees can use their track record on Training Insights as a self-check to keep an eye on their performance during training.

A prospective additional use could be to grant other stakeholders, like the employer I in case of external training facilities, access to insights. They would have the ability to see their employee's progress and prepare and adapt their reintegration into their local production based on the training results.

6.5   Case Studies

The XTEND assistance system for education has been developed in collaboration with partners from both accredited training facilities and educational sciences institutes. To verify and improve its impact on vocational training, the system was tested in different scenarios, two of which are presented in the following chapter. During testing, a deeper understanding of how the system complements the work of trainers and trainees, what challenges were faced and overcome and what features need improvement was gained.

6.5.1   Study 1: Work 4.0

The goal of the first case study is the digitization of real-world applications by the orchestration of work, people and technology in existing processes of work. This use case outlines the joint project "AWARE—Arbeit 4.0" (AWARE - Work 4.0) of the "it's OWL" cluster management [53]. One of the objectives of this project was to pursue the participatory nature and design of digitized work with a particular focus on human-centeredness. For this use case, the XTEND AS was integrated to provide different requirements at the performance and management level, and reduce the challenges of technological system integration with regard to process consistency. Thus, XTEND enabled the availability of knowledge to employees and stakeholders extrinsically and individually. In the context of this project, AR technologies were included in the XTEND AS for an assembly process to support actions by enriching the real workplace with digital information.

During this project, several collaborative workshops were held to inform and integrate the employees with the status and development of the assistance system. XTEND was available to both trainers and trainees with and without experiences on assembly tasks. A usability test [54] and an interview were performed to gather the uncovered problems and discover users' opinions and preferences.

The overall outcome showed that using the XTEND assistance system provides the following advantages:

1. Straightforward guidance for trainees without experiences: Trainees without previous knowledge or experience could complete the assembly task by following the steps.
2. Decreasing the time of the assembly task for the trainees: XTEND could shorten the assembly task and reduce the errors by representing various types of media, e.g. images, videos and AR technology.
3. Improving the comprehension of the process: The AR technology of XTEND helped to understand the assembly steps and detect the target part or tool. This is a significant benefit for complex tasks.
4. Effective evaluation: Trainers evaluated the trainees more effectively based on the gathered data during the training process.

Moreover, XTEND was successfully tested in different training domains during this project. The results proved that the range of XTEND AS applications is not limited to sequential assembly processes.

Nevertheless, creating content for AR was one of the challenges for trainers. Specifically, using the built-in technology framework to anchor individual elements (texts, images) on the intended 3D position was experienced as fairly diffcult by the trainers. It was seemingly hard to position the elements exactly at the position on the object or in the room they wanted them to be. This can be traced back to both their lack of experience using AR and the technique that is required to anchor the elements in the app. As a result, in a later version a new framework was introduced that uses point clouds for a more accurate positioning of AR elements [55].

6.5.2   Study 2: Joint Apprentice Workshop

For the second case study, the XTEND AS was tested in a joint apprentice workshop. This workshop offers standardized training for employees of local SMEs[1] that do not have the capacity for their own on-site workshop. Since joint workshops are a common practice in Germany, the chosen facility is representative for a regular vocational training. In this case, the AS is used to supplement the basic mechanical and electrotechnical training. Specifically, assignments for the first year of training were assisted.

During several workshops, the capabilities of the AS were first explained to the trainers who had the chance to create their prepared exercises afterwards. With increasing repetitions, the trainers got more and more used to the assisted training work ow and down the line they were able to put together new assignments faster with the node graph-based trainer software. Because the management platform was developed with the hierarchical structure of an education organization in mind, setting up classes and courses was perceived as fairly easy once the location of these functions on the web-frontend was understood. Allocating exercise material to the courses could also be done quickly by the trainers. Speaking from trainers' experience using the (brief test of the) AS, following benefits were identified:

1. It is capable to reduce the complexity of single training steps through virtual and digital manuals, which leads to an overall better understanding not only of the exercise but also of the matter at hand.
2. For trainees, the readability of technical drawings improves, when they are digitally augmented with additional, easily accessible information (cf. Fig. 6.5).
3. Through digitally assisted learning techniques, a changed culture of learning can be seen among trainees, with higher motivation to take responsibility for their own study progress and find their own solutions to given challenges. The pursued process and technology competencies in trainees can be reached quicker this way.

However, a few observations and future improvements were made and identified during the study. While using the AS in their training, the trainers found that their personal role pivots from a strict teacher to more of a coach. In this case, the AS takes over the hands-on teaching. The trainer is working with the trainee to overcome challenges when working on assignments on the device or machine and becomes some sort of a mentor trainees only consult when they face a problem they cannot solve themselves.

---

[1] Small and medium-sized enterprises.

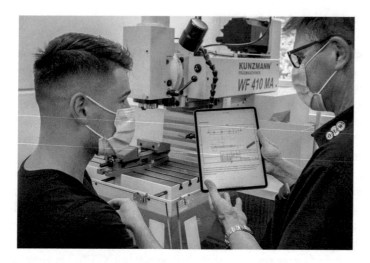

Fig. 6.5 Trainer and trainee discussing a technical drawing shown on the trainee software

A problem of technical nature became apparent when the trainers structured their course material on the management platform. At the time of the study, there was no way of limiting a trainee's access to assignments according to a specific order of completion. That means, the trainers need a way to grant access to subsequent exercises based on the completion or even performance of different assignment that come prior. By setting a specific order of completion for their exercises, trainers can gradually increase the difficulty of the assignments and together with start and due dates, they can keep a fine-grained control over the tasks their trainees are working on. In addition to that, the trainers asked for a way to place clickable links not only on AR-content, but also on pictures and drawings that only show on the screen of the device.

For example, in technical drawings often there is a surface roughness α given using a symbol shown below:

$$\alpha = \sqrt{\dfrac{\textit{turned}}{Rz\ 6.3}} \qquad .$$

To understand what this symbol means, it seems plausible for the user to simply click (or tap on a touchscreen) on this symbol for an explanatory text, rather than "moving away" by scrolling down or opening up a textbook for additional information.

Overall, the assistance system was viewed as very helpful for the training by both trainers and trainees.

6.6   Conclusion and Outlook

An overview of major challenges, limitations, and benefits of digital assistance systems in vocational training is provided in this contribution. Examples of these limitations are: Focusing on one modality (e. g., either in-situ projection, tablet, or augmented reality), inability to collect and analyze data from machines or during the process, and incapability adapting to the trainee's level. Positive effects of employing assistance systems in vocational training include fewer errors, improved quality, offering the right amount of support, increased motivation and speed of learning.

In this context, state-of-the-art systems and their advantages and drawbacks are investigated. A novel, modular assistance system, namely XTEND, is proposed, which tackles the problems of the existing methods. XTEND consists of five modules working and interacting together. The performance of XTEND is evaluated within two case-studies in vocational training.

The outcomes of the case-studies show that employing XTEND in vocational training improved the training level. It provided a better experience and individual support.

The focus of the future work will be on intelligent analytic functions to enhance the automated user adaptations, a greater variety of tools to generate other content for assignments and an improved user experience.

Regarding the user adaptation, currently the XTEND AS implements several discrete levels of user skills. According to the previously set skill level of the user, the amount of detail for assistance is chosen. This can be improved by constantly gauging the speed and precision the user is going through the assignments and assistance. This can be achieved by evaluating the data generated from the trainee software feedback (cf. Sect. 6.4.6) and adapt the assistance based on the results.

In order to expand into different training scenarios, the AS has to support additional types of content. This content is created and incorporated into new assignments using the nodes in the trainer software. New node types, like the clickable links on 2D images mentioned in Sect. 6.5.2, will build the foundation to assist more training scenarios and support further areas of work.

A more intuitive user experience will benefit the AS in its user friendliness and fluency of its workflows. By example of the AR content placement criticized during the first case study (Sect. 6.5.1), improved interaction concepts will increase the technology acceptance among trainers and trainees and streamline creating and working on assignments. This leads to more frequent as well as more efficient use of the software system respectively.

Acknowledgments The authors would like to thank Prof. Carsten Röcker, Sascha Martinetz and Alexander Kuhn for their help. This work was partly funded by the Ministry of Economic Affairs, Innovation, Digitalisation and Energy of the State of North Rhine-Westphalia[2] through The

---

[2] https://www.wirtschaft.nrw/

Technology-Network: Intelligent Technical Systems OstWestfalenLippe.Germany[3] and the Federal Institute for Vocational Education and Training.[4]

References

1. Lasi, H., Fettke, P., Kemper, H. G., Feld, T., & Hoffmann, M. (2014). Industry 4.0. *Business & Information Systems Engineering, 6*(4), 239–242.
2. Lu, Y. (2017). Industry 4.0: A survey on technologies, applications and open research issues. *Journal of Industrial Information Integration, 6,* 1–10.
3. Bröring, A., Fast, A., Büttner, S., Heinz, M., & Röcker, C. (2019). *Smartwatches zur unterstützung von produktionsmitarbeitern*. Mensch und Computer 2019-Usability Professionals.
4. Fellmann, M., Robert, S., Büttner, S., Mucha, H., & Röcker, C. (2017). Towards a framework for assistance systems to support work processes in smart factories. In *International cross-domain conference for machine learning and knowledge extraction* (pp. 59–68). Springer.
5. Haase, T., Weisenburger, N., Termath, W., Frosch, U., Bergmann, D., & Dick, M. (2014). The didactical design of virtual reality based learning environments for maintenance technicians. In *International Conference on Virtual, Augmented and Mixed Reality* (pp. 27–38). Springer.
6. Heinz, M., Büttner, S., & Röcker, C. (2019). Exploring training modes for industrial augmented reality learning. In *Proceedings of the 12th ACM international conference on PErvasive technologies related to assistive environments* (pp. 398–401). ACM.
7. Kohn, V. & Harborth, D. (2018). Augmented reality: A game changing technology for manufacturing processes? In *Twenty-Sixth European Conference on Information System*.
8. Sand, O., Büttner, S., Paelke, V., & Röcker, C. (2016). smART.Assembly–projection-based augmented reality for supporting assembly workers. In *International conference on virtual, augmented and mixed reality* (pp. 643–652). Springer.
9. Mayer-Schönberger, V., & Cukier, K. (2013). *Big data: A revolution that will transform how we live, work, and think*. Houghton Mifflin Harcourt.
10. Li, J., Tao, F., Cheng, Y., & Zhao, L. (2015). Big data in product lifecycle management. *The International Journal of Advanced Manufacturing Technology, 81*(1), 667–684.
11. Fullen, M., Maier, A., Nasarenko, A., Aksu, V., Jenderny, S., & Röcker, C. (2019). Machine learning for assistance systems: Pattern-based approach to online step recognition. In *2019 IEEE 17th International Conference on Industrial Informatics (INDIN)* (Vol. 1, pp. 296–302). IEEE.
12. Lang, D., Wunderlich, P., Heinz, M., Wisniewski, L., Jasperneite, J., Niggemann, O., & Röcker, C. (2018). Assistance system to support troubleshooting of complex industrial systems. In *2018 14th IEEE International Workshop on Factory Communication Systems (WFCS)* (pp. 1–4). IEEE.
13. Steil, J., & Wrede, S. (2019). *Maschinelles Lernen und lernende Assistenzsysteme. Berufe- und Branchen-Screening: Berufsbildung vor neuen Herausforderungen Arbeiten und Lernen mit intelligenten Systemen* (Vol. 14). KI-Chance oder Bedrohung.
14. Hirsch-Kreinsen, H., & Karacic, A. (2019). *Autonome Systeme und Arbeit: Perspektiven, Herausforderungen und Grenzen der Künstlichen Intelligenz in der Arbeitswelt*. transcript Verlag.
15. Manyika, J., Chui, M., Miremadi, M., & Bughin, J. (2017). *A future that works: AI, automation, employment, and productivity* (Vol. 60, pp. 1–135). McKinsey Global Institute Research.

---

[3] https://www.its-owl.com/home

[4] https://www.bibb.de/en/index.php

16. Büttner, S., Sand, O., & Röcker, C. (2017). Exploring design opportunities for intelligent worker assistance: A new approach using projection-based AR and a novel handtracking algorithm. In *European Conference on Ambient Intelligence* (pp. 33–45). Springer.
17. Oestreich, H., Wrede, S., & Wrede, B. (2020). Learning and performing assembly processes: An overview of learning and adaptivity in digital assistance systems for manufacturing. In *Proceedings of the 13th ACM international conference on PErvasive technologies related to assistive environments* (pp. 1–8). ACM.
18. Butollo, F., Jürgens, U., & Krzywdzinski, M. (2019). From lean production to industrie 4.0: More autonomy for employees? In *Digitalization in industry* (pp. 61–80). Springer.
19. Oestreich, H., da Silva Bröker, Y., & Wrede, S. (2021). An adaptive workflow architecture for digital assistance systems. In *The 14th PErvasive technologies related to assistive environments conference* (pp. 177–184). ACM.
20. Sochor, R., Kraus, L., Merkel, L., Braunreuther, S., & Reinhart, G. (2019). Approach to increase worker acceptance of cognitive assistance systems in manual assembly. *Procedia CIRP, 81,* 926–931.
21. Mavrikios, D., Papakostas, N., Mourtzis, D., & Chryssolouris, G. (2013). On industrial learning and training for the factories of the future: A conceptual, cognitive and technology framework. *Journal of Intelligent Manufacturing, 24*(3), 473–485.
22. Heinz, M., Büttner, S., Jenderny, S., & Röcker, C. (2021). Dynamic task allocation based on individual abilities-experiences from developing and operating an inclusive assembly line for workers with and without disabilities. *Proceedings of the ACM on Human-Computer Interaction, 5,* 1–19.
23. XTEND. (2021). Accessed August 08, 2021, from https://www.iosb-ina.fraunhofer.de/en/divisions/Assistance-Systems/Research-topics-projects-and-products/Assistance-system-XTEND.html
24. Tao, W., Lai, Z. H., Leu, M. C., Yin, Z., & Qin, R. (2019). A self-aware and active-guiding training & assistant system for worker-centered intelligent manufacturing. *Manufacturing letters, 21,* 45–49.
25. Büttner, S., Prilla, M., & Röcker, C. (2020). Augmented reality training for industrial assembly work-are projection-based AR assistive systems an appropriate tool for assembly training? In *Proceedings of the 2020 CHI conference on human factors in computing systems* (pp. 1–12). ACM.
26. Dhiman, H., Büttner, S., Röcker, C., & Reisch, R. (2019). Handling work complexity with AR/deep learning. In *Proceedings of the 31st Australian Conference on Human- Computer-Interaction* (pp. 518–522). Springer.
27. Stender, B., Paehr, J., & Jambor, T. N. (2021). Using AR/VR for technical subjects in vocational training–of substancial benefit or just another technical gimmick? In *2021 IEEE Global Engineering Education Conference (EDUCON)* (pp. 557–561). IEEE.
28. Fettke, P. (2008). Business process modeling notation. *WIRTSCHAFTSINFORMATIK, 50*(6), 504–507. https://doi.org/10.1007/s11576-008-0096-z
29. Funk, M., Dingler, T., Cooper, J., & Schmidt, A. (2015). Stop helping me: I'm bored! why assembly assistance needs to be adaptive. In *Adjunct Proceedings of the 2015 ACM International Joint Conference on Pervasive and Ubiquitous Computing and Proceedings of the 2015 ACM International Symposium on Wearable Computers* (pp. 1269–1273). ACM.
30. Hinrichsen, S., & Bornewasser, M. (2019). How to design assembly assistance systems. In *International conference on intelligent human systems integration* (pp. 286–292). Springer.
31. Petzoldt, C., Keiser, D., Beinke, T., & Freitag, M. (2020). Functionalities and implementation of future informational assistance systems for manual assembly. In *International conference on subject-oriented business process management* (pp. 88–109). Springer.

32. Bassendowski, S. L., & Petrucka, P. (2013). Are 20th-century methods of teaching applicable in the 21st century? *British Journal of Educational Technology, 44*(4), 665–667.
33. Cybenko, G., & Brewington, B. (1999). The foundations of information push and pull. In *The mathematics of information coding, extraction and distribution* (pp. 9–30). Springer.
34. Destatis statistisches bundesamt. (2021). Accessed August 08, 2021, from https://www.destatis. de/EN/Themes/Society-Environment/Education-Research-Culture/Vocational-Training/_node. html
35. Elabo: Elabo ausbildungs systeme. (2021). Accessed July 07, 2021, from www.elabo.de
36. Surme: The surgical mentor system. (2019). Accessed July 08, 2021, from https://www. interaktive-technologien.de/projekte/surme
37. Skyware: Skyware connected knowledge. (2015). Accessed July 08, 2021, from https://en. condat.de/solutions/skyware-connected-knowledge
38. ReflektOne. (2015). Accessed August 08, 2021, from https://www.re-flekt.com/de/reflekt-one
39. Raumtänzer. (2017). Accessed August 08, 2021, from https:///www.raumtaenzer.com
40. Festo: Appsist. (2016). Accessed August 08, 2021, from https://www.festo.com/net/de_de/ SupportPortal/Details/417329/PressArticle.aspx
41. Christiani. (2021). Accessed August 08, 2021, from https://www.christiani.de/
42. Ullrich, C., Hauser-Ditz, A., Kreggenfeld, N., Prinz, C., & Igel, C. (2018). Assistenz und wissensvermittlung am beispiel von montage-und instandhaltungstätigkeiten. In *Zukunft der Arbeit–Eine praxisnahe Betrachtung* (pp. 107–122). Springer.
43. Büttner, S., Peda, A., Heinz, M., & Röcker, C. (2020). *Teaching by demonstrating - How smart assistive systems can learn from users* (pp. 153–163). Springer.
44. XTEND. (2021). Accessed August 08, 2021, from https://www.iosb-ina.fraunhofer.de/de/ geschaeftsbereiche/assistenzsysteme/referenzen.html
45. Object management Group: Business process model and notation (bpmn) (03012011). Retrieved from https://www.omg.org/spec/BPMN/2.0
46. Statista. (2021). *Mobile operating systems' market share worldwide from January 2012 to June 2021*. Accessed from https://www.statista.com/statistics/272698/global-market-share-held-by-mobile-operating-systems-since-2009/
47. Büttner, S., Mucha, H., Funk, M., Kosch, T., Aehnelt, M., Robert, S., & Röcker, C. (2017). The design space of augmented and virtual reality applications for assistive environments in manufacturing: A visual approach. In *Proceedings of the 10th international conference on PErvasive technologies related to assistive environments* (pp. 433–440). ACM.
48. Bacca Acosta, J. L., Baldiris Navarro, S. M., Fabregat Gesa, R., Graf, S., et al. (2014). Augmented reality trends in education: A systematic review of research and applications. *Journal of Educational Technology and Society, 17*(4), 133–149.
49. Fehling, C. D., Müller, A., & Aehnelt, M. (2016). Enhancing vocational training with augmented reality. In *Proceedings of the 16th international conference on knowledge technologies and data-driven business*. Springer.
50. Radosavljevic, S., Radosavljevic, V., & Grgurovic, B. (2020). The potential of implementing augmented reality into vocational higher education through mobile learning. *Interactive Learning Environments, 28*(4), 404–418.
51. Kotsifakos, D., Adamopoulos, P., Kotsifakou, P., & Douligeris, C. (2020). Vocational education and training apprenticeship: Using teaching and learning analytics in a learning management system for improved collaboration, individual empowerment and development of apprentices. In *2020 IEEE Global Engineering Education Conference (EDUCON)* (pp. 1775–1782). IEEE.
52. De Lange, P., Neumann, A. T., Nicolaescu, P., & Klamma, R. (2018). An integrated learning analytics approach for virtual vocational training centers. *IJIMAI, 5*(2), 32–38.
53. It's owl cluster management. (2012). Accessed August 08, 2021, from https://www.its-owl.de

54. Nielsen, J. (1994). *Usability engineering*. Morgan Kaufmann.
55. Zhang, L., Van Oosterom, P., & Liu, H. (2020). Visualization of point cloud models in mobile augmented reality using continuous level of detail method. In *The International Archives of Photogrammetry, Remote Sensing and Spatial Information Sciences* (pp. 167–170). Springer.

Designing User-Guidance for eXtendend Reality Interfaces in Industrial Environments

Volker Paelke and Jendrik Bulk

Abstract

Modern industrial environments evolve from collections of physical machines into networked extended reality environments, where in addition to the physical machines digital content becomes increasingly significant. Examples of such digital content include not only the necessary information to control the production, but also digital instructions and product information, information derived from embedded sensors or predictive models, as wells as digital twins of products, machines and entire factories.

This means that user interface designers in such environments have to design interfaces that allow access to and interaction with both the virtual and physical aspects. The design of user interfaces for such eXtendend Reality (XR) environments faces a number of challenges including the change of the implementation platform from a "standardized" PC GUI to I/O devices and paradigms that are suitable for XR. Examples include mobile and wearable devices as well as virtual and augmented reality. There is still a lack of standards and best-practices regarding the design of XR interfaces, especially when interaction and data visualization techniques are required to support work across different platforms.

This chapter focuses on guidance techniques that help users navigate the complex information and interaction tasks in XR environments, as this is an area for which little information is currently available. The chapter explains why different forms of user guidance are required in XR environments, why these pose a challenge for user interface designers, and presents techniques that can be used to aid the design process.

V. Paelke (✉) · J. Bulk
Hochschule Bremen, Human-Computer-Interaction, ZIMT, Bremen, Germany
e-mail: Volker.paelke@hs-bremen.de; Jendrik.Bulk@hs-bremen.de

C. Röcker, S. Büttner (eds.), *Human-Technology Interaction*,
https://doi.org/10.1007/978-3-030-99235-4_7

Keywords

Extended reality · User interface design · Industrial applications

7.1   Introduction: Why Do We Need Guidance Techniques in XR?

Users of visualizations tend to "see" only what they know and expect. It is therefore up to the designer of a visualization to ensure that the user is aware of important information and able to "read" and "understand" it.

This becomes especially challenging with interactive immersive media like virtual and augmented reality because users implicitly control their point of view on the visual content provided, either through head movement or by positioning a hand-held device.

While designers of 2D visuals like diagrams or graphs can easily ensure that essential content is included and visible in the final visual and users of these visuals can easily verify that they have at least "seen" the whole visual, this is not the case in XR environments. If an essential element of the visualization is not within the user's field of view in an XR environment users can easily miss it. There is a whole category of XR puzzle or escape-room games that exploit this as a game mechanism and challenge their users to find information that is "hidden in plain-sight" within an XR environment in order to progress. However, for productive applications in an industrial context this is usually not desirable.

Thus, a need exists to integrate guidance techniques into many XR environments to support users in finding and "seeing" relevant information. In addition to supporting information visualization tasks, guidance is also at the core of many interactive tasks users of industrial XR applications perform, such as route following (being guided towards a spatial location), part picking (being guided to select specific parts) or assistance and instructions for spatial tasks (being guided through specific movements).

7.2   Background

7.2.1   Mixing Realities: What Are AR, VR, MR, XR?

An unfortunate confusion exists with regards to the nomenclature of technologies that combine real and virtual content.

Virtual Reality (VR) is commonly used to describe the technology for computer generated simulated environments that completely substitute the real world. This requires appropriate output devices to create the necessary sensory stimuli. While most existing VR systems focus on the visual and auditory senses, vestibular and proprioceptive senses should also be considered in VR design to avoid simulator sickness and can be used to improve the feeling of presence. Technologies also exist at various stages of development to address the tactile/haptic senses, as well as gustatory and olfactory senses. In the industrial application domain both VR systems using head-mounted-displays (HMDs), and projective VR Systems like CAVEs or Powerwalls are common [1]. Exemplary applications of VR in an industrial context include training, design support, virtual engineering and interaction with the "virtual twin" of a machine, production line or complete plant.

Augmented Reality (AR) extends a real environment by seamlessly integrating virtual elements into it. A definition of AR that is commonly used in the research literature was introduced by Azuma in 1997 [2]. According to this definition an AR system is characterized by:

- The combination of real and virtual elements
- Interactivity in real-time
- Registration in four dimensions (3D + time)

Outside of the research domain much more loosely defined interpretations can be encountered, that might also label systems that use non-real-time combinations (e.g. special effects rendering in movies) or employ virtual elements that are not registered in four dimensions (e.g. text-display on smart-glasses) as "augmented reality". In contrast to common VR use cases which can be used at remote locations, industrial AR application often take place within the production environment, e.g. at assembly workstations or in a warehouse.

Mixed Reality (MR) spans a broader scope of combinations of virtual and real elements. The definition commonly used in the research literature was introduced by Milgram et al. [3]. They examined the relation between VR and AR and defined mixed reality as a spectrum that ranges from the real world without any additions to completely computer-generated virtual reality, with augmented reality employing a part of this spectrum in between, where a predominantly real environment is extended by virtual cues. Notably, the spectrum also gives rise to the definition of "Augmented Virtuality" (AV) as predominantly virtual environments into which real elements are integrated, a common example being virtual TV studios like those used in weather reports into which real presenters are integrated. The broader scope of this MR definition lends itself to the description of systems that integrate different technologies and thus extend beyond the narrow definitions of AR or VR and it has and still is frequently used as such. Unfortunately, in recent years companies (most notably Microsoft) have also used the term "Mixed Reality" to describe a specific type of virtual reality HMD with integrated inside-out tracking. Due to the massive marketing efforts associated with these "Mixed Reality" displays this use is now very common, so that a clarification is always required.

Tangible User Interfaces (TUI) refer to an interaction concept introduced by Ullmer and Ishii [4]. In TUIs the interactions of a user with real-world objects ("tangibles") are tracked (e.g. through embedded sensors) and this information is then used to effect actions in the computer application. In addition to applications that use specially designed tangibles to control an application [5] common industrial use cases now include the tracking of (power-)tools as "tangibles" that are already present at the workplace, so that the progress in a work sequence can be automatically tracked or parameters of the tool (e.g. torque on an impact wrench) can be adjusted automatically. The monitoring of picking tasks for correct type and quantity through appropriate sensors is another example of TUI interaction in industrial environments that uses already present objects as the basis of interaction.

Context-Aware-Interfaces (CAI) aim to simplify interaction by presenting only information and interaction options that are relevant in the current context of use [6]. Industrial systems usually focus on the location and user task as "context", as these are easily accessible, but other variables like personal preferences and intentions can also be used. Simple examples are Location-Based-Services (LBS), that provide a user with information filtered according to the current spatial location.

eXtended Reality (XR) has recently emerged as an umbrella term to describe all forms of combining virtual and real environments. XR can thus be viewed as a generalized term that encompasses AR, VR and MR and also includes techniques that might not fit within the narrow traditional definitions, like TUI, CAI or integrating parts of the interface into the physical (real) environment, e.g. extending a real warehouse shelve with a pick-by-light system based on lasers or projectors.

For the purpose of this chapter we have thus decided to refer to the use of eXtended Reality (XR) as the most generic design space that encompasses the others.

7.2.2 Specific Requirements of Industrial Applications

Guidance tasks in industrial XR environments take place in a variety of application contexts. While designers can draw on previous research from different domains this information is scattered across a wide variety of publications and has usually not been validated in an industrial context. Therefore, designers often have to design and develop guidance techniques from concept to validation before they can be integrated into industrial interfaces. In this section we identify specific requirements of industrial XR environments.

Safety Safety is paramount in industrial XR applications. While entertainment XR applications are commonly used in relatively safe and benign environments, e.g. on the sofa in living rooms, this is not the case for industrial XR applications. Industrial VR applications might be used in separate rooms that can be relatively easy to secure, but many mobile and "on-site" XR applications are used in working industrial sites where hazards like operating machinery, tripping hazards, moving machinery and electrical hazards are present. Ideally an XR system should aim to improve worker safety by actively monitoring potential hazards and providing early warning, but as a minimum industrial "on-site" XR systems must ensure situation awareness of the user, e.g. by ensuring a wide field of view/ peripheral vision, as well as fail-safe operation in the event of a system failure.

Usability Usability is essential for industrial XR applications. While users of an entertainment XR application might be willing to accept certain interaction problems or even regard them as "challenges" that are part of the experience, such problems are unacceptable in industrial XR applications. In training applications, the XR interface should be as transparent or "invisible", as possible, as it would interfere with the intended training results if the users learn to operate the XR interface instead of the underlying task to be trained. In productivity XR applications a certain amount of learning with regards to the XR interface is acceptable, but the system must enable users to achieve their work goals in an effective and efficient way. In addition to usefulness, which can be checked in user tests, user acceptance is another key usability criteria that must be considered in industrial XR applications and that might require elements beyond the system itself, e.g. user participation in the design and introduction or company policies regarding supervision and the use of data.

Creation Effort The creation of XR applications is expensive and time consuming, even compared to the creation of conventional apps. This situation mirrors that of earlier developments in computer graphics: While the potential of computer visualization for communication purposes was recognized early on, widespread use was limited by technological constraints and the fact that only expert programmers had the means to create such graphics. Today, high-level tools and corresponding asset libraries in combination with established content creation work-flows enable the cost-effective creation of computer visualizations, including 3D visualizations and animations.

For XR applications development and authoring tools are approaching a level where widespread creation of customized, application specific content becomes viable. It is now common practice to use game-engines like Unity or Unreal as the basis for XR application development. This works well for applications like games where the still relatively high development costs can be amortised over a large number of instances sold.

The situation is different in most industrial applications, where the number of users and potential customers are limited. The diversity of industrial XR applications leads to design and guidance requirements that are often specific to the use case at hand. In addition, no standards for XR interaction and guidance currently exist, resulting in a lack of code and asset libraries which combined with missing processes, models and expertise for guidance techniques results in time consuming and expensive design processes with often less than satisfactory results. To be viable the development process thus needs to be as streamlined as possible.

7.2.3 Guidance in XR: Why Arrows Are Not Enough

When XR developers notice that the need for guidance either during design or user tests a common quick fix is the introduction of 2D or 3D "arrows". Arrows are popular for a number of reasons:

- First of all, arrows are easy to create both in 2D and 3D and require little in terms of modelling. Most 2D graphics libraries include arrowheads as a standard element for line endings and a 3D arrow is easily created from a cone and a cylinder, which are available as primitives in most 3D toolkits. Arrows are therefore easily available to developers, which makes them attractive for quick prototypes or fixes, without the need to involve graphics designers or 3D modelers.

- Second, arrows have simple geometry that requires little rendering resources and they display well across a wide range of display devices, from optical-see-through displays, over video see-through displays, to pure immersive VR displays.
- Third, arrows are well understood as a visualization for movement or motion paths, as well as pointing out points of interest. Arrows are ubiquitous in classical 2D user interfaces, e.g. as the mouse pointer so that users that are unfamiliar with XR applications can infer their meaning from previous 2D experiences.

However, arrows can also cause a number of problems (see Fig. 7.1):

- Arrows can obscure large sections of the field of view, especially parts that are close to the point of interest, which are usually most relevant.
- 2D arrows in a 3D environment can be unclear or ambiguous with regards to their actual spatial location and orientation.
- 3D arrows can suffer from visibility issues: If 3D arrow orientation is fixed to the content coordinate system, users can view them from orientations that are difficult to understand and interpret, e.g. head on. 3D arrows who's orientation is adapted to the user's point of view avoid this visibility problem but can confuse users because they don't stay fixed.
- Arrows also don't scale well to indicate multiple points of interests or elements/features that are not points (e.g. lines, areas or volumes).
- In long term use arrows are often seen as intrusive or commanding by users.

A key issue with arrows is that they often work "well enough" in early tests of individual features and thus stay in prototypes far longer than initially intended, often up to a point where a replacement would require a substantial redesign. The aim of this chapter is not to replace arrows as a guidance technique "per se", but to provide designers and developers with the necessary toolkit to make informed, conscious choices of guidance techniques.

7.3   Related Work

7.3.1   Guidance Applications in XR

Guidance has been a major feature of augmented reality applications since their inception. When Tom Caudell and David Mizell coined the term "Augmented Reality" in the early 1990s for the technology they were developing at Boeing [7] guidance was at the centre of the proposed applications, including marking positions to drill rivet holes; positioning composite fabrics in a lamination process; and assembling wire harnesses for airplanes.

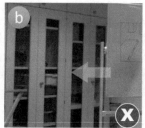

Fig. 7.1 Problems with arrows: (**a**) 2D arrow partly obscures the object of interest and relevant context; (**b**) semi-transparent 2D arrow is ambiguous; (**c**) fixed orientation 3D arrow is difficult to interpret from user's point of view and obscures the object of interest

Azuma's survey of augmented reality in 1997 mentioned guidance as the main application or a central feature in a variety of applications, including maintenance and repair, medical training and surgery, as well as navigation support in military aircraft [2].

The use of guidance techniques in virtual reality also has a long history, e.g. to support navigation in large virtual environments [8] or to provide instructions and assistance in training applications [9].

7.3.2 User Studies of Guidance Techniques

Early publications often describe only the techniques used and sometimes mention design alternatives that were considered. While the lack of evaluation might make judging the suitability of techniques based on such publications more complicated, they can still serve as suitable starting points, e.g. [10] presenting a system using a combination of arrows and virtual objects. More recent publications often include some form of evaluation, e.g. [11] using arrows and virtual objects and [12] using arrows, highlighting and text annotations.

Early papers frequently focus on the introduction of (at the time) novel techniques, e.g. [13] introducing the concept of virtual spotlights and [14] presenting the attention funnel.

More recent publications tend to build on these techniques and include user studies that compare different modifications to the original, e.g. [15] introducing the halo technique as an extension of concentric circles or [16] presenting SWave, an approach that combines elements of the attention funnel with concentric circles and the idea of using animation to convey direction and [17] introducing a spline based extension of lines and arrows and comparing them to highlighting and SWave. A study by [18] introduced the idea of varying visual features based on difficulty.

Studies that evaluate well known techniques are less frequent, but do exist, e.g. [19] evaluating the use of colour, background opacity and concentric circles.

In recent years studies have compared the impact of using different presentation hardware, e.g. comparing conventional paper-based instructions with tablets, projection-based systems and HMDs [20–23].

No generally agreed on framework for structuring the design space of guidance techniques has been established so far, but the need to consider the widely varying requirements of the "same" task at different scopes of interaction has been highlighted by [24] for navigation tasks, where they distinguish coarse and fine navigation. [25] have proposed a classification system for attention guidance techniques in large scale environments, addressing primarily techniques for coarse navigation (e.g. arrows, path, mini-map). The paper by [26] provides an overview of existing work focusing on guidance in cinematic virtual reality, while [27] introduce a visual approach to provide an overview of the design space of XR assistance systems.

The need for user studies to evaluate guidance systems has motivated work towards standardized tasks and evaluation procedures [28, 29]. While no standard has emerged so far, recent work provides a good basis for the design of experiments [30, 31] and data capture and analysis [32].

Publications addressing the perceptual aspects on which guidance techniques are based provide valuable information when existing techniques need to be adapted and modified or when new techniques need to be designed. The paper by [33] provides some general guidelines for the design of augmented reality visualization techniques, including guidance techniques, and [34] identify central perceptual issues in augmented reality.

An important issue regarding user guidance was raised in [35], namely that guidance towards an important, but not critical goal or event, might interrupt more important ongoing tasks and disturb situation awareness, suggesting that designers should carefully balance the "effectiveness" of guidance techniques with the overall task and work environment. An interesting development in this direction are guidance techniques that work on a subliminal level, such as the guidance approach described by [36] which works by slightly blurring less relevant section in the field of view.

Finally, the best user guidance is useless if the corresponding visualization (e.g. added virtual objects in an augmented reality application) is difficult to perceive or confusing to interpret. The book chapter by [37] provides a useful introduction to these aspects with a focus on object visualization.

Thus, while designers of guidance techniques can draw on an increasing amount of published research a number of issues remain:

- The results of published studies are often specific to the hardware and environment used by the authors. While a transfer to different hardware and environments might be possible, the functionality and usability on the actual hardware and in the actual environment should be validated by user tests.
- The test tasks used in published studies might not align well with the tasks encountered in an industrial application context. Even common tasks that are referred to by well-known labels like "picking" or "assembly" can differ significantly in their details, so again a validation is required.
- There is no established framework or process for choosing or designing guidance techniques, leaving it up to the experience of the designer to decide whether existing

techniques can be used, should be adapted or the development of new techniques is required. Given the rapid growth of the field it is unlikely that such experienced designers will always be available.

7.4   Approach

A well-defined process enables effective management of development projects as well as systematic and coherent tool support. A central challenge with the development of guidance techniques in XR applications is that it is usually not possible to establish specific processes for this purpose, so that the design activities must be integrated into the established processes, often from software engineering (e.g. V-model, Scrum) [38] or user-centered-design (e.g. ISO 9241-210) [39].

7.4.1   Design Processes and Process Integration

Established user-oriented design processes can in principle be transferred to XR applications. The ISO 9241-210 standard captures best-practice in the development of interactive systems and formulates a high-level view of a design process that involve users actively in the design process and relies on iterative development with frequent evaluations. The central design activities in this process are:

- To analyze and specify the context of use
- To specify the user and organizational requirements
- To create design solutions
- To evaluate the design solutions against the requirements

The process is iterated until the requirements are met. Differences for the design of XR systems and specifically for guidance techniques arise primarily for individual design activities within these processes. As the design and evaluation typically involves exploratory activities with difficult to predict outcomes it is frequently useful to couple them as loosely as possible with the main project. Common ways to achieve this include design sprints in agile processes like scrum, where sprints are specifically allocated to this exploration. Other options include pre-production phases, where the exploration research is performed prior to the main project, or "micro projects" that are spun out of established processes and later merge their results back into the main project. Later in this chapter we introduce some tools that support the design and evaluation activities with guidance techniques.

A generic, process-independent approach to incorporate the necessary user guidance into XR projects can be subdivided into preliminary activities, that are usually performed outside the guidance specific parts of a project; high-level analysis, that establishes the functional intentions of the guidance techniques; mid-level guidance design that maps

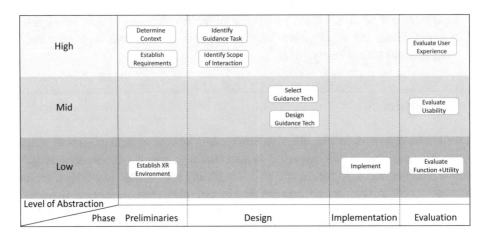

Fig. 7.2 Development activities for selecting and designing guidance techniques

high-level functional intention to concrete guidance techniques; and low-level implementation that is concerned with the technical aspects of implementation such as interaction handling and rendering. The result is then evaluated and the design refined as necessary. Figure 7.2 outlines these activities, organized both by level of abstraction and development phase. When planning a project these activities can be integrated into the established development process.

7.4.2 Design: Activities and Support

In the following subsection we describe the activities identified above in detail and provide suggestions for how these activities can be supported by tools or streamlined to make the design process more efficient and viable. Figure 7.9 shows an extended version of Fig. 7.2 that shows where the supporting tools described in the following subsections can be applied.

Preliminaries: Context and Requirements According to ISO 9241-210 the first step is to analyze and specify the context of use followed by the specification of the user and organizational requirements. As mentioned before, these activities can be integrated into other processes that might use a different layout and nomenclature than ISO 9241. Because the results of these activities are required as a foundation for both the "main" application and the guidance techniques within it, it is useful to perform these tasks first, to establish the main usage context (application environment, users, tasks) and then expand on them with regards to user guidance, where required.

Preliminaries: Establishing the XR Environment In a typical XR project the technological XR environment, especially the devices to be used, will have already been established. However, a co-design approach can be very powerful if possible, e.g. to offset shortcomings of one system (e.g. limited field of view in AR HMDs) with another technology (e.g. pick-by-light). Similar to the context of use and the user requirements the capabilities of the available hardware need to be understood before appropriate guidance techniques can be selected and designed. As use of XR devices become more widespread, evaluation results and experience reports can often be found, otherwise an approach like [40] can be used to evaluate the device at hand or compare potential devices and make an informed decision.

High-Level: Identifying Guidance Tasks Different forms of user guidance can be required depending on the environment, the task and the users of an industrial XR system. While some systems might require only one specific kind of guidance, e.g. a navigation system that guides the user to a location, it is quite common that more complex systems incorporate different guidance tasks, e.g. first guiding the user to a specific location and then providing guidance through an assembly sequence performed at this location. The guidance provided can also differ between users, e.g. an experienced worker might be presented with guidance on what task should be performed, while a trainee might require detailed step-by-step guidance on how to perform the necessary sub-tasks.

Guidance tasks provide a first useful categorization of the guidance required and later allow to browse existing solutions that might be used or adapted. A framework of common guidance tasks in industrial XR systems include:

Spatial Navigation Spatial navigation is a central guidance task in many XR systems ranging form mobile AR applications to large VR environments. As an example, a mobile AR system could support safe movement in factories, especially when robots and autonomous transportation systems are present, by providing navigation guidance in the physical environment. VR applications also frequently require guidance to support effective spatial navigation, e.g. when a large virtual environment is used as a "digital twin" to visualize the current state of production machinery and an event like a maintenance notification requires efficient navigation to the point of interest.

Logical Navigation Logical navigation is a guidance task that is often required in remote assistance and training applications, where users need support when interacting with complex interfaces and visualizations. Similar to help-systems and "wizards" in graphical user interfaces, XR guidance techniques can help users to navigate complex interfaces and information spaces that can encompass both computer-generated ("virtual") and physical ("real") elements.

Picking Picking refers to the process in a warehouse or storage where workers pick specific items to fulfill an order. Picking intrinsically requires guidance so that workers know which items to pick. As picking is one of the most labor and cost intensive activities in a warehouse a lot of effort has been invest in optimizing guidance in this domain, moving from printed pack-lists to XR systems like pick-by-light and pick-by-voice, AR systems and combinations thereof.

Assembly Guidance for assembly tasks can span a wide range: Training systems can provide trainees with detailed step-by-step guidance on how to perform a task; guidance systems for workers in a flexible production environment might limit the guidance to information on what is to be assembled; and remote assistance could focus on one-of real-time support of specific tasks.

Quality Control (QC) QC is another common guidance task that can vary widely according to the specific application: While in automated QC systems XR guidance techniques can be used to visualize and locate (potential) problems that have been identified, XR support of manual inspection task will often provide checklists and references to guide a worker through the process. Combinations of automated and manual approaches are also possible and require a wide range of guidance techniques.

High-Level: Identifying Scopes of Interaction Guidance techniques in XR are required at different scales all the way from the individual workspace to an entire factory. A useful generic first step is to identify the scope in which interaction takes place and guidance is required. As a starting point the following categorization can be used:

At the *micro scale* interaction takes place in a very small area or volume and optical or video magnification techniques are commonly used. Very high precision tracking, high-resolution visualization and magnification are essential requirements at this scope of interaction. Guidance techniques should either work with existing optical or video magnification devices, or a new integrated solution should be considered. XR systems that address micro scale guidance typically use specialized sensors and displays that are integrated into the workplace. Examples include quality control of small components, assembly, maintenance and repair of micro-mechanisms or reworking of SMD circuits.

Workplace scale interaction is typical for work situations where users perform tasks at a stationary workplace, such as a workstation or assembly table or remain at a location for the duration of the task. XR systems at workplace scale require high-precision tracking and high-resolution visualization. Usually workplaces can be adapted to a certain degree to optimize the performance of an XR system, e.g. by installing sensors. The tracking sensors and displays can either be mounted at the workplace or mobile or wearable devices can be used. Typical examples include assembly tasks (stationary workplace for longer time periods) or maintenance tasks (temporary workplace for task duration).

Room scale interaction applies to tasks where users move around physically in a limited, possibly controlled, space while performing a task, e.g. accessing storage shelves. The requirements regarding tracking and visualization accuracy are usually not as high as in workplace tasks, however the spatial volume in which tracking is required is usually much larger and options to adapt the environment are more limited. Mobile or wearable devices are the most common option for room scale XR applications, sometimes combined with additional physical elements, e.g. a pick-by-light system.

Factory scale interaction applies to tasks in which users must navigate large scale environments, either physically as in intralogistics tasks or virtually e.g. in VR systems for factory planning, commissioning or monitoring. Frequently, guidance tasks at factory scale deal with spatial navigation guidance, either along specified pathways or to a specific point of interest. Especially in physical navigation tasks the available tracking accuracy is less than for the other scopes of interaction, as factory scale XR systems cannot usually adapt the environment to the XR system and the system must deal with a wide range of conditions, e.g. mixed in-door/out-door use. Guidance techniques should address this limited accuracy. Typical platforms are mobile and wearable systems for physical navigation and VR systems for the virtual equivalents.

Mid-Level: Selecting and/or Designing Guidance Techniques Having identified the guidance tasks required for an XR system and their respective scopes of interaction the process can continue with the selection or design of appropriate guidance techniques. As a first step it is useful to research potentially suitable techniques for the tasks and review them with regards to the project requirements. Techniques that pass this first filter are than matched to guidance tasks in the system and adapted as required. Guidance tasks for which no suitable technique could be identified require the design and implementation of new techniques. These can frequently be build based on combinations of existing techniques (remix), but sometime a completely new design from scratch is required.

Research and Review As discussed in the related work section an increasing number of studies featuring different kinds of guidance techniques is being published in the scientific research literature. A useful step is to research corresponding digital libraries like the ACM digital library (ACM digital library.)or IEEExplore [41]. However, only a fraction of existing techniques is covered in published research literature. Additional examples can be found in the developer documentation provided by XR hardware/software manufacturers, e.g. Oculus guidelines for VR systems [42] or Microsoft guidelines for Hololens AR systems [43]. While a wide range of different XR projects have published some useful information online there is currently no easy way to find and filter the information.

Assuming that potentially relevant guidance techniques have been identified it is then useful to try to match the available information to some standardized criteria. The following checklist of review features can serve as a starting point, with the list being adaptable to the specific needs of the project:

- Guidance: What kind of guidance task is supported?
- Scope: At which scope of interaction is the guidance technique applied?

- Precision: What level of precision is required in the tracking and what level of precision is achieved by the guidance?
- Guidance elements: What kind of guidance elements are used? E.g. 2D or 3D visual, auditory, tactile or combinations thereof.
- Technology: Are there specific technology requirements? E.g. techniques that rely on video processing might not be applicable to optical-see-through displays; tactile or haptic feedback require specialized hardware.
- Environment: For which environment is the system designed? Does it require special modifications or infrastructure (sensors, markers)? Can it be adapted to the industrial context of use (noise, lighting, EMC)?

Adaptation, Remixing and Designing from Scratch Depending on the results of the research different approaches can be employed:

Adaptation can be used if a suitable technique has been identified. Depending on the materials available (e.g. just a description vs. code) and possibly the license a technique is adapted to the hardware and software environment or reimplemented. While the review should provide some indication of the suitability for the intended task it is still useful to start with a light-weight prototype (see next section on implementation support) with early user tests to validate it, as different environments, technologies and devices might seriously modify usability and functionality compared to the published results.

Remixing combines and modifies elements and ideas of existing solutions to extend their use to different applications and environments or to address specific shortcomings. A rapid iterative design and test approach is essential for developing useful and usable guidance techniques remixing existing ideas, as central aspects of functionality and user experience are difficult to predict and best established using user tests.

Designing from scratch allows to create novel guidance solutions specific to the requirements but involves the highest level of effort and carries a corresponding development risk. Again, a rapid iterative design and test approach is essential to mitigate development risks and ensure usability of the result.

Low Level: Implementation

An appropriate development environment is essential to make the iterative development of testable prototypes viable by providing pre-made components of common features and functionality. The development environment used in a project will usually depend on decisions that are outside of the scope of guidance design, with game-engines like Unity and Unreal becoming increasingly popular, even for "serious" XR applications. Reasons include the widespread support of XR technologies, the availability of pre-made assets, efficient development workflows and the availability of developers that are familiar with these environments.

7.4.3   Support for Evaluation

Evaluation of guidance techniques requires user tests, as there is currently no sufficient knowledge-base to support analytic evaluation techniques. The design of user tests can draw on a wide body of experience, allowing to tailor tests to different goals ranging from simple experiments that only provide rough information for internal design decisions to gathering reliable data as a basis for scientific publications. Tests used in the design of guidance techniques will usually be formative tests, providing information for design decisions without the need for results to be statistically significant. However, it might sometimes be desirable or necessary to conduct more in-depth testing. User tests can be quite costly in terms of money, personal, development resources and time. So called "discount usability" techniques [44] aim to reduce these costs, using the expertise of expert reviewers (e.g. "heuristic evaluation"), simplified design representations (e.g. "paper prototypes") and qualitative tests with small numbers of subjects. However, directly transferring these approaches to XR can be difficult: Heuristics require an in-depth understanding of the domain and ample experience with it on the part of the reviewer, which is in short supply in a rapidly evolving domain as user guidance in industrial XR applications. And paper prototypes struggle to represent the essential combination of virtual and real elements in XR interfaces. Nevertheless, the essence of this approach can be applied to XR evaluation by (re-)using pre-made elements in the tests, test environment, and system under test and tests with small numbers of participants (e.g. 3–5) are of course applicable to XR and allow rapid design decisions and iterations of promising techniques. To reduce test costs our approach uses:

- Pre-made test tasks
- Pre-made test environments
- Test environment building blocks
- Pre-selected variables of interest and corresponding data gathering tools
- Established processes for analysis and decision making

Pre-Made Test Tasks Simplified pre-made test tasks, while not quite as light-weight as paper-prototypes, allow to focus on the guidance techniques of interest and allow to avoid some of the problems that can arise when testing in industrial environments.

Potential issues with industrial tasks in user tests are that the tests have to take occupational health and safety issues into account, especially if (power) tools are involved. Regulations often require previous training, supervision and special insurance, which make it difficult to recruit large and diverse user groups for testing. In addition to addressing safety and security concerns pre-made tasks also allow to address problems of availability (e.g. when a product to be assembled doesn't exist yet) or confidentiality (e.g. when a product has not yet been publicly presented).

It is therefore useful to create pre-made test tasks that are simple and quick to implement, thus allowing to focus on the guidance techniques, and abstract from other complications in XR systems like tracking or networking. Ideally test tasks should be designed to be re-used many times, which then also allows comparison between evaluation results.

For our tests we have created a range of picking and assembly tasks, ranging from micro-scale to workplace-scale using common construction toys such as fischertechnik [45], Lego [46], and Makeblock [47]. As toys these systems are certified as 'safe-to-play' which avoids health and safety issues, and they are relatively inexpensive, which allows for easy replication of tests across multiple instances and sites. All construction toys systems have advantages and disadvantages as a basis for tests tasks: Makeblocks allows to create sturdy structures, but requires some tools for assembly, while Lego and fischertechnik can be assembled without tools. Lego is widely available and offers an immense range of different blocks, while fischertechnik allows to create a wide range of functional objects from a more limited set of blocks. Figure 7.3 shows an example of an assembly test task using fischertechnik. For a more in-depth discussion we refer the reader to [48].

The use of construction toys is by no means obligatory and other existing objects (for picking tasks) or disassembled products (for assembly tasks) can be used as the basis for pre-made test tasks. Based on our experiences in user-test conducted over a 8-year period a few guidelines can be summarized for good test tasks:

- While the guidance requirements of a test task should be similar to the corresponding real-world task, the test task can differ, which can increase re-use of test tasks.
- Good test tasks can be scaled from simple to complex. E.g. from a pick task with a small set of options to a large set of options, or from a 2-part assembly to a complex hierarchical assembly.
- Unless there are specific reasons (e.g. in training applications) to handle very small or large physical objects, the objects used in test tasks should be easily "grabbable" (e.g. from 15 to 100 mm).
- Physical components used should be robust, without sharp edges and individually replaceable.
- 3D models of all physical components used in the test should be readily available to create instructions and XR visualizations.
- "Interesting" tasks can help test participants to understand a task and keep them motivated longer. E.g. in a picking task, users might have to perform large numbers of pick orders—explaining that the components from each pick order assemble into complete assemblies, showing the final assembly or letting users assemble one of these themselves at the end of the test can be helpful to engage test participants.
- Regarding the practicality of organizing tests, adjusting them as designs change, and reusing elements between projects a smaller set of versatile components is more useful than a large set of specialized components.

Fig. 7.3 Example pre-made test task: assembly test task using fischertechnik

Pre-Made Test Environments The situation with test environments mirrors that of test tasks: Conducting tests at the actual work-place can be impractical due to safety, security or confidentiality reasons or the corresponding workplace might simply not be available for tests. Pre-made test environments represent the essential aspects of the work environment from the guidance perspective, match the scope of interaction and spatial layout, while abstracting and simplifying other aspects.

Pre-made test environments and the elements therein can be viewed as the equivalent of film sets and props in film production. To support tests across a wide range of XR applications the "set" and "props" should be pre-made both in physical form to enable AR style interfaces, and as virtual 3D models to support VR style interfaces, as well as the creation of visualizations.

Figure 7.4 shows an example of a small-scale picking test environment. The test "prop" represents the scope of interaction (on the worksurface) and spatial location of objects. It also provides additional elements like the marker to simplify 3D tracking and thus abstracts considerations of the technical implementation that might exist in the real workspace (markers might not be possible, lighting might be an issue). Such props can be rapidly produced, modified and adjust to the specific test context. Props based on rapid-prototyping techniques (using laser-cut plywood in our example) are easily moved and stored between tests. A small pick test environment like the one presented here is also useful as the basis for assembly tasks as it provides a simple and well-structured way to provide the necessary elements for an assembly task under test.

Figure 7.5 shows an example of a larger picking test environment "set". Again, the main feature required is simply a representation of the scope of interaction (workspace) and the spatial layout of the elements to be picked, while the actual storage elements can be abstracted and are represented by cheap generic containers, here labelled with a large amount of tracking markers to ensure proper tracking during the tests.

Figure 7.6 shows the larger picking test environment "set" being used in a test of guidance techniques for picking.

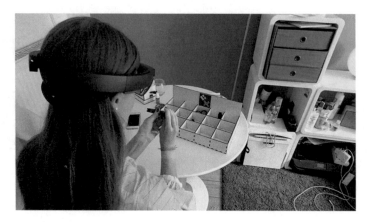

Fig. 7.4 Example pre-made test environment: small-scale picking test environment

Fig. 7.5 Example pre-made test environment: larger picking test environment "set" at "workplace" scale

Test Environment Building Blocks Some tasks, especially navigation tasks, are difficult to represent by a pre-made physical test environment. The sheer physical size required often prevents the creation of a representative "set" and use of the real environment is often

Fig. 7.6 Example use of the larger picking test environment "set" in a guidance test

precluded by concerns of safety and potential interference with the ongoing activities taking place. One option in these cases can be to use a virtual environment as the test environment and integrate the XR techniques as their VR equivalents. To make the creation of such virtual environments viable techniques for the rapid construction of environments are required that allow to adjust the environment to the specific test task, e.g. creating tasks with different navigational complexity. Obviously, the same setup can also be used to prototype and test VR techniques. Figure 7.7 shows an example of our testbed environment editor based on Unity.

The testbed uses a modular approach to both the construction of navigation and locomotion techniques and virtual (test) environments. Building on the modular structures provided in the Unity game engine our testbed allows to quickly create 3D indoor and outdoor environments that can then be used to test and evaluate guidance and locomotion techniques. The right side of Fig. 7.7 shows a selection of predefined spatial elements from which such environments can be build up using drag/drop, the left side shows several different test environments, ranging from outdoor environments with buildings for large scale navigation to indoor navigation with detailed guidance requirements. Figure 7.8 shows an example of an industrial navigation test environment created with the tool.

In addition to the construction of test environments we also provide a set of ready-to-use guidance techniques for navigation, e.g. 2D-displays like Minimaps, or 3D solutions like a 3D compass. This set of techniques can be used in two ways: As a baseline reference for user tests against which the performance of other techniques can be compared and as a basis for the development and adaptation of new techniques. Because the techniques are provided as standard Unity prefabs they can be easily modified and extended, simplifying the development of application specific adaptations. Over time developers can create and expand the library of "proven" baseline techniques by adding their own new techniques to the library.

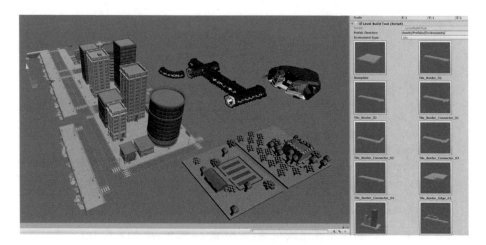

Fig. 7.7 Example test environment building blocks: Navigation guidance testbed creation using our customized Unity editor interface

Fig. 7.8 Industrial navigation test environment created using the testbed editor

Pre Selected Variables of Interest in User Tests and Corresponding Data Gathering Tools In addition to (re-)using pre-made test tasks and environments the (re-)use of test materials and questionnaires provides another avenue to reduce the cost of XR user guidance tests. While a key focus of XR guidance technique user tests, especially in early stages, will be on the identification of interaction problems it will often also be important to gather additional data, either by measurement or through user feedback reports.

After completing a test, users will usually be asked to provide feedback on aspects such as usability, comfort and ergonomics. There are established questionnaires for these individual aspects, but their combined scope is often too large for test users to complete them all. It is therefore essential that designers carefully decide which information is really required and limit the set of questionnaires accordingly. If a wide coverage of feedback topics is essential, a viable approach can be the creation of integrated customized questionnaires that integrate topics of different domains and trade depth for breath. It is important to note that the results from such custom questionnaires can't be compared to results from standardized questionnaires and that it is essential to validated custom questionnaires before use. The following list provides an overview of measures commonly used in XR user tests. To streamline the process and simplify both the tests and the analysis of the results it is essential to be selective and carefully consider which of these (or additional) measures are essential to the task under study:

- Completion time of a task
- Numbers of errors made during a task
- Numbers of errors corrected during a task
- Numbers of errors persisting in the final result
- Measures of user fatigue
- Measures of user attention/distraction
- Measures of user's cognitive task load
- Collection of data on cybersickness symptoms
- Feedback on technological acceptance of the hardware used
- Feedback on the perceived utility of the system
- Feedback on the perceived usability of the system
- Feedback on the perceived user experience of the system

Established Process for Analysis and Decision Making This subject is closely related to the preceding one and they should be considered together. Before conducting tests with users and gathering data, a clear and well-defined process for the analysis of the resulting data and the way it is going to be used for decision making should be established. This includes a process to address feedback on interaction problems, as these should receive priority in getting resolved.

7.5   Reflection and Future Work

Guidance technique are media elements that are not present in reality, but added as artificial extensions or modifications in an XR application. The requirement for user guidance arises from the great amount of freedom and control that users of XR applications enjoy. While the possibility to freely explore can improve the user experience, this freedom can also be a

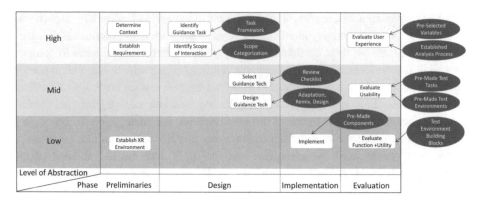

Fig. 7.9 Supporting tools for individual development activities (in dark bubbles)

major usability problem if users don't recognize interaction possibilities or have no clear idea of what to look for. The solution to this problem is to guide users by providing explicit cues in the XR environment that recommend future actions. Guidance in 2D user interfaces is typically provided by standardized interaction components (widgets) that rely on the user's ability to recognize and use familiar elements. Virtual environments in the entertainment and narration domains often use scoring feedback, storytelling techniques and virtual characters to provide guidance, but these techniques are difficult to apply in the industrial domain. In this chapter we have focused on a more general approach that augments XR environments with explicit techniques that serve to guide the viewer.

As reported a significant body of literature exists that describes such explicit guidance techniques, with some publications including an evaluation or comparison to other techniques. However, not all existing guidance techniques are described in publications and it would be useful to provide support in order to simplify the activities in the "mid-level" design activities, such as "Research and review", "Match and adapt" and "Remix and design". In this chapter we have described some techniques and tools that can be used to integrate the necessary design and evaluation activities into established development processes. Figure 7.9 illustrates which design activities can be addressed by which technique or tool. While these techniques help to make the integration of customized user guidance into XR applications more viable there is still a large opportunity to further improve the process through additional support.

Possible ways to provide such support at the "Research and review" level in the future could include the creation of standardized description formats for common properties of guidance techniques and simplified ways to both "publish" and "search" guidance techniques, e.g. by means of a public repository or database. Similar to the publication of graphics "gems" [49] or UI "pattern" libraries, e.g. [50], such repositories could provide a platform to publish and share designs and experiences with guidance techniques that might not warrant the publication of a scientific publication.

At the "Match and adapt" level a growing body of techniques are becoming available in open source repositories, however their reuse is often complicated by the use of rapidly changing and platform specific APIs. Just like XR in general, the design, development and distribution of XR guidance techniques would thus benefit from standardized, platform independent content description formats.

Finally, at the "Remix and design" level libraries of common elements in combination with specialized tools for their combination and extension are required. 3D game engines like Unity and Unreal provide not only a useful basis for the implementation of XR systems, but can also be extended with custom elements and user interfaces that allow to create such high-level tools (as in our 3D navigation testbed) without the need to reimplement the functionality already available.

Concluding, based on recent and ongoing technological improvements in XR base technologies like displays, tracking and rendering the key challenges in industrial XR applications increasingly move from the technological challenges of making XR systems work in an industrial environment to designing industrial XR applications that effectively support users in their work tasks. The design and implementation of guidance techniques forms an essential part of such systems. In this chapter we have described approaches that help to make the design and implementations of such techniques effective and efficient and have identified some open challenges to further optimize the design, reuse and evaluation of XR guidance techniques.

References

1. Dörner, R., Broll, W., Grimm, P., & Jung, B. (2016). Virtual reality und augmented reality (VR/AR). *Informatik-Spektrum, 39*(1), 30–37. https://doi.org/10.1007/s00287-014-0838-9
2. Azuma, R. T. (1997). A survey of augmented reality'. *Presence: Teleoperators and Virtual Environments, 6*(4).
3. Milgram, P., Takemura, H., Utsumi, A., & Kishino, F. (1995). Augmented reality: A class of displays on the reality-virtuality continuum. In *Paper presented at the Telemanipulator and Telepresence Technologies* (Vol. 2351, p. 282). IEEE. https://doi.org/10.1117/12.197321
4. Ishii, H., & Ullmer, B. (1998). Tangible bits: Towards seamless interfaces between people, bits and atoms. In *Conference on Human Factors in Computing Systems - Proceedings*. IEEE. https://doi.org/10.1145/258549.258715
5. Paelke, V., Nebe, K., & Geiger, C. (2012). Multi-modal, multi-touch interaction with maps in disaster management applications. *International Archives of the Photogrammetry, Remote Sensing and Spatial Information Sciences, XXXIX-B8*, XXII.
6. Moran, T. P., & Dourish, P. (2001). Introduction to this special issue on context-aware computing. *Null, 16*(2–4), 87–95. https://doi.org/10.1207/S15327051HCI16234_01
7. Caudell, T. P., & Mizell, D. W. (1992). Augmented reality: An application of heads-up display technology to manual manufacturing processes. In *Proceedings of the Twenty-Fifth Hawaii International Conference on System Sciences* (pp. 659–669). ACM. https://doi.org/10.1109/HICSS.1992.183317

8. Darken, R. P. (1995). Wayfinding in large-scale virtual worlds. In *Paper presented at the conference companion on human factors in computing systems, Denver, Colorado, USA* (pp. 45–46). ACM. https://doi.org/10.1145/223355.223419

9. Boud, A. C., Haniff, D. J., Baber, C., & Steiner, S. J. (1999). Virtual reality and augmented reality as a training tool for assembly tasks. In *Paper presented at the 1999 IEEE international conference on information visualization* (pp. 32–36). IEEE. https://doi.org/10.1109/IV.1999.781532

10. Reiners, D., Stricker, D., Klinker, G., Müller, S., & Igd, F. (1999, November). Augmented reality for construction tasks: Doorlock assembly. In *Proceedings of the first International Workshop on Augmented Reality (IWAR'98), San Francisco* (pp. 31–46). A K Peters.

11. Tang, A., Owen, C., Biocca, F., & Mou, W. (2003). Comparative effectiveness of augmented reality in object assembly. *In Paper presented at the Proceedings of the SIGCHI Conference on Human Factors in Computing Systems, Ft. Lauderdale, Florida, USA*, pp. 73–80. https://doi.org/10.1145/642611.642626

12. Henderson, S., & Feiner, S. (2011). Exploring the benefits of augmented reality documentation for maintenance and repair. *IEEE Transactions on Visualization and Computer Graphics, 17*(10), 1355–1368. https://doi.org/10.1109/TVCG.2010.245

13. Khan, A., Matejka, J., Fitzmaurice, G., & Kurtenbach, G. (2005). Spotlight: Directing users' attention on large displays. In *Paper presented at the Proceedings of the SIGCHI Conference on Human Factors in Computing Systems, Portland, Oregon, USA* (pp. 791–798). IEEE. https://doi.org/10.1145/1054972.1055082

14. Biocca, F., Owen, C., Tang, A., & Bohil, C. (2007). Attention issues in spatial information systems: Directing mobile users' visual attention using augmented reality. *Journal of Management Information Systems, 23*(4), 163–184. https://doi.org/10.2753/MIS0742-1222230408

15. Baudisch, P., & Rosenholtz, R. (2003). Halo: A technique for visualizing off-screen objects. In *Paper presented at the Proceedings of the SIGCHI Conference on Human Factors in Computing Systems, Ft. Lauderdale, Florida, USA* (pp. 481–488). ACM. https://doi.org/10.1145/642611.642695

16. Renner, P., & Pfeiffer, T. (2017). Attention guiding techniques using peripheral vision and eye tracking for feedback in augmented-reality-based assistance systems. In *Paper presented at the 2017 IEEE Symposium on 3D User Interfaces (3DUI)* (pp. 186–194). IEEE. https://doi.org/10.1109/3DUI.2017.7893338

17. Renner, P., Blattgerste, J., & Pfeiffer, T. (2018). A path-based attention guiding technique for assembly environments with target occlusions. In *Paper presented at the 2018 IEEE Conference on Virtual Reality and 3D User Interfaces (VR)* (pp. 671–672). IEEE. https://doi.org/10.1109/VR.2018.8446127

18. Radkowski, R., Herrema, J., & Oliver, J. H. (2015). Augmented reality-based manual assembly support with visual features for different degrees of difficulty. *International Journal of Human-Computer Interaction, 31*, 337–349.

19. Imbert, J., Hodgetts, H. M., Parise, R., Vachon, F., Dehais, F., & Tremblay, S. (2014). Attentional costs and failures in air traffic control notifications. *Ergonomics, 57*(12), 1817–1832. https://doi.org/10.1080/00140139.2014.952680

20. Blattgerste, J., Strenge, B., Renner, P., Pfeiffer, T., & Essig, K. (2017). Comparing conventional and augmented reality instructions for manual assembly tasks. In *Proceedings of PETRA'17* (pp. 75–82). Association for Computing Machinery. https://doi.org/10.1145/3056540.3056547

21. Büttner, S., Funk, M., Sand, O., & Röcker, C. (2016). Using head-mounted displays and in-situ projection for assistive systems: A comparison. In *Proceedings of 9th ACM international conference on pervasive technologies related to assistive environments* (p. 44). ACM.

22. Funk, M., Kosch, T., & Schmidt, A. (2016). Interactive worker assistance: Comparing the effects of in-situ projection, head-mounted displays, tablet, and paper instructions. In *Proceedings of 2016 ACM international joint conference on pervasive and ubiquitous computing* (Vol. 2016, pp. 934–939). ACM.

23. Korn, O., Schmidt, A., & Hörz, T. (2013). The potentials of in-situ-projection for augmented workplaces in production: A study with impaired persons. In *Extended Abstracts CHI'13* (pp. 979–984). Association for Computing Machinery. https://doi.org/10.1145/2468356.2468531

24. Hein, P., Bernhagen, M., & Bullinger, A. C. (2019). Improving visual attention guiding by differentiation between fine and coarse navigation. In *Paper presented at the 11th International Conference on Virtual Worlds and Games for Serious Applications, VS-Games 2019, Vienna, Austria, September 4–6, 2019* (pp. 1–4). ACM. https://doi.org/10.1109/VS-Games.2019.8864539

25. Renner, P., & Pfeiffer, T. (2020). AR-glasses-based attention guiding for complex environments: Requirements, classification and evaluation. In *Paper presented at the Proceedings of the 13th ACM International Conference on Pervasive Technologies Related to Assistive Environments, Corfu, Greece*. IEEE. https://doi.org/10.1145/3389189.3389198

26. Rothe, S., Buschek, D., & Hußmann, H. (2019). Guidance in cinematic virtual reality-taxonomy, research status and challenges. *Multimodal Technologies and Interaction, 3*(1). https://doi.org/10.3390/mti3010019

27. Büttner, S., Mucha, H., Funk, M., Kosch, T., Aehnelt, M., Robert, S., & Röcker, C. (2017). The design space of augmented and virtual reality applications for assistive environments in manufacturing: A visual approach. In *Proceedings of 10th International conference on pervasive technologies related to assistive environments* (pp. 433–440). ACM.

28. Funk, M., Kosch, T., Greenwald, S. W., & Schmidt, A. (2015). A benchmark for interactive augmented reality instructions for assembly tasks. In *Proceedings of 14th International Conference on Mobile and Ubiquitous Multimedia (MUM'15)* (pp. 253–257). Association for Computing Machinery. https://doi.org/10.1145/2836041.2836067

29. Hou, L., Wang, X., & Truijens, M. (2015). Using augmented reality to facilitate piping assembly: An experiment-based evaluation. *Journal of Computing in Civil Engineering, 29*, 1. (2015). https://doi.org/10.1061/(ASCE)CP.1943-5487.0000344

30. Blattner, J., Wolfartsberger, J., Lindorfer, R., Froschauer, R., Pimminger, S., & Kurschl, W. (2021). A standardized approach to evaluate assistive systems for manual assembly tasks in industry. In *Proceedings of Conference on Learning Factories (CLF)*. ACM. https://doi.org/10.2139/ssrn.3858632

31. Illing, J., Klinke, P., Grünefeld, U., Pfingsthorn, M., & Heuten, W. (2020). Time is money! Evaluating augmented reality instructions for time-critical assembly tasks. In *19th international conference on mobile and ubiquitous multimedia*. ACM.

32. Büttner, S., Prilla, M., & Röcker, C. (2020). Augmented reality training for industrial assembly work - Are projection-based AR assistive systems an appropriate tool for assembly training? In *Proceedings of CHI'20* (pp. 1–12). Association for Computing Machinery. https://doi.org/10.1145/3313831.3376720

33. Furmanski, C., Azuma, R., & Daily, M. (2002). Augmented-reality visualizations guided by cognition: Perceptual heuristics for combining visible and obscured information. In *Paper presented at the Proceedings. International Symposium on Mixed and Augmented Reality* (pp. 215–320). ACM. https://doi.org/10.1109/ISMAR.2002.1115091

34. Kruijff, E., Swan, J. E., & Feiner, S. (2010). Perceptual issues in augmented reality revisited. In *Paper presented at the 2010 IEEE international symposium on mixed and augmented reality* (pp. 3–12). IEEE. https://doi.org/10.1109/ISMAR.2010.5643530

35. Roda, C. (2010). Attention support in digital environments. nine questions to be addressed. *New Ideas in Psychology, 28*(3), 354–364. https://doi.org/10.1016/j.newideapsych.2009.09.010

36. Hata, H., Koike, H., & Sato, Y. (2016). Visual guidance with unnoticed blur effect. In *Paper presented at the Proceedings of the International Working Conference on Advanced Visual Interfaces, Bari, Italy* (pp. 28–35). ACM. https://doi.org/10.1145/2909132.2909254

37. Kalkofen, D., Sandor, C., White, S., & Schmalstieg, D. (2011). *Visualization techniques for augmented reality* (pp. 65–98). IEEE. https://doi.org/10.1007/978-1-4614-0064-6_3

38. Schwaber, K. & Sutherland, J. (2020). *The scrum guide*. Retrieved November 9, 2021, from https://scrumguides.org/docs/scrumguide/v2020/2020-Scrum-Guide-US.pdf

39. ISO 9241-210:2019(en). Retrieved from https://www.iso.org/standard/77520.html

40. Dhiman, H., Martinez, S., Paelke, V., & Röcker, C. (2018). Head-mounted displays in industrial AR-applications: Ready for prime time? In *Proceedings of HCI International 2018, Las Vegas, NV, USA*. ACM. https://doi.org/10.1007/978-3-319-91716-0_6

41. IEEE Xplore. Retrieved from https://ieeexplore.ieee.org/

42. Oculus developers guide. Retrieved from https://developer.oculus.com/learn/

43. Microsoft guidelines for hololens AR systems. Retrieved from https://docs.microsoft.com/en-us/windows/mixed-reality/design/app-patterns-landingpage

44. Nielsen, J. (1989). Usability engineering at a discount. In *Paper presented at the proceedings of the third international conference on human-computer interaction on designing and using human-computer interfaces and knowledge based systems* (2nd ed., pp. 394–401). IEEE.

45. Fischertechnik. Retrieved from https://www.fischertechnik.de/de-de/produkte

46. Lego. Retrieved from https://www.lego.com/

47. Makeblock. Retrieved from https://www.makeblock.com/

48. Paelke, V., Röcker, C., & Bulk, J. (2019). A test platform for the evaluation of augmented reality head mounted displays in industrial applications. In *Proceedings of advances in manufacturing, production management and process* (pp. 25–35). IEEE.

49. Glassner, A. (1990). In A. S. Glassner (Ed.), *Graphics gems*. Academic Press.

50. UI pattern library. Retrieved from http://ui-patterns.com/

Lenssembly: Authoring Assembly Instructions in Augmented Reality Using Programming-by-Demonstration

8

Thomas Kosch, Pascal Knierim, Mareike Kritzler, Daniel Beicht, and Florian Michahelles

Abstract

Managing the knowledge of assembly workers is crucial due to the valuable personal expertise of collected information over time that is hard to articulate. Unfortunately, the accumulated knowledge disappears when workers leave the company. Methods to record and transfer assembly knowledge between workers rarely exist due to the time-consuming documentation of assembly steps. This paper presents Lenssembly, a mobile augmented reality system utilizing programming-by-demonstration to record, detect, and generate assembly instruction sequences using a head-mounted display. The assembly instructions are automatically detected using a neural network, preventing the need for manual documentation and time-intensive content creation for each assembly step.

T. Kosch (✉)
Utrecht University, Utrecht, the Netherlands

HU Berlin, Berlin, Germany
e-mail: t.a.kosch@uu.nl

P. Knierim
Bundeswehr University Munich, Munich, Germany
e-mail: pascal.knierim@unibw.de

M. Kritzler
Siemens Corporate Technology, Berkeley, CA, USA
e-mail: mareike.kritzler@siemens.com

D. Beicht
LMU Munich, Munich, Germany

F. Michahelles
TU Wien, Vienna, Austria
e-mail: florian.michahelles@tuwien.ac.at

© The Author(s), under exclusive license to Springer Nature Switzerland AG 2023
C. Röcker, S. Büttner (eds.), *Human-Technology Interaction*,
https://doi.org/10.1007/978-3-030-99235-4_8

In a user study (N = 12) with two different assembly tasks, participants favored the recording functionality of Lenssembly while conducting fewer errors and perceiving less task load than traditional paper instructions. We discuss the implications of our results and conclude how technologies create repositories for storing and transferring expert worker knowledge.

Keywords

Augmented reality · Artificial Intelligence · Programming-by-demonstration · Industry 4.0

8.1   Introduction

Assembly knowledge preservation has become a relevant factor in manual production lines. Lot sizes become smaller, which is attributed to a decrease in mass production and an increase in personalized production for individual customer needs [1]. Despite an increasing degree of automation, manual assembly workers are the driving force to maintain the trend of individual mass production [2]. Gone are the times where pure rote learning of instruction steps was sufficient to perform the manual assembly. Instead, individual production at assembly lines which are expected to rise [3]. Current practices require workers to memorize assembly instructions on-demand, requesting junior and senior workers to adapt their assembly procedures frequently to fulfill the unique product assembly requirements. The assembly knowledge is passed from a senior worker to a junior worker in verbal or written form. Common instruction modalities include verbal communication or printed instructions [4]. Such instruction modalities do not scale well with small sizes and frequently require senior workers to teach new instructions.

However, the senior workers' time is precious, and they may leave the company, effectively taking their accumulated expert knowledge with them. The current practice is to reverse engineer the assembly process since no documentation standards for assembly procedures exist. Here, junior workers are confronted with constant variations of assembly instructions that may lead to an increase in working memory demand and error rates [5–7]. The issues above are counterproductive towards an effective knowledge transfer between junior and senior workers.

Assistive systems at workplaces have been researched to ameliorate this effect [8]. Such assistive systems may use external displays [9], pick-by-light systems [10], or Augmented Reality (AR) systems, such as projections or Head-Mounted Displays (HMDs) [11] to support workers during their assembly. The feasibility of these systems has been the subject of various evaluations [12, 13] attesting to the positive effects of interactive worker assistance. However, one reason preventing assistive assembly systems from entering the production lines of enterprises is the extended complexity of recording and assembly instructions generation. Here, the initial configuration and the authoring of assembly instructions are time-consuming. Moreover, it requires senior workers proficient with the assembly procedure to prepare them alongside their working function in a company. Those requirements might result in an early decline of assistive assembly systems, although these systems offer promising solutions to convey assembly knowledge and optimize knowledge management.

This paper presents Lenssembly, an HMD-based prototype recording, generating, and displaying assembly instructions using a programming-by-demonstrating approach. Workers can record assembly instructions during assembly procedures through image data received from the integrated cameras. The assembly steps and associated objects are detected using a neural network translating the assembly instruction sequence into a spatial holographic representation. Lenssembly supports the modular recording of assembly steps, which can be stored to create an assembly instruction repository. Junior workers can then utilize this repository to view and assemble manufacturing procedures. We evaluate Lenssembly in a user study with two assembly tasks and 12 participants, comparing the usability of the teaching procedure and the assembly performance between Lenssembly and traditional paper instructions. We find that all participants could create assembly instructions and perform all assembly tasks successfully using Lenssembly. Furthermore, our results revealed that using Lenssembly contributes to a lower error rate and perceived task load compared to paper instructions. Finally, interviews revealed that Lenssembly is a suitable tool to preserve the assembly procedures of senior workers, which junior workers can use to learn new assembly procedures.

8.2   Contribution Statement

The contribution of our work is threefold: We (1) present Lenssembly, an AR-based assistive system that records and replays assembly instructions using a neural network through a user-driven programming-by-demonstration approach. We (2) conduct a user study (N = 12) with two different assembly tasks to evaluate the efficiency and usability of Lenssembly compared to traditional paper instructions. Finally, we (3) discuss how ubiquitous technologies benefit from our approach to populate knowledge repositories hosting assembly procedures for effective knowledge transfer between workers. We are confident that our work paves the way for pervasive knowledge documentation passed to other users utilizing ubiquitous technologies.

8.3   Related Work

In the following section, we first introduce related research regarding AR-supported assembly guidance, followed by an outline of assembly authoring and different object and action.

8.3.1 Augmented Reality Supported Assembly Guidance

The use of AR for assembly guidance instead of other assembly instruction modalities was already the subject of previous research. Boud et al. [14] presented the idea of using AR for manual assembly tasks. Specifically for the effectiveness of assembly task guidance, Boud et al. [14] compared five guidance methods, including conventional engineering drawings, immersive VR, and context-free AR. In a user study where participants had to conduct an assembly task using one of the different methods, they showed that VR and AR were outperforming the traditional 2D engineering drawings. The AR system was further rated as the most effective method. Interestingly, the assembly tasks were up to three times faster using AR compared to the VR methods and more than eight times faster than the traditional engineering drawings.

A series of studies showed the positive effects of augmented worker assistance. Henderson et al. [15], and Tang et al. [16] studied the advantages of AR compared to traditional knowledge and assembly instruction transfer methods, finding that AR-based guidance significantly reduced the task completion times, number of errors, and cognitive workload. Nilsson and Johansson [17] investigated the acceptance of AR instructions and confirmed the users' preference for AR supported instructions over traditional learning methods.

Researchers began to evaluate functional prototypes to evaluate augmented assembly instructions. A hand-held assembly system was presented by Billinghurst et al. [18]. A mobile phone is used as a see-through display to view complex models and detailed assembly instructions. Their study shows that AR with overlayed animations resulted in the lowest task completion time than static AR. Westerfield et al. [19] translated this concept to HMDs in a motherboard assembly scenario. Their research concludes that AR assembly guidance can significantly improve learning success. This hints towards a learning effect when lessons are frequently repeated in an interactive learning environment [20, 21].

Bannat et al. [22] presented how projection-based systems can support assembly workers. A camera and a projector mounted above the table displays the assembly instruction sequence. A camera is used to track the user's actions and verify the assembly. Inspired by the approach from Bannat et al., several projection-based systems were presented that improve the assembly performance at stationary and mobile workplaces for workers [12, 23–25] and workers with cognitive impairments [9, 26]. Furthermore, Kosch et al. [6, 27, 28] showed through the use of electroencephalography how the worker's cognitive workload is effectively reduced using in-situ AR. Hence, assistive assembly systems can be implicitly benchmarked in addition to already established assembly system performance metrics [29, 30]. Büttner et al. [31] investigated training duration and learning effects of AR-driven assembly support systems. Their findings conclude that participants show a faster assembly performance when using AR-based assistive technologies in the first 24 h. Their results imply that, while assistive technologies improve the assembly performance during initial training, there may be no improvements for long-term use.

8.3.2 Assembly Authoring, Object, and Action Recognition

The construction and order of assembly steps require a worker (i.e., the author) to demonstrate those steps while a system recognizes these actions. Marker-based approaches were initially used to accomplish object detection and tracking. Molineros et al. [32] addressed the sensing problem for object tracking and connection detection with the help of encoded markers. Each assembly object had a marker that was uniquely recognized and tracked. A previously computed assembly graph is used as a basis for the representation of all possible states of parts and feature descriptors. Computer vision algorithms are used to identify, track, and verify the assemblies and attachments in real-time using the marker. Gupta et al. [33] demonstrated an assembly authoring and guiding system for Lego Duplo bricks. The system uses color and depth information from a Kinect camera for object tracking and action inference. The functionality of the system is restricted to a table surface with four marked regions. The user can freely move the assembly object within the play area during the guidance and authoring task, where the object is tracked. A virtual replica is displayed on a monitor showing the assembly object in the same pose as the in-hand physical model. Before a new Lego Duplo brick is added to the physical model, it must first be placed in the "add box", where the size and color of the model are matched against a virtual model. If the attachment is not correctly recognized, the user can restart the detection process by placing his hand in a recheck box. Finally, objects can be removed from the physical model by placing the detached object in the remove box. For the alignment of the digital replica with the physical model, a transformation between the camera point cloud and the virtual point cloud is calculated with an iterative closest point algorithm. New attachments and detachments are detected by calculating a belief distribution value over several frames, describing addition and removal candidates, since single poses (i.e., single frames) contain insufficient information. This approach can be combined with recent studies that utilize eye tracking to understand how assembly instruction information is accessed by the individual user [34].

Neural networks have become a popular method to estimate object and their states for further processing in AR. Su et al. [35] proposed a method to estimate the pose of objects for AR applications. They use a neural network to detect the pose and state of objects for further use in AR applications. Roitberg et al. [36] presented a vision-based approach for recognizing human activities in industrial environments using deep learning. The system enables the recognition of hand activities like selecting points and objects grasping and coarser activities such as assembly and processing of workpieces. Multimodality is used to create a single output by combining multiple sensors. For this, data from a Kinect v2 for human skeleton tracking, an Asus Xtion Pro for object recognition, and a Leap Motion for detailed hand skeleton tracking are fused with a self-developed control framework for data synchronizing. Activities are classified in multiple abstraction levels using hierarchical hidden Markov models after feature selection and dimensionality reduction. Agrawala et al. [37] proposed design principles on how to display assembly instruction sequences for users. Consequently, Büttner et al. presented a system that records assembly instructions of Lego Duplo bricks using an RGB and depth sensor [38, 39]. The prototype records the user's actions and visualizes the assembly instructions to the user. However, the assembly instruction reconstruction is limited to a stationary setting and relies on complementary object colors to enable a robust recognition. While the added depth sensor provides a more robust detection based on the objects' 3D constitution, it may not work reliably for similar objects with similar colors.

Neural networks have gained an increase in popularity for object and action detection recently. Shinde et al. [40] present an approach where human actions are recognized using the YOLO object detection framework. Therefore, they trained the object detector with images consisting of appropriate labeled actions. When an action label is detected with a given threshold over multiple frames, the associated action is classified for this period. An object detection algorithm for action classification allowed the classification of activities with a small number of images. It showed that even single images could be sufficient to recognize the action. The work of Bhattacharya et al. [41] introduced an approach for generating AR work instructions by expert demonstration. The execution of the procedure is first recorded and then processed to create the learning environment. For this purpose, the system consists of two parts: the demonstration and refinement phases. In the demonstration phase, a static near-range 3D sensor captures a predefined surface. New steps are detected by continuously checking whether a hand is in the image. The current point cloud is compared with a previous one and checked for a new object when the hand disappears from the image. Subsequently, an algorithm tries to identify the movement of the newly detected object to create an accurate animation. Finally, the refinement phase allows us to modify the recorded steps and add additional text, images, and videos. However, this approach is limited to stationary settings as a near-range sensor must be placed in a fixed position. Recently, Kong et al. [42] presented TutorialLens, an AR authoring and demonstration system to create instructions for the operation of user interfaces. Users can author interface tutorials using a computer vision-based demonstration recording the 3D coordinates of finger positions. TutorialLens reproduces these finger movements for users who are not proficient with the user interface through a see-through device. Users are then hinted at the correct interaction via AR.

Previous research showed how interactive worker assistance could be used to increase assembly efficiency and productivity. However, at the same time, assembly instructions for those assistive systems need to be created, which is a laborious task. Here, past research looked into activity and action recognition to ease the authoring of the assembly instructions. However, previous work required manual interaction with an assistive system to teach assembly instruction procedures or was functional under controlled conditions. We close this gap by presenting Lenssembly. This mobile assembly instruction authoring system ubiquitously records the worker's interaction with workpieces to automatically

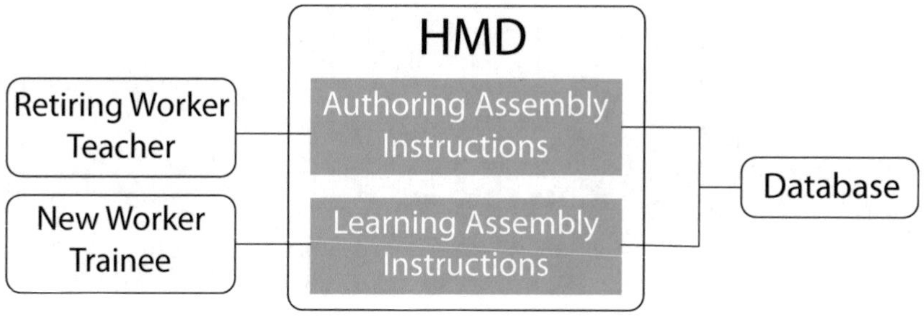

Fig. 8.1 Application *s*tructure of Lenssembly. An Authoring Mode enables to create and author new assembly instruction procedures. The learning procedure uses the recorded instructions to transfer this knowledge to new workers via AR. A Playback Mode displays the assembly instructions in AR and verifies if the assembly is conducted correctly

generate assembly instruction sequences, where objects are detected using a neural network.

8.4   Lenssembly: An Assembly Authoring and Playback System

Lenssembly runs as a mobile setup on a Microsoft HoloLens as HMD. Here, Lenssembly features an Authoring Mode and Playback Mode to capture assembly steps of a senior worker and replay them to trainees or junior workers. Figure 8.1 depicts the concept of Lenssembly. Both, the Authoring Mode and Playback Mode use the front-facing camera of an HMD (i.e., a HoloLens) to capture changes on the workplace to learn the order of new assembly steps (i.e., in the Authoring Mode) or show a digital representation of the next assembly step (i.e., in the Playback Mode). Objects and the worker's hands are detected using a trained neural network. We use the YOLOv3 [43] algorithm to detect and track objects. YOLOv3 applies a single neural network to the full image, making it suitable for the efficient real-time detection of objects. Objects must be annotated beforehand in pictures fed into the neural network, resulting in a model embedded into applications. The following section describe the details of the Authoring Mode and Playback Mode.

8.4.1   Authoring Mode: Expert Authoring and Recording Systems

The Authoring Mode allows expert and senior workers to record the order of assembly steps using an automatically "programming-by-demonstration" approach. The recording system allows workers to author new assembly task workflows and store them persistently in a data model by demonstrating the assembly steps in a workplace. This includes assembly steps and hands, tools, and actions that require attachments between several components. The Authoring Mode system utilizes a neural network to detect the sequence

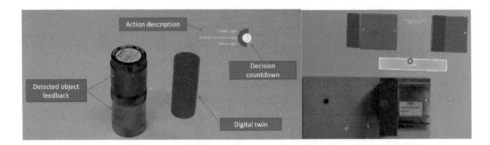

Fig. 8.2 Procedure step suggestion process for a detected connected object in the Authoring Mode.
Left: A connection between the yellow and green object was detected and is added to the assembly
procedure after the decision timer expires. **Right:** Conflict resolution menu models two detected
actions for the same objects. The user must choose one of the two actions by clicking on the
corresponding model

of the assembly steps and worker actions. The user can manually proceed or wait until a
decision countdown expires, upon which the current assembly instruction step is saved.
The worker receives continuous feedback about the learned assembly procedure and
displayed digital representation. The previously prepared digital twin is saved into the
app and displayed upon detecting the associated real-world object. All objects are captured
using the point-of-view camera of the HoloLens. Figure 8.2 depicts an illustration of the
Authoring Mode view.

The worker can choose an existing assembly task or create a new one upon starting
Lenssembly. A virtual keyboard or voice commands are used throughout the whole
training procedure to provide text input. The worker can create a new work procedure
for each step and perform the respective assembly. Lenssembly provides constant feedback
on how the final representation will look like for the trainee. The process enables experts to
see what the system detected and to intervene if necessary. First, the procedure step is
always described as a triplet in the upper right corner of the HMD. In the case of an
attachment action, a digital twin is created next to the associated physical objects of the
action. To create a digital twin, we use pre-generated 3D models of the individual captured
objects utilizing the CAD models to create a 3D representation in AR. The 3D
representations of the digital twins are prepared beforehand and displayed when the
associated objects' constellations are detected. A five-second timer is started upon
detecting an object. The assembly procedure is saved when the timer expires, and subsequently, the next step can be performed—the timer restarts when corrections to the
assembly are made. The current assembly step training procedure is canceled when the
front-facing camera loses the tracking of the objects. If multiple actions are detected during
the same assembly step, the expert worker can review and correct the assembly steps using
a conflict resolution menu (see Fig. 8.2).

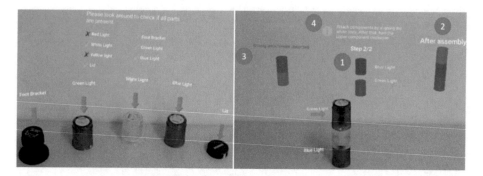

Fig. 8.3 Lenssembly checks the available parts prior the assembly. **Left:** A checklist is displayed and an arrow with a description denotes the part. The assembly begins when all parts are available. **Right:** Lenssembly displays (1) the current procedure step, (2) the pre-trained result, (3) potential wrong actions, and (4) textual information messages that support the assembly

8.4.2 Playback Mode: Trainee Replay and Learning System

Previously trained assembly instructions can be replayed using a Playback Mode functionality. Lenssembly displays all authored assembly procedures for selection to the user. A feedback location (i.e., the worktable) has to be selected by the user to begin with the assembly process using a drag and drop gesture. Then, a part list appears that visually checks if the assembly parts are available in the workplace (see Fig. 8.3). The assembly starts afterward. A previously prepared 3D rendering of a CAD model is rendered next to the physical assembly object. A description above the object provides additional details about how to perform the assembly. Generated animations are played to display how attachments must be performed (e.g., through a predefined blinking arrow displayed above the detected object). Also, arrows depict which objects have to be attached. The assembled objects turn green, and the application continues with the next assembly step when a correct assembly is detected. Wrong actions are displayed on the left side of the workplace to inform the user about potential corrections of their assembly. Again, all objects are tracked using the integrated point-of-view camera of the HoloLens. Figure 8.3) displays how the Playback Mode is displayed to the user.

8.5   Evaluation of Lenssembly Through a User Study

We conduct an evaluation of the Authoring Mode and Playback Mode system of Lenssembly. We explain the used assembly tasks, provide details about the trained model to detect workpieces and describe the methodology.

Fig. 8.4 The signaling column (**left**) and PLC task (**right**) that were used to evaluate Lenssembly. Both assembly tasks require a screwdriver

8.5.1 Assembly Tasks

We selected two assembly scenarios to evaluate Lenssembly in a user-centric study. First, we use YoloV3 to train a neural network that detects the assembly components, tools, and worker's hands. Second, we utilize the assembly of a signaling column and programmable logic controller for the user study.

Signaling Column Assembly The signaling column assembly task consists of seven components that are stacked on top of each other (see Fig. 8.4). At the lower end is a bracket, and the upper back is closed by a lid. Every component, except the screwdriver, has one or two attachment points located on the top or bottom of the object where other fitting parts can be attached. The bracket can be screwed on to fasten the signaling column. This assembly task was chosen because of the various possible results and procedures that can lead to the same goal. Actions, such as the connections between the components, the worker's hand, and the screwdriver, need to be detected by Lenssembly. We have selected the signaling column assembly task since it can be varied through the order of the attachments without changing the assembly task itself.

Programmable Logic Controller Assembly The programmable logic controller (PLC) task contains five hardware components (see Fig. 8.4) that require the worker to mount several components on a mounting rail. The components are placed next to each other. The PLC task is more complex than the signaling column task since the modules placed next to each other and on the components cannot be exchanged. In addition, it requires the frequent use of the screwdriver since all modules need to be fastened.

8.5.2 Data Set Collection and Model Training

We recorded videos of all assembly workpieces in the first step to acquiring image data of the objects. Videos were made from different angles and directions to retrieve a diverse set of images. Furthermore, all videos were taken against diverse backgrounds and under other lighting conditions. We recorded the videos with 30 frames per second. We extracted every fifth frame of the videos for the labeling process to get a good distribution of multi-angled images. We manually removed blurred images and replaced them with adjacent frames from the video. We ensured that all objects in the data set were represented approximately equally often during the data collection. In total, we extracted over 5600 frames from the recorded videos. Each captured frame contained one or multiple objects. The data set included 17 classes from the two previously introduced assembly tasks. A list of all classes with their associated assembly task can be found in Table 8.1. Tools (i.e., a screwdriver) and hands are listed extra since they are part of both tasks.

Data annotation and labeling is the process of labeling data for supervised learning machine learning methods. This involves the object localization inside a frame and their respective manual labeling. We manually labeled the objects with bounding boxes in the image using the open-source tool OpenLabeling. OpenLabeling generates the annotation and associated labels in the appropriate YOLOv3 textfile format. Darknet2 is used as an open-source implementation for YOLOv3. Each file stores information about all bounding boxes of the corresponding image, describing their respective class id and position (i.e., xCenter, yCenter, width, and height), and uses normalized pixel coordinates between 0 and 1. Finally, we achieved a good distribution with only the Hand class having significantly more pictures~ (see Table 8.1). After generating the basic truth data, the annotated data set was split into a train and a test data set. We divided the data into 85–15 splits using a random-based approach. YOLOv3 resized the input images to 608x608 pixels. We used the previous labeled images for training, resulting in 11,467 training instances. The number of filters was set to 66 in the YOLOv3 configuration file. The final model was trained for 176k iterations until a training loss of less than 0.1 was reached, resulting in a loss of 0.06.

8.5.3 Methodology

We evaluate Lenssembly using a mixed-method study design opting for a within-between-subject design. We divided the participants into two groups. The first group (N = 6) was asked to assemble the signaling column and PLC unit using the Playback Mode of Lenssembly. Afterward, the participants filled a NASA-TLX questionnaire and custom Likert scales accompanied by a semi-structured interview. The second group (N = 6) was asked to assemble the signaling column and PLC unit using paper instructions. The participants were asked to fill a NASA-TLX questionnaire and custom Likert scales similar to the first group. In addition to the assembly, the second group was asked to build the

Table 8.1 Final label distribution of the workpieces, tools, and hands

Class	Instances	Assembly task
Yellow light	659	Signaling column
Red light	684	Signaling column
Green light	693	Signaling column
White light	663	Signaling column
Blue light	663	Signaling column
Lid	686	Signaling column
Bracket	674	Signaling column
Load current supply (lid closed)	689	PLC
CPU (lid closed)	660	PLC
Digital input/output module (lid closed)	655	PLC
Mounting rail	580	PLC
Load current supply (lid opened)	641	PLC
CPU (lid opened)	650	PLC
Digital input/output module (lid opened)	665	PLC
U-connector	632	PLC
Screwdriver	667	Tool
Hand	906	Tool
Total	11,467	

signaling column using the Playback Mode to gain additional qualitative feedback about the mode. Finally, we conducted a semi-structured interview. We counterbalanced the assembly conditions (i.e., assembly of signaling column and PLC unit) according to the balanced Latin square. After the assembly, all participants of both groups were invited to train and test a new assembling procedure of a signaling column using the Authoring Mode to gain additional qualitative feedback.

We compare the assembly efficiency of Lenssembly using paper instructions during the participant's assembly using the Playback Mode. Printed paper instructions are a standardized modality to convey assembly instructions that have been used in research to compare novel assembly instruction modalities [44]. The paper instructions included one instruction step per page. The page has to be flipped to continue with the following assembly instruction. Pictures were generated using available CAD models of the corresponding workpieces. Also, both the recorded assembly procedures in AR and the paper instructions contained written text to support the assembly. We used the Microsoft HoloLens as HMD. Figure 8.5) the study workplace. We outline the independent and dependent variables in the following.

Independent Variables We employ the assembly instruction modalities as a factor with two levels (i.e., AR with Lenssembly and paper instructions). Participants were either assembling the signaling column task or PLC task using Lenssembly or using paper

Fig. 8.5 Left: Study setup with the signaling column and PLC assembly task. Printed paper instructions were used as baseline. The finally assembled **(middle)** signaling column and **(right)** PLC unit

instructions depending on the group. In addition, all participants were teaching the signaling column using the Authoring Mode of Lenssembly after their assembly.

Dependent Variables We measure the task completion time for every single assembly step during assembly with the paper instructions and Playback Mode of Lenssembly. We measured the number of errors made for each assembly task. An error is always counted when the assembly step itself or the mounting of parts was conducted erroneously (e.g., wrong order of attachments). Here, we subdivide the errors between independent and dependent errors. We specify independent errors as wrong attachment steps or missing assembly objects. For example, independent errors are subsequently performed without the user's dependent error being recognized by the user. Each participant was asked to fill in a NASA Task Load Index (NASA-TLX) [45] questionnaire after each assembly procedure. Afterward, we conducted semi-structured interviews to gain additional feedback about the Playback Mode and Authoring Mode.

8.5.4 Procedure

We greeted the participants and provided them with a written description of the study to ground each participant regarding their intention. The participants were familiarized with the study setup after they provided their informed consent. Furthermore, our participants provided their demographic data, previous AR experience, and knowledge about manual assembly. A 1-min video an introduction of the Playback Mode of Lenssembly. It demonstrated the basic UI elements, controls, and the execution of hand tool actions. The study started afterward with the assembly using either Lenssembly or paper instructions according to the counterbalanced order.

In the first part of the study, our participants started with assembling the first workpiece after putting on the HoloLens or viewing the printed paper instruction. Here, the participant was either asked to assemble the workpiece (see Fig. 8.5 for the fully assembled result). For the signaling column, the participants were attaching the subparts of the signaling column

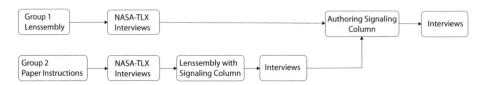

Fig. 8.6 Illustrated study procedure. We use mixed within-between-subject design where both groups experience the assembly using either Lenssembly or paper instructions. All participants author a signaling column after the assembly using the Authoring Mode. Interviews were held between the sessions

until it resulted in the whole signaling column that consisted of seven objects and six assembly steps. On the other hand, the PLC task consists of five objects and fifteen procedure steps that frequently involve tool actions. In addition, participants filled a NASA-TLX questionnaire after each assembly. Finally, all participants were authoring a signaling column using the Authoring Mode.

The second part of the study focused on the Authoring Mode feature of Lenssembly. First, the participants were asked to train an assembly procedure of the signaling column. Again, an introductory video explained how the Authoring Mode works, how it suggests detected actions, and how they can intervene to correct the system if necessary. Next, the participants were invited to build the signaling column in their fashion due to the modular components. The participants were asked to test their training procedures afterward. Finally, a semi-structured interview was conducted that examined the usability and user acceptance of the Authoring Mode. Figure 8.6 depicts the study procedure.

8.5.5 Participants

We recruited 12 volunteers (three female, nine male) to participate in the user study via mailing lists. The participants' age ranged between 18 and 55 years. Four participants had no previous experience with AR, where six participants reported rare experience with AR. Two participants were using AR frequently. Six participants had prior experience with the HoloLens. None of the participants had previous experience with manual assembly.

8.6 Results

In the following, we report the results of the task completion time, the number of errors, and subjectively perceived task load.

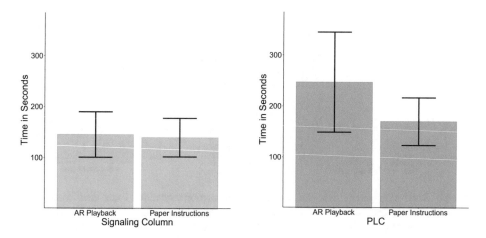

Fig. 8.7 Task completion times for the signaling column (**left**) and the PLC assembly (**right**). Paper instructions require less time for complex assembly procedures. The error bars depict the standard error

8.6.1   Task Completion Time

Paper instructions required less time compared to instructions displayed in the Playback Mode when assembling the signaling column (paper instructions: $M = 140.00$s, $SD = 37.80$s, AR-based instructions: $M = 145.55$s, $SD = 44.49$s). For the PLC task, participants require less time using paper instructions ($M = 170.00$s, $SD = 46.84$s) compared to the Playback Mode ($M = 247.00$s, $SD = 98.33$s). Potential reasons for longer task completion times using the AR-based instructions can be that participants were unused to the HMD itself and the potential waiting times between the assembly steps to display the next instruction. Figure 8.7 illustrates the mean task completion times.

8.6.2   Number of Errors and Task Load

We counted the number of errors conducted with paper and the employed AR-based instructions during the Playback Mode. We describe the number of independent and dependent errors. The documentation method for the signaling column task had 0.83 errors and 1.00 independent errors on average, resulting in 1.83 errors. No error was recorded from participants using the Playback Mode. For the PLC assembly task, we measured an averaged independent error rate of 0.33, with no independent error occurring during Playback Mode (overall error rate: $M = 0.33$). However, the documentation manual had an independent error rate of 1.83 and a dependent error rate of 0.33 (overall error rate: $M = 2.16$). Figure 8.8 shows the averaged number of errors. Furthermore, we analyze the NASA-TLX questionnaire to quantify the participants' subjectively perceived task load.

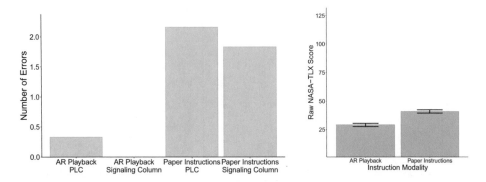

Fig. 8.8 Left: Averaged number of errors for both instruction modalities. Paper instructions elicit a larger number of paper instructions compared to the Playback Mode. **Right:** Raw NASA-TLX scores for both instruction modalities. The Playback Mode requires less workload compared to paper instructions

We find a higher level of workload for paper instructions (M = 40.84, SD = 1.44) compared to instructions displayed in the Playback Mode (M = 29, SD = 1.42). Figure 8.8 shows the TLX scores between the Playback Mode and paper instructions.

8.6.3   Qualitative Results

The following section examines the findings of the semi-structured interviews. Most participants liked that they had their hands free during the assembly and did not have to occupy their hands with physical instructions. However, one participant (P9) stated that it is hard to do the actual assembly when handling the AR application and the assembly task. The participant referred to the tool actions, where the search for tools "[...] leads to abnormal head movements." (P9). This is mainly attributed to the narrow field of view of the HoloLens. However, we see this as a technical limitation that will be resolved in the future. Asking about potential helpful features that are present in the application, eight participants stated that they liked the arrow feedback (P1, P2, P5–P7, P10–P12). Two participants added that it helped them to identify the objects even if you do not know what they look like (P7, P12). This was useful in the PLC assembly task. However, one participant found there was sometimes "[...] too much information with the arrows" (P3) which led to confusion. P7 liked that you can look at the rendered 3D models from all perspectives and that you can even look into them, providing an experience which would not be possible with documentation or videos. The participants also liked the fact that the application tells you when a task is completed. One participant stated that "It makes you feel good. You can tick off the task, continue, and know that the previous step was correct." (P10). Some participants were disturbed that certain information was often not in the field of view of the assembly environment. We asked what worked well and what caused problems during the learning process. Four participants stated that they had problems carrying out the screwing action because their hands were not recognized (P2, P3, P9, P10). Two participants explained that they often did not see the visual feedback when the system successfully recognized an object. Thus, they missed the next step in the process (P3, P9), resulting in higher task completion times. One participant stated that this is probably the case because the confirmation text on the HMD lies on a different level than the physical objects (P3). The participant also stated that "[. . .] if you concentrate on the assembly and your eyes are focused on it, you simply overlook information on the HMD." (P3). We asked if they liked that the system checks for the correctly assembled objects before proceeding to the next step. The participants liked the concept since it detects and catches errors early on. One participant thought it is essential for work where a certain level of safety must be maintained (P11). We continued to ask technical questions about the HMD. First, we asked what features of the HoloLens would need to be improved to increase the application's usability. Ten participants stated that the narrow field of view is one of the main limitations (P1–P8, P11, P12). Five participants had problems with the resolution as well as the contrast and found it hard to distinguish between certain colors.

Finally, we were asking questions about the perceived utility and usefulness of the Authoring Mode. Most of the participants perceived the recording process as intuitive. One participant stated that "[...] this is the next step in learning assembly tasks after documentations and video instructions." (P3). All participants stated that they either wanted to set the time for the recording countdown manually or use a dedicated gesture, button, or voice command. One participant would like to disassemble objects later in the recording to undo previously recorded procedure steps (P9). Overall, all participants could imagine using an HMD application similar to the developed one in the future to store their assembly knowledge persistently.

8.7   Discussion

Can we use ubiquitous technologies to store and preserve assembly knowledge? The presented user study results show that our participants could learn the recorded assembly tasks and train new assembly procedures using our system. This section discusses the results, addresses the remaining challenges, and lays out the vision of user-generated knowledge repositories.

8.7.1   Lenssembly Requires More Time than Paper Instructions

While the signaling column assembly task did not differ in completion times between the AR application and the documentation, the PLC task showed a descriptive difference. However, paper instructions outperformed AR-based instructions in terms of task completion times during the PLC task. One reason was the tool actions, where the tool has to be registered by the HMD, hence taking more time than using the tool right away. Furthermore, some users intuitively grabbed the screwdriver without looking at the tool to perform the pickup action, ignoring the previous demonstration of handling these actions that led to the need to register the tool again. This is consistent with the collected qualitative statements, where participants stated thfat the special performance of tool actions could lead to unusual head movements. Also, some participants had problems performing hand tool actions since their hands were not reliably detected. One reason could be that the data set contained hand images from people with a similar handshape. A more diverse set of hand images can lead to a more robust model. The additional actions that should be performed with the tools were partially forgotten and led to problems during execution, contributing to longer execution times.

Although the overall assembly with AR took longer than paper instructions, participants were faster going through the checklist for the available parts. The participants stated that the arrows and named labels at the location of physical objects are more convenient than searching for objects using images in the paper instructions. Verifying the availability of the required assembly parts was favored by the participants. However, another issue that arose during the assembly was the narrow field of view of the camera, preventing a successful detection of the performed assembly. We believe this can be circumvented by adding additional wearable cameras around the HMD or improvements to the existing camera. Participants also complained that the field of view is limited. We see this as a technical limitation that is likely to be resolved with future releases of HMDs (e.g., the HoloLens 2).

8.7.2   Lenssembly Elicits Fewer Errors and Less Task Load

We observed that text written in the paper instructions was more likely to be ignored than visually represented in AR. We attribute the higher number of errors for paper instructions compared to AR-based instructions to these observations. The participants confirmed this, who either did not read textual representations of assembly instructions or missed them. Our results imply that if time is not a critical factor, AR-based instructions lead to fewer errors, hence maintaining a quality level of the assembled product compared to paper-based instructions. Contrary to this, paper-based instructions can be used if time is a critical component. An analogy can be drawn toward the perceived task load. It is advisable to use AR-based instructions when the user's perceived workload should be kept minimal. For example, safety-critical assembly scenarios benefit from the Playback Mode used by junior workers.

Finally, we believe Lenssembly is successful in conveying assembly instructions to junior workers. Overall, the results show that the learning environment with automatic object detection and feedback generation led to successful assembly procedures for all participants. It reduces the error rate compared to printed documentation by providing feedback and preventing dependent errors. The current restrictions lead to the same or slower completion times compared to documentation. However, most participants stated that learning with Lenssembly was more engaging and that after the assembly, instruction was known. The HMD is not needed anymore after completing several assembly trials. We believe that Lenssembly provides an entry into the assembly of complex workpieces, which is not needed anymore after the worker gets used to the assembly procedures.

8.7.3 Recording Assembly Instructions

The evaluation of the recording subsystem showed that the participants were able to record a custom signaling column assembly procedure. All participants were able to record their assembly procedures after a short introduction. Similarities between colors, such as red and yellow, were challenging to distinguish for the deployed classifier. However, since the error only occurred in one room, the problem could have been different lighting conditions. Most participants built the signaling column with all available components and wanted to see their automatically generated learning environment afterward.

8.7.4 Limitations

The presented implementation and the study design are prone to certain limitations. The neural network can only detect an object that has been trained before. This requires laborious labeling of data which was performed manually during this study. Furthermore, this could have caused some noise in the training data since the bounding boxes do not cover the exact shape of the assembly workpieces. However, 3D representations of workpieces are usually available as a CAD model. We believe that these models can be exploited to generate a neural network for the parts and the respective assembly order, hence obviating the need for manual labeling. Another limitation affecting the study is the limited field of view of the HoloLens itself. However, this is a technical limitation that will be improved through future developments. A final fundamental limitation is the manual preparation and association of the digital twins towards the detected objects. The arrangement of the different hologram parts has to be stored into Lenssembly directly to be displayed upon detection of the assembled object. However, we are confident that future work can use the detected object model to generate the digital twin on the fly. Finally, we acknowledge that our sample partly represents persons who have experience with the HoloLens. This experience can skew our results in favor of the HoloLens. However, we are confident workers will achieve a similar performance after a brief settling-in period.

8.7.5   Future Work

Our research provides the first step of creating a knowledge repository that is not limited to the interactive representation of assembly instructions alone. In the first step, we want to utilize existing 3D CAD models to automatically generate a neural network that can detect single workpieces in an assembly procedure to create assembly instructions automatically. Furthermore, we envision creating a user-generated repository that conveys assembly instructions enriched by individual assembly styles that optimize previously known assembly procedures. Also, we will investigate potential learning effects that emerge during the use of AR-based instructions. Here, we are interested in differences in using assembly instruction in different modalities, including the required length of use and assembly performance. Finally, we will incorporate the suggestions made by the participants with a subsequent evaluation of learning effects for various assembly scenarios.

8.8   Conclusion

This paper presents Lenssembly, an AR-based system that enables the knowledge transfer of assembly instructions between senior and junior workers. We developed an AR application for an HMD that enables the flexible authoring of assembly instructions. Workers can record their assembly procedure using a programming-by-demonstration approach, reducing the complexity of recording assembly instructions while generating instructions simultaneously. As a result, the laborious content creation of assembly instructions is no longer required. In a user study with 12 participants that evaluates the Authoring Mode and Playback Mode, we find a reduced number of assembly errors and self-reported task load when using Lenssembly compared to paper instructions. However, Lenssembly elicited a higher task completion time compared to paper instructions. Interviews revealed that training new assembly procedures was a pleasant experience that all participants have successfully conducted. We conclude that Lenssembly eases the authoring of further assembly instructions for assistive systems and supports junior workers to learn new assembly instruction procedures. Our work describes how recent advances in artificial intelligence can be used to preserve assembly knowledge by creating user-generated assembly sequences that others can retrieve for learning purposes. We are confident that our work paves the way for future usable assembly systems that automatically generate assembly instructions.

References

1. Wang, Y., Ma, H. S., Yang, J. H., & Wang, K. S. (2017). Industry 4.0: A way from mass customization to mass personalization production. *Advances in Manufacturing, 5*(4), 311320. https://doi.org/10.1007/s40436-017-0204-7

2. Fellmann, M., Robert, S., Büttner, S., Mucha, H., & Röcker, C. (2017). Towards a framework for assistance systems to support work processes in smart factories. In A. Holzinger, P. Kieseberg, A. M. Tjoa, & E. Weippl (Eds.), *Machine learning and knowledge extraction* (p. 5968). Springer.
3. Yin, Y., Stecke, K. E., & Li, D. (2018). The evolution of production systems from industry 2.0 through industry 4.0. *International Journal of Production Research, 56*(1–2), 848861. https://doi.org/10.1080/00207543.2017.1403664
4. Gjeldum, N., Salah, B., Aljinovic, A., & Khan, S. (2020). Utilization of Industry 4.0 related equipment in assembly line balancing procedure. *Processes, 8*(7), 864. https://doi.org/10.3390/pr8070864
5. Brettel, M., Friederichsen, N., Keller, M., & Rosenberg, M. (2014). How virtualization, decentralization and network building change the manufacturing landscape: An industry 4.0 perspective. *International Journal of Mechanical, Industrial Science and Engineering, 8*(1), 3744.
6. Kosch, T., Funk, M., Schmidt, A., & Chuang, L. L. (2018). Identifying cognitive assistance with mobile electroencephalography: A case study with in-situ projections for manual assembly. *Proceedings of the ACM on Human Computer Interaction, 2*, 11. https://doi.org/10.1145/3229093
7. Rüßmann, M., Lorenz, M., Gerbert, P., Waldner, M., Justus, J., Engel, P., & Harnisch, M. (2015). Industry 4.0: The future of productivity and growth in manufacturing industries. *Boston Consulting Group, 9*(1), 5489.
8. Büttner, S., Mucha, H., Funk, M., Kosch, T., Aehnelt, M., Robert, S., & Röcker, C. (2017). The design space of augmented and virtual reality applications for assistive environments in manufacturing: A visual approach. In *Proceedings of the 10th ACM international conference on PErvasive technologies related to assistive environments*. ACM. https://doi.org/10.1145/3056540.3076193
9. Funk, M., Mayer, S., & Schmidt, A. (2015). Using in-situ projection to support cognitively impaired workers at the workplace. In *Proceedings of the 17th international ACM SIGACCESS conference on computers & accessibility*. ACM. https://doi.org/10.1145/2700648.2809853
10. Guo, A., Raghu, S., Xie, X., Ismail, S., Luo, X., Simoneau, J., Gilliland, S., Baumann, H., Southern, C., & Starner, T. (2014). A comparison of order picking assisted by head-up display (HUD), cart-mounted display (CMD), light, and paper pick list. In *Proceedings of the 2014 ACM International Symposium on Wearable Computers, ISWC'14* (p. 7178). Association for Computing Machinery. https://doi.org/10.1145/2634317.2634321
11. Büttner, S., Funk, M., Sand, O., & Röcker, C. (2016). Using head-mounted displays and in-situ projection for assistive systems: A comparison. In *Proceedings of the 9th ACM international conference on PErvasive technologies related to assistive environments, PETRA'16*. Association for Computing Machinery. https://doi.org/10.1145/2910674.2910679
12. Funk, M., Baechler, A., Baechler, L., Korn, O., Krieger, C., Heidenreich, T., & Schmidt, A. (2015). Comparing projected in-situ feedback at the manual assembly workplace with impaired workers. In *Proceedings of the 8th international conference on PErvasive technologies related to assistive environments*. ACM. https://doi.org/10.1145/2769493.2769496
13. Zheng, X. S., Foucault, C., Matos da Silva, P., Dasari, S., Yang, T., & Goose, S. (2015). Eye-wearable technology for machine maintenance: Effects of display position and hands-free operation. In *Proceedings of the 33rd annual ACM conference on human factors in computing systems, CHI'15* (p. 21252134). Association for Computing Machinery. https://doi.org/10.1145/2702123.2702305
14. Boud, A. C., Haniff, D. J., Baber, C., & Steiner, S. (1999). Virtual reality and augmented reality as a training tool for assembly tasks. In *1999 IEEE International Conference on Information Visualization (Cat. No. PR00210)* (p. 3236). IEEE.

15. Henderson, S. J., & Feiner, S. K. (2011). Augmented reality in the psychomotor phase of a procedural task. In *2011 10th IEEE International Symposium on Mixed and Augmented Reality* (p. 191200). IEEE.

16. Tang, A., Owen, C., Biocca, F., & Mou, W. (2003). Comparative effectiveness of augmented reality in object assembly. In *Proceedings of the SIGCHI conference on Human factors in computing systems* (p. 7380). ACM.

17. Nilsson, S., & Johansson, B. (2008). Acceptance of augmented reality instructions in a real work setting. In *CHI'08 extended abstracts on Human factors in computing systems* (p. 20252032). ACM.

18. Billinghurst, M., Hakkarainen, M., & Woodward, C. (2008). Augmented assembly using a mobile phone. In *Proceedings of the 7th international conference on mobile title suppressed due to excessive length 21 and ubiquitous multimedia* (p. 8487). Association for Computing Machinery. https://doi.org/10.1145/1543137.1543153

19. Westerfield, G., Mitrovic, A., & Billinghurst, M. (2015). Intelligent augmented reality training for motherboard assembly. *International Journal of Artificial Intelligence in Education, 25*(1), 157172. https://doi.org/10.1007/s40593-014-0032-x

20. Lindgren, R., & Johnson-Glenberg, M. (2013). Emboldened by embodiment: Six precepts for research on embodied learning and mixed reality. *Educational Researcher, 42*(8), 445452. https://doi.org/10.3102/0013189X13511661

21. Werrlich, S., Daniel, A., Ginger, A., Nguyen, P., & Notni, G. (2018). Comparing HMD-based and paper-based training. In *2018 IEEE International Symposium on Mixed and Augmented Reality (ISMAR)* (p. 134142). ACM. https://doi.org/10.1109/ISMAR.2018.00046

22. Bannat, A., Wallhoff, F., Rigoll, G., Friesdorf, F., Bubb, H., Stork, S., Müller, H. J., Schubö, A., Wiesbeck, M., Zäh, M. F., et al. (2008). Towards optimal worker assistance: A framework for adaptive selection and presentation of assembly instructions. In *Proceedings of the 1st international workshop on cognition for technical systems*. Cotesys.

23. Funk, M., Bächler, A., Bächler, L., Kosch, T., Heidenreich, T., & Schmidt, A. (2017). Working with augmented reality? A long-term analysis of in-situ instructions at the assembly workplace. In *Proceedings of the 10th ACM international conference on PErvasive technologies related to assistive environments*. ACM. https://doi.org/10.1145/3056540.3056548

24. Funk, M., Kosch, T., Kettner, R., Korn, O., & Schmidt, A. (2016). motioneap: An overview of 4 years of combining industrial assembly with augmented reality for Industry 4.0. In *Proceedings of the 16th international conference on knowledge technologies and datadriven business* (p. 4). ACM.

25. Funk, M., Sahami Shirazi, A., Mayer, S., Lischke, L., & Schmidt, A. (2015). Pick from here! - An interactive mobile cart using in-situ projection for order picking. In *Proceedings of the 2015 ACM international joint conference on pervasive and ubiquitous computing*. ACM. https://doi.org/10.1145/2750858.2804268

26. Kosch, T., Kettner, R., Funk, M., & Schmidt, A. (2016). Comparing tactile, auditory, and visual assembly error-feedback for workers with cognitive impairments. In *Proceedings of the 18th international ACM SIGACCESS conference on computers & accessibility*. ACM. https://doi.org/10.1145/2982142.2982157

27. Kosch, T., & Chuang, L. (2018). Investigating the impact of assistive technologies on working memory load in manual assembly through electroencephalography. In *2nd international neuroergonomics conference: The brain at work and in everyday life*. Frontiers Research Foundation.

28. Kosch, T., Karolus, J., Ha, H., & Schmidt, A. (2019). Your skin resists: Exploring electrodermal activity as workload indicator during manual assembly. In *Proceedings of the ACM SIGCHI*

Symposium on Engineering Interactive Computing Systems, EICS'19. ACM. https://doi.org/10. 1145/3319499.3328230

29. Funk, M., Kosch, T., Greenwald, S. W., & Schmidt, A. (2015). A benchmark for interactive augmented reality instructions for assembly tasks. In *Proceedings of the 14th international conference on mobile and ubiquitous multimedia.* ACM. https://doi.org/10.1145/2836041. 2836067

30. Kosch, T., Abdelrahman, Y., Funk, M., & Schmidt, A. (2017). One size does not fit all - Challenges of providing interactive worker assistance in industrial settings. In *Proceedings of the 2017 ACM international joint conference on pervasive and ubiquitous computing.* ACM. https://doi.org/10.1145/3123024.3124395

31. Büttner, S., Prilla, M., & Röcker, C. (2020). Augmented reality training for industrial assembly work - Are projection-based AR assistive systems an appropriate tool for assembly training? In *Proceedings of the 2020 CHI Conference on Human Factors in Computing Systems, CHI'20* (p. 112). Association for Computing Machinery. https://doi.org/10.1145/3313831.3376720

32. Molineros, J. M. (2001). *Computer vision and augmented reality for guiding assembly.* Springer.

33. Gupta, A., Fox, D., Curless, B., & Cohen, M. (2012). Duplotrack: A real-time system for authoring and guiding Duplo block assembly. In *Proceedings of the 25th annual ACM symposium on user interface software and technology* (p. 389402). ACM.

34. Heinz, M., Büttner, S., & Röcker, C. (2020). Exploring users' eye movements when using projection-based assembly assistive systems. In N. Streitz & S. Konomi (Eds.), *Distributed, ambient and pervasive interactions* (p. 259272). Springer.

35. Su, Y., Rambach, J., Minaskan, N., Lesur, P., Pagani, A., & Stricker, D. (2019). Deep multistate object pose estimation for augmented reality assembly. In *2019 IEEE International Symposium on Mixed and Augmented Reality Adjunct (ISMAR-Adjunct)* (p. 222227). ACM. https://doi.org/ 10.1109/ISMAR-Adjunct.2019.00-42

36. Roitberg, A., Somani, N., Perzylo, A., Rickert, M., & Knoll, A. (2015). Multimodal human activity recognition for industrial manufacturing processes in robotic workcells. In *Proceedings of the 2015 ACM on International Conference on Multimodal Interaction* (p. 259266). ACM.

37. Agrawala, M., Phan, D., Heiser, J., Haymaker, J., Klingner, J., Hanrahan, P., & Tversky, B. (2003). Designing effective step-by-step assembly instructions. *ACM Transactions on Graphics, 22*(3), 828837. https://doi.org/10.1145/882262.882352

38. Büttner, S., Peda, A., Heinz, M., & Röcker, C. (2020). Teaching by demonstrating how smart assistive systems can learn from users. In N. Streitz & S. Konomi (Eds.), *Distributed, ambient and pervasive interactions* (p. 153163). Springer.

39. Funk, M., Lischke, L., Mayer, S., Shirazi, A. S., & Schmidt, A. (2018). *Teach me how! Interactive assembly instructions using demonstration and in-situ projection* (p. 4973). Springer. https://doi.org/10.1007/978-981-10-6404-34

40. Shinde, S., Kothari, A., & Gupta, V. (2018). Yolo based human action recognition and localization. *Procedia Computer Science, 133*, 831838.

41. Bhattacharya, B., & Winer, E. H. (2019). Augmented reality via expert demonstration authoring (AREDA). *Computers in Industry, 105*, 6179.

42. Kong, J., Sabha, D., Bigham, J. P., Pavel, A., & Guo, A. (2021). Tutoriallens: Authoring interactive augmented reality tutorials through narration and demonstration. In *Title suppressed due to excessive length 23 Symposium on Spatial User Interaction. SUI'21.* Association for Computing Machinery. https://doi.org/10.1145/3485279.3485289

43. Redmon, J., & Farhadi, A. (2018). *Yolov3: An incremental improvement.* ACM.

44. Blattgerste, J., Strenge, B., Renner, P., Pfeiffer, T., & Essig, K. (2017). Comparing conventional and augmented reality instructions for manual assembly tasks. In *Proceedings of the 10th International Conference on PErvasive Technologies Related to Assistive Environments,*

PETRA'17 (p. 7582). Association for Computing Machinery. https://doi.org/10.1145/3056540.3056547

45. Hart, S. G., & Staveland, L. E. (1988). Development of NASA-TLX (task load index): Results of empirical and theoretical research. In P. A. Hancock & N. Meshkati (Eds.), *Human mental workload, advances in psychology* (Vol. 52, pp. 139–183). North-Holland. https://doi.org/10.1016/S0166-4115(08)62386-9

Escaping the Holodeck: Designing Virtual Environments for Real Organizations

9

Alyssa Rumsey and Christopher A. Le Dantec

Abstract

The dominate narrative surrounding Industry 4.0 encourages manufacturing companies to adopt and adapt new forms of smart technologies which include purpose-built sensor platforms, advanced data capture and analytic capabilities, and Augmented Reality (AR) and Virtual Reality (VR). Recognizing that technology fundamentally changes the nature of work, we need to understand how these kinds of tools are affecting organizational structures and skill requirements in order to design technologies that work at work. We conducted a mixed methods VR study which facilitated access to participant observation and interviews (n = 21) on a commercially available VR toolset at a US-based global aviation manufacturer. The VR environment was deeply integrated into the manufacturer's existing enterprise stack which introduced a whole set of new utility and usability constraints beyond the environment itself that need to be considered when developing useful VR. Our findings provide insights into the impact of VR on human performance augmentation and skill acquisition revealing larger infrastructural challenges and design characteristics which need to be addressed during the development and implementation stages of new digital technologies in industrial workplace settings.

Keywords

Empirical study · Qualitative methods · Virtual reality · Future of work

A. Rumsey (✉) · C. A. Le Dantec
Georgia Institute of Technology, Atlanta, GA, USA
e-mail: arumsey3@gatech.edu; ledantec@gatech.edu

© The Author(s), under exclusive license to Springer Nature Switzerland AG 2023
C. Röcker, S. Büttner (eds.), *Human-Technology Interaction*,
https://doi.org/10.1007/978-3-030-99235-4_9

9.1   Introduction

The discipline of human computer interaction has a well-established history of studying technologies at work. Early contributions to the field illustrated the impact of groupware technologies like email on office practices which eliminated entire job fields and upset power dynamics [1]. In decades since that initial research, we have experienced the consequences of mobile devices, increasing communication expectations, and further erosion of the boundaries around traditional office place settings. Recognizing that applications like Lotus Notes resulted in organizational change, we need to understand how VR will alter organizational structures and workplace practices to create technology that supports and extends worker abilities [2–4]. Without an understanding of the organizational structure and divisions of labor, Industry 4.0 technologies stand to reinforce structures of control and increase the complexity of organizational processes, missing the opportunity to reimagine a future of work that is more worker centric. This is especially important with the proliferation of new, lower-cost consumer grade devices in industrial settings which have not been developed within the backdrop of office environments.

Consumer grade VR is no longer just an avenue of entertainment; it is being used to replicate the real world in professional fields changing the nature of work. Unlike prior forms of VR—CAVES or Powerwalls—which were permanent, expensive installations, consumer head-mounted VR—like products from Oculus and HTC—offer an easier-to-set-up, low-cost alternative. These devices are more accessible and create new inroads into industrial workplaces where bearing the cost and disruption of a CAVE or Powerwall was previously prohibitive. However, in these industrial settings, these kinds of VR devices are no longer relegated to single individual users or white-collar professionals. They affect the entire organization and supply chain, requiring changes in workforce talent and skills, and the establishment of new processes and procedures [5, 6].

Manufacturing is one industry experimenting with these new forms of consumer technologies in industrial settings offering an inroad to the complexities of VR design and use for precision-oriented work. While VR in manufacturing has long established roots, applied work in this area often occurs on purpose-built platforms meant to demonstrate a specific capability, or in short-term observations and pilot studies where the extent of use within regular work routines is implicit [7–9]. Contrary to the growing outlook that VR works at work, costly issues of integration with existing enterprise systems and a lack of translation between the virtual and established routines of work prevent the full adoption and use of VR in enterprise organizations. To realize the benefits of VR in these kinds of settings it is critical that we understand how to design immersive environments that enable real work to actually get done in a more immersive way.

Current VR interfaces and techniques allow for embodied interaction, but these techniques do not necessarily offer the same precision and detailed sophistication as their 2D counterparts. There is a gap between new forms of interaction with controllers and headsets and the long-established workarounds and routines based on stationary desktop and mouse practices that necessitate reevaluating how we design virtual environments to make VR not only usable but also useful in the real world. The kind of embodiment and immersion emphasized in the current VR user experiences are tailored for entertainment [10, 11], but overlook the complexities of integration and technology change required to achieve the adoption and use of these kinds of new technologies in organizations [12]. Yet these are key for an industrial use-case that involve precision-oriented work. To ease the adoption of new tech like VR in enterprise organizations, we need to design experiences that extend and support workers' skills, and that build on the specific and unique affordances of immersive environments.

Issues facing consumer grade VR are not limited to usability design characteristics within the virtual environment, but also how the virtual integrates into workplaces altering decision-making practices and workforce skills [9]. Taken together, these areas demonstrate the need to understand and address usability concerns to design technologies that support and extend workers abilities. Otherwise, we risk engendering the future workforce with the same notions of power and authority that result in one group taking on more work when it does not benefit them [13, 14].

To better comprehend the technical, usability, and organizational constraints of VR in the workplace, we conducted a qualitative VR study at a US-based aerospace manufacturer. The VR environment in our study was deeply integrated into the manufacturer's existing enterprise stack which introduced a whole set of new utility and usability constraints beyond the VR environment itself that need to be considered when developing useful VR for organizations. Our findings begin to unpack the kinds of support and expectations that employees have for VR at work, moving beyond just usability requirements. The virtual introduces a whole new kind of immersion to an already immersive workplace environment, resulting in an entirely new type of work product that needs to be integrated into routines and structures. We need to address how this category of smart technologies moves between development and consumer applications into real organizations that require transitioning into and out of the Holodeck.

The main contributions of our work draw attention to the incongruities between how the virtual is made, created, and used for precision-oriented work common to organizational settings. The lack of well known "good VR UI design" practices for industry contributed to VR development that was based on assumptions of intuitiveness in the virtual world. This led to the absence of adequate VR onboarding practices and discounted the complexity of transitioning work into the virtual. Further, participants had improper initial beliefs of what the VR environment was capable of simulating (e.g. haptics, virtual object collisions). Mismatches in expectations transcended to the inability to present and disseminate information from users in VR to other non-VR stakeholders. Adopting these kinds of platforms changes the way data is produced and disseminated across the organization effecting job roles and processes and procedures. To enable organizational problem solving these sorts of tools have to be flexible enough to adapt to the complexities of workplace practices and support and preserve tacit knowledge. We need to reconsider the affordances and interactions necessary for precision-oriented work to be accomplished in VR if these tools to be adopted in organizational settings.

9.2   Related Work

The history of VR in manufacturing provides a backdrop to understand the limitations and opportunities for VR development and use. These kinds of technologies have been built under the guise of streamlining work to deliver cost savings [15]. With the uptake of consumer grade VR more ad hoc use cases are being developed by third party software providers outside of the context of use which could have serious consequences for the future of work[ers]. To design technology for manufacturing entails a deep understanding of workplace practices that enables flexibility to respond to the subtle and ongoing change vital for long term integration [12]. Accounting for all the complexities of organizations requires taking a more holistic approach to the design of virtual systems for industrial use [16]. In a previous study we looked specifically at how VR implementation and use in manufacturing necessitate changes in routines of work, decision making, and job skills for adoption to occur within large organizations [6]. By recognizing the influence of context and organizational studies on the adoption and use of VR, we can begin to identify how these shape the design considerations of virtual environments for the workplace.

9.2.1   Immersive Environments in Manufacturing

Manufacturing industries have explored the use of VR for well over two decades [7, 9]. It has been viewed as the next logical step away from the complexities of desktop-based computer-aided design (CAD) tools which "impair collaboration and direct interaction" [17, 18]. In manufacturing, design tools are detail-oriented tools used to digitally represent a future product. This level of precision has made these kinds of tools hard to learn requiring intensive training with steep learning curves making it extremely difficult for non-expert users [17]. Virtual reality presents the opportunity to improve the human computer interface for complex product modeling and expand access to novice users, but it has to contend with being both detail oriented and interactive [17].

Cave Automatic Virtual Environments (CAVEs) were some of the first forms of VR in manufacturing organizations [19, 20]. These projector-based systems were geared towards increasing collaboration and embracing ideas of ubiquitous computing, moving away from work being tied to a desktop. CAVE environments provided a higher image resolution and a larger viewing area for detail-oriented design work. However, the inability to share or describe the experience to others presented difficulty for adoption in the workplace and was counter intuitive to increasing collaboration. These kinds of VR systems were limited by the same constraints as traditional desktops, relegated to a single user in control, a finite location, and a select group of expert users plus exorbitant setup costs and physical accessibility requirements.

The affordability of consumer-grade devices has led to the revival of VR in manufacturing along with the promise of delivering operational efficiencies. Studies have shown the potential of consumer head-mounted VR to deliver savings by helping users spot design errors sooner and manipulate objects easier than traditional desktop applications [21]. Being immersed is also a way to encourage collaboration across teams and reduce training time required to use 3D modeling systems [22–24]. However, these are the results from controlled lab environments and custom-built VR applications that are siloed self-enclosed systems meant to demonstrate a VR capability. It remains unclear if these outcomes can be realized long-term for consumer-grade hardware when implemented in real organizations with the complexities of enterprise system integration.

In the context of manufacturing, there are two main approaches to developing VR solutions. First, companies which were already deploying industrial-level modeling tools like Siemens and Dassault Systèmes have introduced VR-as-a-plugin for their existing applications [25, 26]. This type of VR arrangement relies on user familiarity with previous desktop versions of the software. In these situations, design guidelines are already predefined by existing software which can constrain the immersive experience. Legacy system adaptations already place the designer at a disadvantage because they are "under more constraints of compatibility in that changes have to be minimized to avoid disrupting users...making fresh substantially different approaches impossible" [27]. An area of research that needs more attention is the creation of VR tools for developers to overcome technical barriers for coding complex immersive environments.

The alternative approach put forth by new applications such as Gravity Sketch [28] and Mindesk [29] are standalone VR-only tools. These kinds of applications tend to focus more on simplifying the CAD toolset, making them more suited to create immersive experiences from the onset. When compared to the VR-as-a-plugin approach, these tools suffer from the lack of compatibility with long established legacy software and standards which are often integral to daily operations at large enterprises. These solutions are designed to replace existing systems putting a huge cost burden on the manufacturer let alone reskilling the workforce. Due to the changes required to integrate such standalone tools into organizations, these kinds of new technologies are often poorly adopted and can lead to increased negative perceptions of technology in the workplace [1, 2].

On the one hand there are tabla rasa systems that are focused on ease of use and on the other, bolt-on additions that are rooted in integration and compatibility. Standalone VR really takes advantage of the affordances of the virtual to deliver an immersive experience. All users are treated as novices and changes to processes and procedures for CAD modeling are built into how the user interacts with the environment. Whereas VR as a plugin assumes that transitioning existing 2D interface elements into the 3D with minimal modifications will still result in a 'better' experience. VR is treated as if it is endowed with natural affordances that do not have to be carefully articulated. Both overlook how existing practices are shaped and informed by routines and physical office spaces outside of the 2D interface. Creating useful VR requires identifying what needs to be brought over from the ease-of-use approach to orient and [re]skill professionals as they move from their familiar environments into VR. At work the product of VR is not just about the experience it needs to enable a whole set of interactions finetuned to the work being completed and communicated across the organization.

9.2.2 Designing for Context of Use

There are different goals when using VR to accomplish certain kinds of work like precision engineering that demand different kinds of interactions. Early work in VR recognized that synthetic virtual worlds needed to be comprised of key elements including immersion, interaction, and presence to facilitate an effective virtual experience [10, 11]. These parameters helped propel the development of VR and orient the field around the in-VR user experience. As VR continued to progress design parameters expanded to include: interaction usability, engagement usability, multimodal system output usability, and side effects usability [10]. These considerations emphasize requirements like wayfinding, navigation, and object manipulation, as well as visual, auditory and haptic outputs needed to increase presence and immersion [10]. All of these features are geared towards reducing system intricacies through natural interaction during the in-VR experience. This approach to VR differs significantly from the engineering product design process for manufacturing which entails detailed information be displayed about the product and components like material composition and load factors.

A key part of performing engineering analysis for manufacturing is the ability to replicate the real world to evaluate human posture and positioning for ergonomic assessments. In the desktop platforms this has been approached by creating models of the human body that can be manipulated. In VR, this has been attempted through whole body participation, wearing a body suit to become your own avatar—giving an entirely new meaning to suiting-up for work [30]. While this does demonstrate the capability of VR to offer a full body visceral experience, it produces a large amount of new data that now needs to be understood and manipulated. Despite the fact that the system maybe usable that does not mean it is useful in the context of work. Fundamentally, this kind of system requires a high level of trust in the virtual and results in new skill requirements and job roles affecting the organizational structure and changing the nature of work. Trying to directly replicate the real world and human body in VR demonstrates technical strides at the sacrifice of professional standards of practice and communication.

Looking at VR specifically geared towards workplace design reviews and product development shows how the affordances of VR are being adapted to deliver an experience. Interactivity work in VR for product design is exploring the benefits of combining 2D and 3D virtual sketching to support creativity during the design process [31]. Simplifying the virtual environment to produce seamless shape rendering does not address the kinds of detailed information and precision controls necessary for engineering analysis. Other areas of focus are attempting to address collaboration with projection-based CAVE systems by creating multi-viewpoint interaction to enable multiple users to control the virtual environment while coexisting in the same space for design reviews [32]. But as work is being pushed into more remote settings, relying on colocation seems implausible. It remains unclear if these kinds of advanced tools are flexible enough to adapt to the complexities of workplace practices and support and preserve tacit knowledge.

Even studies of the effect of VR on task performance remain focused on the ease of execution and system design neglecting the complexities of workplace integration that we know to be critical for technology adoption and use [10, 33]. One study has drawn attention to the mental effort of VR on task performance illustrating that walking in VR increased mental effort and decreased task performance [34]. While this was a simple lab experiment, we can imagine how mental effort may increase when the demands of meeting a production schedule are introduced to the context of using VR. Unlike 2D displays, VR allows for different ways of interacting that are context dependent and require reconsidering the kinds of affordances and development practices driving workplace VR technologies.

9.2.3 Designing Immersive Environments for Organizations

The implementation and use of VR in manufacturing has thus far revolved around technology offering solutions to problems of inefficiency in people and processes to make products faster and cheaper resulting in more profitable organizations. This framing emphasizes that the people performing the work are not fully capable and are a problem that needs to be fixed. Adopting this problem-solution model enables designers to develop tools in a vacuum because they have been given a specific problem definition. The paradox of this perspective is that the solution doesn't actually fix the problem but can engender new ones. With the pervasiveness of consumer-grade VR, users and workplaces now have to confront issues of privacy based on VR data that can reveal underlying cognitive or mobility issues [35]. It is no longer just about the objects being worked on but about the people doing the work. We also do not know the lasting safety and health impacts on long term VR users [36, 37]. Even Oculus does not recommend long duration use of more than 30-minute intervals without taking a break [38]. We can conceive the tradeoff for VR use resulting in more outsourced labor who cannot refuse the job or the creation of new physical and mental requirements for workers as stringent as becoming an astronaut. The flow of data through VR for work dispels the idea of an alternate reality and reminds us of the consequences these experiences can have in the real world.

To begin to grapple with the "dynamic relationships between people, products, social activities, and the context that surrounds a system" we need to take a systems-based approach to the design of technology in organizations [16]. If we apply the lens of serving, one of the critical constructs of product service ecologies, then VR should serve the people and the organization allowing us to contend with multiple user groups [16]. Central to this approach is the recognition that the designer should also be internal to the system being developed [16]. It is from this systems-based vantage point that we can draw on the field of organizational studies to see the influence that design and technology have on corporations. The relevance of this work is growing especially as we see an increase in remote work that is creating new grounds for technology development.

Approaching the design of VR from the organizational studies perspective recognizes the implications of technology on the distribution of work and the dynamics of power and control [1, 14]. We know from extensive field studies in this area that for technology adoption to occur change is required from either the people, the processes, or the technology [4]. To account for the complexities of technology in organizations design must occur within the context of use; the realities of change occur simultaneously and subtly over long periods of time [39]. This is especially true in large organizations where work occurs across distributed teams and multiple user groups rely on the same technology to accomplish different kinds of tasks [40]. To extend the applicability of VR in organizations, it needs to act as a common platform for communication across different user groups this relies on identifying and designing for the particularities of each user group [6].

Usability concerns for VR in industry go beyond the basic design parameters that are unique to virtual environments. One of the particularities of VR in manufacturing is the level of detail and precision required to perform work. Consumer VR comes with the expectation of entertainment, a fluid visual experience, but this conflicts with the functionality of performing engineering analysis tasks. The "expectation is that products will automatically have usability 'engineered in'" [41]. To make the virtual environment visually engaging requires specialized skills and significant time investment. In the workplace these kinds of development skills have not been part of traditional job requirements. Additionally, this type of behind-the-scenes work is not prioritized in manufacturing which is fundamentally geared towards delivering a tangible product or outcome [6]. By centering the design of VR within practice, we can begin to understand how to create both usable and useful virtual environments.

9.3   Context and Research Methods

Our study of the use of VR for engineering analysis was conducted at a global aviation manufacturing company headquartered in the United States. We worked directly with the human factors and maintainability team at the manufacturer to explore the capabilities of a new VR toolset. The VR capabilities were available on a standard industry-wide software platform used for viewing and analyzing 3D product models. The department's interest in using VR was to reduce the time spent conducting engineering assessments and to identify design errors earlier in the product development cycle as a path toward cost savings and process efficiency. Realizing these goals relied on understanding and transferring of a very complex set of tasks into the virtual environment. Traditionally, these tasks were executed on the desktop version of the software by human factors engineers who had established routines and deep expertise.

The role of the human factors and maintainability team was to represent the mechanic throughout the product design process to prevent human error during assembly and repair, and to protect employee health and safety. During new product development, the team was responsible for ensuring that mechanics would be able to physically access components with their hands and tools in the field or on the manufacturing shop floor. It was essential to make certain that the product can actually be assembled in the real world by real people.

The software platform used in our study was how these types of ergonomic and line-of-sight assessments were evaluated with 3D CAD models prior to the production of a physical product.

Working with the human factors team, we selected a single line removable unit as the main focus of the VR study. This particular line removable unit was a simple 3D model attached to a larger product assembly in the virtual environment. Only a tool sweep and path removal analysis were required to complete a full human factors and maintainability assessment on the part. A tool sweep analysis ensures there is enough room between parts for the movement of the hand tools that mechanics use. Similarly, a removal path analysis creates a designated "keep-out zone" so that each part had enough clearance to be removed and replaced. These analyses were critical to guaranteeing that products could actually be assembled and serviced by mechanics in the field. We used this guiding principle 'thinking like a mechanic' to inform our design decisions when customizing the VR toolset because that was the ultimate role the human factors and maintainability team played during the product design lifecycle.

9.3.1 Customizing the VR Environment

Translating existing work tasks into the virtual environment required significant customization of the VR toolset. Out-of-the-box, the VR environment was barren–imported part files were left floating in a gridded space. Navigation relied on only one controller and the VR navigation menu which was comprised of multiple 'pages' of icons displayed in a circle. The menu was hierarchical, i.e., icon selections would reveal sub-menus. The feature set was derived from earlier versions of the software that had been developed for CAVE environments and brought with them assumptions about how a person would interact with that environment (e.g. single mouse, menu driven). The software developers of the VR toolkit made explicit decisions to bring over these elements from the 2D environment partly because of limitations of the software but expressively to help participant transition to the 3D environment. It is possible that starting from scratch we would have designed something entirely different, but we wanted to work clearly within the system that was already being used by the human factors team and accessible across the organization and supply chain. The adaptation of VR to engineering analysis allowed us to better understand how to take existing processes and procedures embedded in a large organization and morph them into the virtual environment.

We tailored the layout of the environment by creating five workstations, akin to the manufacturing shop floor embodying our "think like a mechanic" design principle (see Fig. 9.1). We added a floor, enabled shadows and reflections, color coded part files and material types to create a more realistic setting for the analysis. We also imported realistic objects including a workbench, pegboard and tools such as hammers and wrenches. These additions were necessary to perform tool sweeps using the Attach/Release mode because users needed access to hand tools to attach them to the line removable unit in a way that a

Fig. 9.1 (Top) Test environment customized for our study; (Bottom) Detailed screenshot from usability testing

mechanic would actually use the tool in the field. For example, a user might Attach/Release a virtual wrench to a bolt on an engine model to see if enough space was provided for a mechanic to unscrew the bolt and maneuver tools freely. Finally, we included work instructions billboards at each station shown in Fig. 9.2 to help guide participants since there were no directions provided in the VR environment. The work instructions were stationary items in the environment requiring users to move between the workbench, the product model, and the work instructions similar to the shop floor. Each of the five workstations corresponded to one task that needed to be completed by users. Each task was paired with one icon on the modified VR navigation menu. The tasks included:

- TASK 1: Get familiar with the Immersive Menu
- TASK 2: Navigate the space and using Teleport Mode
- TASK 3: Perform a tool sweep with Attach/Release Mode
- TASK 4: Annotate the part with Markup Mode
- TASK 5: Perform part removal with Path Planning Model

We customized the navigation menu for the human factors team by removing icons that were unrelated to their work tasks informed by field observations and interviews. We mapped existing processes—tool sweep and removal path analyses—to the preset

"Path Planning" and "Attach/Release" modes available on the VR navigation menu. We also included "Teleport" and "Markup" modes because they were affordances specific to the VR environment. Teleportation was the main way to navigate the space. Markup allowed users to create 3D annotations on parts for design reviews. Additionally, to create a more visual trail that users could follow to navigate through the immersive menu, we color coded the trail of icons. For example, all the icons that needed to be selected to perform path planning had a yellow background and when they were selected the background would turn a darker shade of yellow.

To match the kinds of emersion needed for a compelling and functional VR experience, we customized the VR toolset based on existing workplace practices identified through our fieldwork. By contextualizing the setting for VR use, we were able to explore how VR would change the nature of work, as discussed in our previous publication [6], and inform design considerations that need to be addressed for the use of VR in large organizations.

9.3.2 Data Analysis and Procedures

To gain an understanding of the impact of VR on workplace practices, we conducted a mixed methods qualitative study relying on multiple modes of inquiry—surveys, interviews, field observations—over the course of 6 months. We used a grounded approach to analyze all interview transcripts (n = 21) and field notes. The transcripts were coded iteratively applying Charmaz's method of open coding which allowed us to build our themes over each stage of the project from kickoff to VR development and final usage [42].

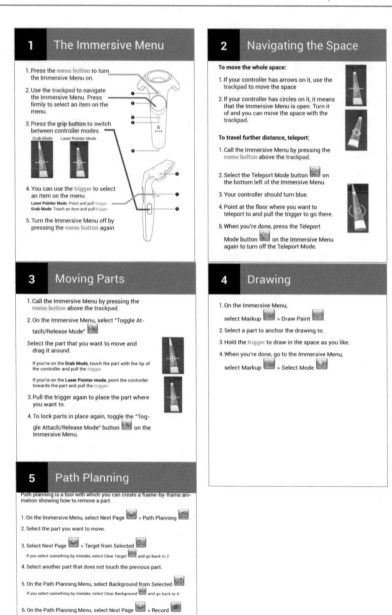

Fig. 9.2 Work instructions

An initial week-long technology assessment trip was crucial to documenting existing work practices. We interviewed the core team members (n = 5) during this time to detail expectations and gain an understanding of how the human factors team used the 3D modeling software to perform engineering analysis tasks. We also established relationships with the external VR developers and internal IT professionals during this trip to the manufacturer's headquarters. Key requirements like note taking, exporting/sharing files, snapshots, personalized toolbars were all features routinely used by the human factors and maintainability group to accomplish their work tasks. We also identified organizational themes related to flexibility, technical lag, communication, expertise, workforce skills. These initial interviews and field observations served as the basis for the customization of VR environment.

We customized the VR environment at our university research lab using an HTC Vive headset and dedicated VR desktop. To facilitate the customization process, we had screen share sessions and ongoing conversations with the third party VR developers and IT professionals at the manufacturer. As a research team, we met weekly to track and review our design customization progress leading up the to the deployment of the VR environment. Through this process we became expert users of the VR toolset in order to customize it for the human factors team. All field notes were compiled and coded iteratively during the development phase as a part of our fieldwork.

The final stage of the VR study included testing the customized VR toolset with the human factors and maintainability team at the manufacturer. All participants (n = 16) successfully completed the experience consisting of a demographic survey, a Steam VR tutorial, and the custom VR experience, followed by a 45–60 min semi-structured debrief interview. Each participant session lasted approximately 2 h per person and occurred in March 2019 at the manufacturer's headquarters. All VR sessions were screen recorded using OBS Studio. A research team member actively monitored each session using a mirrored screen view displayed on an external monitor. Participants—15 male, 1 female—were recruited for the VR study by the human factors team and were all employed by the aviation manufacturer. VR experience ranged from one person being very experienced to a majority of participants having no experience. All participants self-reported previous knowledge conducting engineering analysis tasks. The focus of the VR experience was to have participants reflect on how the VR application emulated existing work tasks.

9.4   Findings

VR has a long history of development and testing within manufacturing that has continued to build off of the same ideas of technology delivering time and cost savings often at the expense of the workforce [14]. To create usable and useful technologies in the context of work we need to design technologies that support workers while still meeting organizational goals. We begin to unpack these opportunities by drawing attention to how knowledge transfer occurs and the particularities of context of use along with the complexities of organizational and technical system integration that is critical for adoption to occur within large organizations [2, 43]. More broadly our work highlights the need for VR interface guidelines specifically for organizations where work is geared towards precision.

Our approach to understanding VR in the workplace was not to conduct a traditional usability study but examine how to transition existing practices into VR in ways that support embodied expertise of trained staff, complex workflows, and organizational structures—all required to successfully get work done. These kinds of infrastructures present really challenging sets of constraints for an already difficult task involving a complicated software environment. Our findings demonstrate what it means to design and deploy VR in this setting. Pointing out the affordances of a task-based tool and identifying pathways for how to transition VR into organizational settings. From the individual user experience to new environmental expectations, disruptions in embodiment point to the kinds of support needed apart from entertainment driven virtual experiences. These include making design trade-offs for VR in organizations, balancing user needs with organizational priorities and the technical capabilities of the hardware and software. Decisions that are all highly dependent on context and require evaluating existing IT systems and processes. Ultimately, VR necessitates confronting larger issues of communication and information dissemination across the organization. This suggests the need to create VR design practices for industry that address not only usability constraints but also support knowledge translation both into and back out of the Holodeck.

9.4.1 Disrupting Workplace Norms

Creating an embodied experience in VR is critical to leveraging the affordances of the virtual environment like interactivity but workplace VR necessitates reexamining how embodiment is upheld when transitioning from 2D to 3D environments. The use of VR at work revealed breakdowns in embodiment based on design assumptions about the user interface. The developers mapped the 2D software onto the 3D environment viewing this level of consistency as a way to encourage users to engage with VR. This approach to adapting VR for work, ignored the fact that work was already embodied especially considering a specialized software program had been learned and workarounds established to perform engineering analysis tasks in 2D. We built strategies into the VR environment to try and support the transition between 2D and 3D based on existing workplace practices. However, disruptions in embodiment still occurred because the underlying icons and menu structure did not provide enough cues to participants.

The VR developers assumed that familiarity with the desktop software would result in smooth transitions to the 3D environment requiring less user interface support. Yet, by taking the hierarchical menu structure and icons out of context, users didn't recognize them, putting the VR environment in direct conflict with established norms and expectations of use. When participants were asked if they recognized the navigation menu icons most stated that they did not. According P10, "I'm so used to seeing the buttons at this size, right? I guess I got use to seeing them and maybe I don't even know what they look like." As an expert user, P10 was so used to performing these kinds of engineering analysis tasks that he was doing it by rote memory. Even though we reduced the number of icons available on the navigation menu, it was still overly complicated for users and familiarity with the menu ribbons on the desktop software did not help users assimilate. As emphasized by P5, "You saw me fiddling around with those menus. They [engineers and mechanics] would do that for 5 s and be like, this is like a waste of time. I'm going to throw this headset in the garbage." P5 volunteered to participate in the study but could not imagine his peers voluntarily incorporating VR into their work routines. The pressure of getting work done and learning new tools was at odds with one another. Adopting VR for work requires careful articulation and tradeoffs between existing practices and features that is not as straight forward as mapping information architectures from 2D to 3D.

Difficulties with the navigation menu interface were also compounded with trouble locating the menu. When participants tried to make the navigation menu appear they pushed the menu button on the controller with their arms down by their sides, resulting in the menu appearing outside their field of view. The habit of holding the controller at the side of your body was compared by P6 to using a mouse at a computer station. It was extremely difficult to get users to understand that wherever the controller was pointed was where the menu would appear—depth matters in 3D. The action of holding the controller in front of the VR headset to see the navigation menu is a simplistic illustration of the kinds of assumptions about physicality that did not transition to the workplace. Spatial awareness is an affordance assumed to be one of the benefits motivating VR use especially at an individual consumption level but, at an organizational level, users are also faced with learning a new set of practices (that they may not have even opted into) requiring even more support in the virtual environment.

The work instructions and layout were meant to bridge the gap between virtual reality and the real-world engineering analysis tasks. We selected the five workstation layout based on discussions with the human factors team to replicate the manufacturing assembly environment where final product testing would occur. However, users were hesitant to walk between the stations, the workbench, and the product model. When trying to recall what action to take next, P11 sighed and said, "Remembering all the instructions on these sheets is a little bit difficult." Yet, without the work instructions and additional objects users would have been left in the dark without any cues as to what to do in the virtual environment. When standing in front of the 3D model, P11 continued by saying "But right now I'm probably relying a lot on what you're telling me to select." Verbal guidance increased significantly because the layout and work instructions relied on ease of move-ment in the virtual space that was not natural for users irrespective of their experience level in VR. Nevertheless, VR goes against office practices where work occurs sitting down and stationary. These are learned customs and VR introduces new ones that require both transitioning existing and developing new rituals. Longer term in-vivo studies of the use

Fig. 9.3 Screenshot from Steam VR Tutorial. Image: CC-BY Jeffrey Grubb

of VR are needed to further understand how learning curves effect adoption and contribute to changes in workplace practices.

9.4.2 Conflicting Realities

Maintaining embodiment in VR moves beyond just visual sign posting by adding another kind of immersion to an already immersive workplace environment. VR introduces a whole new layer of interaction that comes with expectations about physical space, mobility, and manipulation. Participants thought they were going to be able to step into a whole new world that was not only easier to navigate but also more compelling than the existing desktop software which would result in immediate time and cost savings as emphasized by management and reiterated by participants. However, ideas of entertainment and intuitiveness, commonly associated with VR, conflicted with the kinds of detail-oriented work that the human factors and maintainability team needed to perform.

The stark contrast between the Steam VR tutorial and the engineering analysis VR experience highlighted the different performance expectations for using VR at work. During the tutorial experience it was common that participants would interact directly with the robot instructor (see Fig. 9.3). Talking and commenting on the action requests out loud, participants made statements like "come on buddy" and "what's next I'm ready for ya." Users were rapidly mashing the buttons and by the end of the lesson and many made statements expressing their excitement and desire to stay in the tutorial longer. The entertainment-focused training wizard produced by Steam underscored the gap in production quality and interaction between the more bare-bones environment rolled out with the industry VR toolset.

The VR tasks selected for the study were chosen because they replicated the kinds of work the human factors team performed regularly on the desktop software. Transitioning this work into virtual reality engaged entirely different sensibilities than those needed to perform the same operations using the desktop software. Participants expected that the laws of physics would be upheld when performing precision-oriented tasks like part manipulation and path planning. As P14 expressed, "I'm picking up a part, but am I in the part? Am I holding onto the part? I'm putting on the wrench on. Oops, now I just busted right through the other side." P2 stressed that "When we're doing the grabbing the wrench or [moving] the parts, people aren't going to like that because everything we do is tactile. You've got to be able to touch it." Participants even discussed that not only should VR provide feedback for both grabbing parts and part collisions, but it should also be able to replicate the solid material properties of the components. While the desktop modeling software does not have tactile or haptic feedback, these were viewed as requirements for VR to be useful. Expectations of practicality and accuracy in engineering needed to be built into the environment.

Additionally, for analysis tasks that required moving parts around, participants found the lack of precision to be a cause for concern. P7 stated "When I was trying to put that one [part] together, it was perfect, and it wasn't getting close. What does it need?" Coordinate data that reinforced part placement in the desktop software was not available in the virtual world. P16 resolved "If you don't have a perfectly steady hand, you can't really just snap to the part." P16 felt embarrassed that he wasn't able to pick up these features of VR as quickly as he wanted. He blamed this on his age and inability to have "steady hands". In contrast, P14 highlighted the more natural method of direct 3D manipulation by stating, "'Oh, I don't like that part there, let's put it here.' It's a three second move. Not file, transition, select part, X axis, scroll, scroll, alt, reopen [application], load part. Right?" However, this ease of movement is at the sacrifice of precision and it remains unclear how these actions would be saved or accessed by engineers, once the VR headset is removed [44].

Part of the virtual experience is that you can push the boundaries of what is real but engineering demands exacting real-world scenarios. In many VR use cases, the fact that physics is allowed to work differently is viewed as a benefit because problems can be approached more creatively. In engineering that is not always the desired case. The more detail-oriented work tasks required a level of precision that demanded more information and control. The realities of completing engineering analysis tasks in the virtual exposed the difficulties of transitioning entertainment focused tools—consumer grade VR—into structured workplace organizations as a result of clashing design values.

9.4.3   Getting Lost in Translation

The adoption and use of VR tools in the workplace relies heavily on how well these kinds of experiences represent and convey the work that users are trying to accomplish. This includes supporting embodiment through the design of user interface elements based on workplace practices and addressing affordances of VR like physicality while also ensuring that the in- and out- put of VR can speak to the organization. How information and models were shared between users and across the larger organization was imperative to executing the work of human factors and maintainability engineers. There was a clear tension between the VR experience which was focused on the individual compared to the actual workplace practices which were collaborative. The software developers did not consider how knowledge work needed to escape the holodeck for consideration and decision making.

The software developers kept emphasizing that they were creating the VR tool for an "ideal customer" to use during design reviews. However, our observations revealed that the software platform at the aerospace manufacturer was primarily used by individual engineers and designers. Whereas design reviews happened at a managerial level. PowerPoint was used to facilitate decision making in these types of meetings typically held in conference room settings with multiple stakeholders. It was a rare occurrence to use the existing desktop software platform "live" because conference rooms were not equipped with the necessary hardware to run the program. Many participants also reported that the desktop version was unreliable often crashing and freezing during use. According to P5 introducing VR into these review meetings, "would be way harder to do than just put it in a PowerPoint that they can print out." This perspective was support by P5 who stated, "I don't know how to deliver that type of content to customers." The single-player nature of the VR headset made it hard for users to communicate their experience with people outside of the virtual environment which was a key feature of how the existing platform was being used to make design decisions. Using VR at work means that these kinds of systems need to be able to communicate outwards to coworkers and colleagues because decision making in organizations does not happen individually. VR must serve as a nexus for communication.

To bridge the gap between stakeholders and the work that would happen in the VR environment, the human factors and maintainability engineers saw themselves performing this knowledge translation role for others across the organization. They discussed this in terms of creating a centralized VR team to be able to own the process and disseminate best practices. P15 expressed, "I think a lot of that VR hardware and software would have to become centralized that may start pulling it back towards the maintainability organization." When envisioning how VR would be taught P6 shared that "As soon as you're asking for someone who isn't as familiar with it to set it up and spend their time to do that and maybe they're having an issue and trying to solve that, I think you run into even more of the pushback." However, a centralized VR team and dedicated space constrain the affordances of consumer-grade VR like its ability to be mobile. Also, recall that the stationary nature of CAVE environments was one of the factors that contributed to their decline. VR risks centralizing work that was previously easily distributed.

The manufacturer recognized that our study was the first step to integrating VR into the workplace. After our testing concluded, it took the human factors team another 6 months to acquire VR capable hardware and the appropriate software permissions to be able to load the immersive environment on company owned equipment. We coached them up to the point where they were able to successfully launch the VR plugin to an empty gridded environment. After accomplishing such large organizational hurdles, the user experience was not conducive to accomplishing real work. Not only did the usability of the application need to be improved, but the ability to communicate results and pathways for learning were missing component critical to ensuring the acceptance and use of VR within the organization.

9.5 Discussion

From the manufacturer's perspective the goal of our study was to understand the use cases of VR that could deliver organizational efficiencies. The first step towards this goal required understanding what it would take to transition current workplace practices into the virtual. Management initially viewed the VR plugin as a direct translation of the existing desktop software which was supported by the developers who described the toolkit as having the same features and icons as the 2D version. However, direct translation of desktop tooling and interactions actually disrupted the virtual experience illustrating was actually a series of different kinds of translations that needed to occur for VR to be effective at getting work done and circulating the product of work across the organization. A collection of translations are necessary not only for incorporating VR into existing workflows but also for transitioning that work back out from the 3D environment to the office environment. This relies on designing VR in the context of existing and legacy enterprise IT infrastructures that acknowledge the constraints and design tradeoffs necessary to support the transition of knowledge and professional [re]skilling developed in other tool sets or environments. The advantage of the VR-as-a-plugin is that new capabilities can be added atop legacy systems and offer an inroad for large-scale roll out of new capabilities rather than be contained as one-off point solutions. The challenge presented here is combining the benefits of VR with tacit knowledge of legacy software users to create immersive systems that work for workers and the organization.

9.5.1 Translating from 2D to 3D

There is an adage in user interface design that, when supporting users who have developed expertise in a given tool, "consistency is better than better." Which is to say maintaining expectations and functionality is crucial to enabling productivity in new versions of software. The consistency of the visual presentation of tools in our VR experience through familiar icons, menu structures and the shopfloor environment was meant to provide that bridge; however, the shift to the virtual environment was enough to disrupt the context altogether and so the larger set of cues that would help the participants orient and complete tasks were missing. The research team's verbal feedback provided during user testing became much more important for supporting both task-based instruction and creating any kind of immersive experience for participants.

Traditionally, immersion is about constructing a whole-body sensory experience that transports users to a new reality [45]. This often comes in the form of visual graphics that capture imagination and draw users into the experience even deeper [45]. Escapism is bound up with immersion and is part of the narrative experience but at work we need to be aware of the boundaries that have been carefully crafted between different disciplines and professional roles. Immersion for workplace VR is one that maintains transparency and creates consistency by mimicking reality like the laws of physics. While VR captured the imagination of participants like P the nature of the device isolated them from the rest of their job responsibilities. VR was less about the traditional immersive experience and more a way to provide visibility and detailed information. These discrepancies led to disruptions in embodiment demonstrating the mismatch between precision work and the way VR had been authored and created.

The assumption made by the VR developers was that immersive was interchangeable with intuitive. The developers assumed the transition to VR would be easy because the icons and menus remained the same as in the desktop version. They were just transitioning existing capabilities into the 3-D environment. These decisions ultimately impacted the user experience and limited the effect of the virtual toolset. As previously stated by P10"I'm so used to seeing the buttons at this size, right? I guess I got used to seeing them and maybe I don't even know what they look like." Removing the icons from their usual context disrupted the translation of information from the 2D interface to the 3D environment. Participants did not recognize the icons and the hierarchical menu structure convoluted the experience. Working within these parameters came with its own set of constraints when customizing the virtual environment but it reflects real world experience deploying and using VR-as-a-plugin solutions.

Further, participants were lost in the environment because they were used to orienting themselves based on a whole set of interrelated people, platforms, and objects. We typically use a variety of artefacts when doing work: we have a whole workspace with notes, perhaps multiple screens, post-its, manuals, a whole mix of digital and physical reference material that get used and referenced even when working on one task. VR puts a hard wall between that larger environment and the task at hand. The virtual instruction billboards were developed to help provide context for the work tasks but were confusing because we tried to recreate a standardized work environment when in reality everyone has individual ways of reminding themselves about what they do. This implies that all of these individual cues need to be re-created in the VR space for it to be a productive place to work—again, shifting away from entertainment experiences where removing reminders of the real world is what enables suspension of disbelief and emersion. Part of being productive comes from all the ways we offload cognition into artifacts and people in our environment giving us the ability to take action and make decisions [46, 47].

In the virtual action comes through our ability to explore and navigate the environment key to creating a sense of agency. In immersive experiences agency is about creating autonomy by giving participants the ability to choose whether or not they act [45]. The verbal guidance provided by the research team exemplified a lack of agency drawing attention to the need for additional forms of navigational support. This is especially important for VR use in organizations because unlike VR for recreation, users may not have the time or the desire to go and learn VR best practices. We know from prior studies that the often-invisible work of learning and making VR work is not viewed as a high priority by management [omitted for review]. As previously stated by P6 "you run into even more of the pushback" when asking people to figure out how to use new tools on company time. VR creates completely new ways of organizing and doing work—we are reminded of P14 who talked about how easy it was to manipulate objects. But, in order to capitalize on the full potential of VR, we need to support agency by incorporating more guidance and learning mechanisms into workplace VR platforms. What became clear through running our study is that people need to be taught how to take advantage of the unique capacities of 3D space in workplace settings.

There are multiple possible channels to use within a VR space that go beyond what we're accustomed to on the desktop. One of the benefits of VR is the different methods of instruction that can be provided in an immersive world [48–50]. Applying this to engineering analysis tasks, we can imagine that the viewpoint of the human factors engineers is different from the design engineers and the maintenance technicians who all use the same software to perform different tasks. Each person maybe looking at the same virtual product model, but the experience reveals different kinds of information based on the users professional role. Users could select between multiple forms of instruction (auditory, visual, haptic) and be able to learn and adapt to the environment on their own. However, developing transformative experiences requires understanding the particularities of work, the kinds of artifacts relied upon in the larger environment, and figuring out which of those need to be reproduced (and how to reproduce them).

It is not enough to have a 3D model and a VR headset, these kinds of experiences need a variety of support to both make a convincing environment (object collision, physics, navigation) and one that facilitates the detail work being done. As we experienced, people had higher expectations for VR to uphold the laws of physics and present more information about components that was not offered in the desktop computer platform. P14 expected that components would act like solid objects and restrict user movement so that you could not "bust through the other side." In practice making those objects "solid" requires a lot of behind-the-scenes development. Further, this kind of development is different from entertainment environments that are not focused on supporting precision interactions or whose precision is tuned to be entertaining and not high tolerance engineered parts. As industrial applications of VR expand it is our responsibility as designers and developers to create convincing environments that can be adapted by industry and support workers through the accurate translation of tasks into VR pointing to the need to reevaluating the kinds of authoring tools available for workplace VR.

9.5.2   Translating from 3D Back to 2D

Addressing usability and the in-VR experience to support the transition of work into the virtual is only one piece of integrating these kinds of tools into organizational settings. A key component missing from the VR toolset was the ability to share information with other people and other systems—even when built atop those other systems as the plug-in was in the case of our study. While learning how to navigate and manipulate the VR environment for participants was the first hurdle, translating that work back out of the environment proved to be even more critical for successfully using VR at work. It is just as important to understand how work needs to escape the immersive as it is to understand how to enter and orient to complete that same work within the virtual environment.

Currently, this problem space has primarily been discussed from the technical vantage point of data exchange between systems not considering the organizational procedures or processes that would need to change as a result [9]. Translating data encompasses more than just figuring out how tessellation information can be shared between CAD systems and the VR environment. We know the way data is produced shapes social practices and influences how data sharing happens in organizations [51]. This is directly tied to people's job roles which affect their relationships with data [51]. Vertesi and Dourish argue that this is a part of the larger data economy that needs to be investigated more if we are to design collaborative infrastructures that support people and established norms and values assigned to data [51]. Moving from manipulating data in the 2D desktop, with coordinate controls and finetuned keystrokes, into the VR space where you are manipulating objects results in a substantial difference in how data is both created and disseminated, and we see this chasm in our work.

As we worked with the software developers to build out support for the study, we realized there was a substantial gap between what they imagined as the core output of the tool and how the tool needed to perform within the context of the full organization. There was not a shared technological framing of VR that we know is key to the adoption and use of new technologies in organizations [2]. This was made clear in how results from the human-factors assessment would need to come out of the immersive environment and circulate within the engineering and design organizations. While problems might be identified by individuals in the VR environment, design reviews happened at a managerial level as problems were escalated. The mode of communication within these reviews was through PowerPoint slide decks that documented and facilitated decision making during the review. As previously stated by P5, PowerPoint was an easier alternative to VR in conference room settings with multiple stakeholders. We would point out this is more broadly the case when collaborative decision making is not always in-person or co-located. Immersive environments have to reconcile the fact that so much work happens collaboratively via platforms like PowerPoint.

The outcome of work in VR must be a product that can travel between disciplines and established organizational routines and structures. Additionally, the kind of immersion one might want for analyzing and working with 3D CAD models is different from the kind of immersion one might need to support collaboration. Here immersion is less about separation and more about permeability because information needs to flow into and out of the virtual environment. VR acts not only as a producer of data but as a kind of data filter determining what stays behind and what moves on to the larger enterprise system. Form factor matters both in-vr and out of VR. Moving representations across these different environments, whether conceived in VR or not, is an important problem for VR as a tool that enables individual and organizational problem solving.

The translation work that needs to occur between the organization and machine presents an opportunity to consider VR as a boundary object when designing immersive experiences for work. Boundary objects perform the translation between multiple social groups in manufacturing [40, 52], and framing the issue as a boundary object provides a way to link the individual VR experience to a group-level experience where knowledge and insight carry across distinct representations. By treating VR as a boundary object, we can incorporate tacit knowledge and work towards creating a communication language across all stakeholders. This would enable VR to become the true nexus of communication and an integral part of how work gets actually gets accomplished.

9.6   Conclusion

As more consumer-grade technologies like VR become pervasive in workplace settings, it is essential that we design experiences that both are productive and meaningful for workers. The use of VR for engineering analysis is only one example of the expanding impact of these types of new technologies on workplace practices. Taken outside of their intended use, these consumer devices deserve more attention when implemented in industrial settings. Our study of one such setting yielded information about designing for complex work tasks and the often-undervalued work of translation essential for VR to be adopted in organizations. One approach to designing these immersive environments is treating these tools as boundary objects to develop a communication language between the undoubtedly multiple end users across an organization. VR has the potential to open up new work domains and enable unskilled labor to fill existing jobs but to realize these capabilities designers must recognize the importance of understanding workplace practices.

In the field of HCI this is particularly relevant work as we see technology permeate into new workplaces and create entirely new kinds of jobs in a gig economy. A vehicle can now be your office and your boss can be a smart phone equipped with an algorithm [53]. A far cry from traditional workplaces yet these devices mediate the ways in which work is organized and accomplished similar to early organizational studies. Machine learning and artificial intelligence are already recognized as being sources of bias outcomes that affect hiring practices and influence decision making [54, 55]. We need to design technologies with a deep understanding of how work gets accomplished in practice rooted within the structure of the organization or risk reinforcing hierarchies of control or worse, creating new ones. This demands that we revitalize organizational studies as a critical component of design ecologies to deal with the mounting complexities of technology implementation and use.

References

1. Grudin, J. (1994). Groupware and social dynamics: Eight challenges for developers. *Communications of ACM, 37*, 92–105.
2. Orlikowski, W. J. (1993). Learning from notes: Organizational issues in groupware implementation. *Inf Soc, 9*, 237–250. https://doi.org/10.1080/01972243.1993.9960143
3. Barley, S. R. (1986). Technology as an occasion for structuring: Evidence from observations of CT scanners and the Social Order of Radiology Departments. *Administrative Science Quarterly, 31*, 78–108.
4. Janson M., Brown A., & Cecez-Kecmanovic D. (2006). *Interweaving groupware implementation and organization culture.*
5. Rumsey, A., & Le Dantec, C. A. (2019). Clearing the smoke: The changing identities and work in firefighting. In *Proceedings of the 2019 ACM Designing Interactive Systems Conference (DIS 2019)* (pp. 581–592). ACM.
6. Rumsey, A., & Le Dantec, C. A. (2020). Manufacturing change: The impact of virtual environments on real organizations. In *Proceedings of the 2020 ACM CHI Conference on Human Factors in Computing Systems (CHI'20)*. Springer.
7. Berg, L. P., & Vance, J. M. *Industry use of virtual reality in product design and manufacturing: A survey*. Springer. https://doi.org/10.1007/s10055-016-0293-9
8. Berg, L. P., & Vance, J. M. (2017). An industry case study: Investigating early design decision making in virtual reality. *Journal of Computing and Information Science in Engineering, 17*, 1–7. https://doi.org/10.1115/1.4034267
9. Seth, A., Vance, J. M., & Oliver, J. H. (2011). Virtual reality for assembly methods prototyping: A review. *Virtual Reality, 15*, 5–20. https://doi.org/10.1007/s10055-009-0153-y
10. Stanney, K. M., Mollaghasemi, M., Reeves, L., Breaux, R., & Graeber, D. A. (2003). *Usability engineering of virtual environments (VEs): Identifying multiple criteria that drive effective VE system design.*
11. Gupta, R., & Zeltzer, D. (1995). Prototyping and design for assembly analysis using multimodal virtual environments. *ASME Database Symposium, 4485*, 887–903.
12. Orlikowski, W. J. (1995). *Improvising organizational transformation over time: A situated change perspective*. Massachusetts Institute of Technology.
13. Kling, R. (1991). Cooperation, coordination and control in computer-supported work. *Communications of the ACM, 34*, 83–88. https://doi.org/10.1145/125319.125396
14. Grudin, J. (1988). Why CSCW applications fail: Problems in the design and evaluation of organizational interfaces. In *Proceedings of the 1988 ACM conference on Computer-supported cooperative work (CSCW'88)* (pp. 85–93). ACM.
15. Gill, S. A., & Ruddle, R. A. (1998). Using virtual humans to solve real ergonomic design problems. In *IEEE International Conference on Simulation'98* (pp. 223–229). Springer.
16. Forlizzi, J. (2013). *The product service ecology: Using a systems approach in design*. Springer.
17. Arrighi, P. A., & Mougenot, C. (2019). Towards user empowerment in product design: A mixed reality tool for interactive virtual prototyping. *Journal of Intelligent Manufacturing, 30*, 743–754. https://doi.org/10.1007/s10845-016-1276-0

18. Sidharta, R., Oliver, J., & Sannier, A. (2006). Augmented reality tangible interface for distributed design review. In *Proceedings - Computer Graphics, Imaging and Visualisation: Techniques and Applications, CGIV'06* (pp. 464–470). IEEE.
19. Lehner, V. D., & DeFanti, T. A. (1997). Distributed virtual reality: Supporting remote collaboration in vehicle design. *IEEE Computer Graphics and Applications, 17*, 13–17. https://doi.org/10.1109/38.574654
20. Purschke, F., Schulze, M., & Zimmermann, P. Virtual reality-new methods for improving and accelerating the development process in vehicle styling and design. In *Proceedings. Computer Graphics International (Cat. No.98EX149)* (pp. 789–797). IEEE Computer Society.
21. Satter, K., & Butler, A. (2015). Competitive usability analysis of immersive virtual environments in engineering design review. *Journal of Computing and Information Science in Engineering, 15*. https://doi.org/10.1115/1.4029750
22. Wolfartsberger, J. (2019). Analyzing the potential of virtual reality for engineering design review. *Automation in Construction, 104*, 27–37. https://doi.org/10.1016/j.autcon.2019.03.018
23. Bordegoni, M., & Caruso, G. (2012). Mixed reality distributed platform for collaborative design review of automotive interiors. *Virtual and Physical Prototyping, 7*, 243–259. https://doi.org/10.1080/17452759.2012.721605
24. Freeman, I. J., Salmon, J. L., & Coburn, J. Q. (2016). CAD integration in virtual reality design reviews for improved engineering model interaction. In *Volume 11: Systems, design, and complexity*. American Society of Mechanical Engineers.
25. Siemens Industry Software Inc. *Virtual reality with PLM*. Accessed January 28, 2020, from https://www.plm.automation.siemens.com/global/en/products/collaboration/virtual-reality.html
26. Hillner, B. (2019). *Solidworks. New VR mode in eDrawings desktop professional 2019*. Accessed January 28, 2020, from https://blogs.solidworks.com/solidworksblog/2019/07/new-vr-mode-in-edrawings-desktop-professional-2019.html
27. Gould, J. D., & Lewis, C. (1983). Designing for usability-key principles and what designers think. *Conference on Human Factors in Computing Systems - Proceedings, 28*, 50–53. https://doi.org/10.1145/800045.801579
28. Gravity Sketch Gravity Sketch. Accessed January 28, 2020, from https://www.gravitysketch.com/
29. Mindesk Inc. *Mindesk – Real time VR CAD*. Accessed January 28, 2020, from https://mindeskvr.com/
30. Baskaran, S., Akhavan Niaki, F., Tomaszewski, M., Gill, J. S., Chen, Y., Jia, Y., Mears, L., & Krovi, V. (2019). Digital human and robot simulation in automotive assembly using Siemens Tecnomatix process simulate: A feasibility study. In *Proceedings of SME North American Manufacturing Research Conference (NAMRC47)* (p. 155). Springer.
31. Arora, R., Kazi, R. H., Anderson, F., Grossman, T., Singh, K., & Fitzmaurice, G. (2017). Experimental evaluation of sketching on surfaces in VR. In *Conference on Human Factors in Computing Systems - Proceedings* (pp. 5643–5654). Association for Computing Machinery.
32. Simon, A. (2006). First-person experience and usability of co-located interaction in a projection-based virtual environment. *ACM Symposium on Virtual Reality Software and Technology VRST, 2006*, 23–30. https://doi.org/10.1145/1101616.1101622
33. Orlikowski, W. J., & Gash, D. C. (1994). Technological frames: Making sense of information technology in organizations. *ACM Transactions on Information and System Security, 12*, 174–207.
34. Luong, T., Martin, N., Argelaguet, F., & Lecuyer, A. (2019). Studying the mental effort in virtual versus real environments. In *26th IEEE Conference on Virtual Reality and 3D User Interfaces, VR 2019 – Proceedings*. IEEE.

35. Outlaw, J. & Persky, S. (2019). *The hidden risk of virtual reality - And what to do about it.* World Economic Forum. Accessed September 14, 2020, from https://www.weforum.org/agenda/2019/08/the-hidden-risk-of-virtual-reality-and-what-to-do-about-it/
36. Penumudi, S. A., Kuppam, V. A., Kim, J. H., & Hwang, J. (2020). The effects of target location on musculoskeletal load, task performance, and subjective discomfort during virtual reality interactions. *Applied Ergonomics, 84*, 103010. https://doi.org/10.1016/j.apergo.2019.103010
37. Lavoie, R., Main, K., King, C., & King, D. (2020). Virtual experience, real consequences: The potential negative emotional consequences of virtual reality gameplay. *Virtual Real.* https://doi.org/10.1007/s10055-020-00440-y
38. Oculus Health and Safety Warnings.
39. Orlikowski, W. J. (1992). The duality of technology: Rethinking the concept of technology in organizations. *Organization Science, 3*, 398–427.
40. Bechky, B. A. (2003). Sharing meaning across occupational communities: The transformation of understanding on a production floor. *Organization Science, 14*, 312–330. https://doi.org/10.2307/4135139
41. Thomas, P., & Macredie, R. D. (2002). *Introduction to the new usability* (Vol. 9, pp. 69–73). Academic Press.
42. Charmaz, K. (2014). *Constructing grounded theory* (2nd ed.). Sage.
43. Feldman, M. S., & Orlikowski, W. J. (2011). Theorizing practice and practicing theory. *Organization Science, 5*, 1240–1253. https://doi.org/10.1103/PhysRevD.82.045002
44. Wu, P., Qi, M., Gao, L., Zou, W., Miao, Q., & Liu, L. L. (2019). Research on the virtual reality synchronization of workshop digital twin. In *Proceedings of 2019 IEEE 8th Joint International Information Technology and Artificial Intelligence Conference, ITAIC 2019* (pp. 875–879). https://doi.org/10.1109/ITAIC.2019.8785552
45. Murray, J. (1998). *Hamlet on the Holodeck* (pp. 97–125). The MIT Press.
46. Hollan, J., Hutchins, E., & Kirsh, D. (2000). Distributed cognition toward a new foundation for human-computer interaction research. *ACM Transactions on Computer-Human Interaction, 2000*, 174–196.
47. Hutchins, E. (1995). *How a cockpit remembers its speeds.* Wiley.
48. Kumaravel, B. T., Nguyen, C., DiVerdi, S., & Hartmann, B. (2019). TutoriVR: A video-based tutorial system for design applications in virtual reality. In *Conference on Human Factors in Computing Systems - Proceedings.* Association for Computing Machinery.
49. Mohr, P., Mandl, D., Tatzgern, M., Veas, E., Schmalstieg, D., & Kalkofen, D. (2017). Retargeting video tutorials showing tools with surface contact to augmented reality. In *Conference on Human Factors in Computing Systems - Proceedings* (pp. 6547–6558). Association for Computing Machinery.
50. White, S., Lister, L., & Feiner, S. (2007). Visual hints for tangible gestures in augmented reality. In *2007 6th IEEE and ACM International Symposium on Mixed and Augmented Reality, ISMAR.* ACM.
51. Vertesi, J., & Dourish, P. (2011). The value of data: Considering the context of production in data economies. In *Proceedings of the ACM Conference on Computer-supported cooperative work CSCW* (pp. 533–542). ACM. https://doi.org/10.1145/1958824.1958906
52. Star, S. L., & Griesemer, J. R. (1989). Institutional ecology, 'translations' and boundary objects: Amateurs and professionals in Berkeley's Museum of Vertebrate Zoology, 1907-39. *Social Studies of Science, 19*(3), 387–420.
53. Roxenblat, A. (2018). *Uberland: How algorithms are rewriting the rules of work.* University of California Press.

54. Raghavan, M., Barocas, S., Kleinberg, J., & Levy, K. (2020). Mitigating bias in algorithmic hiring: Evaluating claims and practices. In *Proceedings of the 2020 conference on fairness, accountability, and transparency* (pp. 469–481). IEEE.
55. Sako, M. (2020). Artificial intelligence and the future of professional work. *Communications of the ACM, 63*, 25–27. https://doi.org/10.1145/3382743

Gamification in Industrial Production: An Overview, Best Practices, and Design Recommendations

10

Oliver Korn 📟

Abstract

This work describes gamification as a path to increase both productivity and motivation of persons working in industrial production. While gamification has been established in pedagogy or health more than two decades ago, its transgression to the industrial domain started around the year 2010. A discussion of production-specific requirements and the psychological background provide an overview on production-oriented gamified solutions in recent years. We look at how gamification designs evolved to minimize distraction while maximizing acceptance. Based on three best practices, we describe ways to neatly integrate gamification into workflows, use context-awareness to augment work and adapt the challenge-level to keep users in a state of flow. Furthermore, we investigate ways to further increase acceptance by creating user-specific "bottom-up" gamification designs, like custom agents and branded gamification. The overview concludes with design recommendations tailored for the production domain.

Keywords

Gamification · Production · Assistive technology

10.1   Introduction

Gamification is an "umbrella term for the use of video game elements in non-gaming systems to improve user experience (UX) and user engagement" [1]. Of course, gamification is not new—there are several predecessors in the education domain, like

O. Korn (📧)
Affective & Cognitive Institute, Offenburg University, Offenburg, Germany
e-mail: oliver.korn@acm.org

"edutainment" or "serious games". Also, gamification is increasingly used in the health domain [2]. This history is discussed in Sect. 10.2.

The question is, why gamification should be introduced in industrial production at all? With the domain's notorious obsession for quality, would gamification not be perceived as an inadequate distraction? Ten years ago, I would have agreed—gamification was considered inadequate for production environments by most stakeholders. However, with the success of mobile devices, intuitive and "playful" interaction became common in society and spread to other platforms and domains. Playful interaction and attractive design are no longer considered "nice to have"—instead, their absence is noted even in applications in the production domain. Especially members of "Generation Y" [3] and "Generation Z" [4] are skeptic and reluctant towards applications which do not reflect current design standards and aesthetics. This leads to a continuous spread of attractive and often also "gameful" design attributes into various work contexts. While at first primarily areas with social components like call centers [5] or human resource management [6] were covered, soon other domains were explored.

For a long time, industrial applications in production were resistant to such developments: classic systems for production assistance were purely functional (see Fig. 10.1). Human Machine Interaction (HMI) regularly does not live up to the solutions developed in the broader field of Human Computer Interaction (HCI). A Fraunhofer study on HMI from 2011 states that "from the variety of modern interaction techniques only touch screens found their way to machine interfaces in production environments" [7]—and even a recent article still states that "daily work routines in manufacturing companies are rarely considered [for gamification] resulting in underutilized potential in production environments" [8]. While there sure is potential, at least the scientific community's interest has improved considerably in both gamification of production and gamification of logistics, as a review paper shows [9].

To summarize, making industrial work more attractive or introducing fun has not been a development goal for decades and even today such elements are considered unusual—instead, quality, safety, and consistency are paradigm. The industrial domain's gradual consideration of soft factors like motivation and user experience started around the year 2010: in 2011 the technology research agency Gartner rather optimistically predicted that 70% of the global 2000 businesses will manage at least one "gamified" application or system already by 2014 [10]. Moreover, several studies (see Sect. 10.3) showed that gamification not only heightens motivation but also increases work speed—which makes it profitable for manual production.

The combination of changing societal expectations, an increased interest in the potential benefits of gamification, and the technological advances in motion recognition (see Sect. 10.2.1) resulted in a series of work on gamification of production environments which will be discussed in this article. Starting with the background, we describe the special requirements of the production domain (Sect. 10.2.1), advances in motion recognition, as well as developments in the concept of gamification and emotion recognition (Sect. 10.2.2). In Sect. 10.3, we describe the development of gamified assistive systems for industrial production starting with the first concepts 2012 until recent developments in 2020. We then describe three best practices (Sect. 10.4) and close with a section on design recommendations.

10.2   The Background

10.2.1 The Production Domain

As explained in the introduction, for decades industrial applications in production lacked behind the "cool stuff" developed in the broader field of human computer interaction (HCI). These discrepancies may partly be a result of the traditional distance between mechanical engineering and information technology, two disciplines which even today struggle to move closer together. Another hindrance is the paradigm of production companies to produce "zero errors" or "total quality". This goal is reflected in the concept of "Poka-Yoke", where systems are designed to be "fail-safe" or "mistake-proof". This is mostly realized by organizational methods or physical appliances, like a machine that can only hold a part if it is assembled correctly.

Indeed, while in HCI users typically consciously interact with an interface, in industrial production the main target of a user's interaction is not an interface but rather the current work component or the machine. While human machine interaction (HMI) plays an important role when steering advanced machines by computerized numerical control (CNC), processes requiring human interaction like assembly are less digitized. According to the authoritative guideline VDI 2860 Technology for Assembly and Handling [11] the central assembly activities are joining, handling, fitting, controlling and auxiliary functions like labeling. Interacting with assistive systems is not one of them, although it might be considered an "auxiliary function". Although elements like instructions or even technical details can be shown on screens close to the working area, the typical view is that these must be used with caution as they might distract the user's focus from the "actual" work.

Accordingly, computer-based assistive systems in production focused on instructional guidance and the control of work results rather than motivational support. Typical systems described or controlled the steps in the work process, often by using images or videos illustrating how to place or assemble a specific part (see Fig. 10.1).

Nevertheless, given the comparatively good measurability of work in production both on the process level and on the results level, it is surprising that gamified assistive systems have been dormant for so long. Indeed, recognizing states, for example in order to inform Enterprise Resource Planning Systems (ERP) or Manufacturing Execution Systems (MES), has long been a goal in the production domain [12].

Probably, this reluctance roots in the production domain's demand for "total quality": as long as there was no way to make sure that the process (for example assembly) was not compromised by adding gamification elements, or even by changing to new interaction

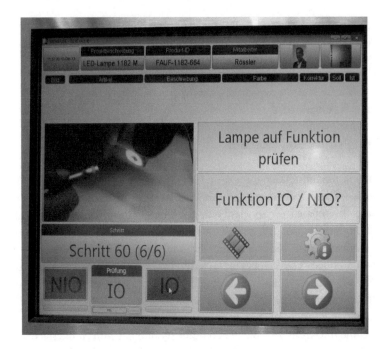

Fig. 10.1 Due to the total quality paradigm, assistive systems in production have been focusing on ensuring that work steps are correct rather than fun during the work process

methods, the industry preferred to stay on the safe side. Thus, for the production industry to accept innovations in interaction, there were several special requirements [13]:

1. implicit interaction, resulting in the sub-requirement motion recognition
2. (augmented reality, for example by) projection of information
3. error detection, resulting in the sub-requirement of recognizing fingers or objects

While the projection of information had already been established in the year 2000 [14] and adapted for production environments [15], real-time recognition of human motion was new for this domain. Accordingly, the industry started to become interested as soon as the first technologies were introduced which allowed a continuous analysis of work processes. One of these early systems called "QualityAssist" by the company Sarissa was based on ultrasonic waves emitted by "trackers" on gloves the assembly worker had to wear and received by sensors mounted above the assembly unit. It compared workers' motions with pre-stored motion sequences in real-time (Fig. 10.2).

However, this way of tracking was rather expensive and at the same time cumbersome as the workers had to wear gloves. While the form of interaction was new to production environments, the feedback was still presented traditionally by a small touch screen indicating the next steps and highlighting detected errors.

Fig. 10.2 The system "QualityAssist" by Sarissa was based on ultrasonic waves. It was one of the first systems allowing continuous process tracking

This changed in late 2010 with the launch of Microsoft Kinect: the first low-price solution allowing real-time interactions on consumer hardware while being able to handle a full range of human body shapes and sizes undergoing motions [16]. The sensor drew these capabilities from the 3D-depth-of-field-sensor and the IR-technology developed by the Israeli company PrimeSense, which was bought by Apple in 2013. Already in 2012, a system using the Kinect for natural interaction was introduced [17]. The same article describes both the use of projection to minimize distraction as well as gamification to increase motivation, especially during repetitive work tasks.

10.2.2 Gamification and Flow

The idea to use games to promote "serious" purposes came up in pedagogy where motivating pupils to learn has always been a key challenge. There are several attempts to integrate gamification into work processes, especially in the service sector. Reeves & Read describe "ingredients" to gamify work and meticulously map existing game elements like avatars, leaderboards, leveling and reputation to general business processes [18]. Another approach is the Playful Experiences framework (PLEX) which lists ways to improve the playful aspects of user experience [19] by looking at 22 experience types like "humor" or "subversion". Based on these types, certain "design patterns" are recommended.

Beyond frameworks, there is much work on how gamification is to be put into practice. Generic design elements to do that include "points, badges, leaderboards, progress bars, performance graphs, quests, meaningful stories, avatars and profile development" [20]. In an article on "Designing a System for Playful Coached Learning in the STEM Curriculum"

[21] gamification methods for technology-oriented activities are described in more detail. The authors explain ways to integrate some of the elements mentioned above as well as the concept of "extra lives" and the pyramid design developed specifically for production-oriented tasks (described in Sect. 10.3). Furthermore, they point out that it is important to re-establish a real-world connection to further motivation in the long-term, for example by rewarding the first place of the leader board or every person who achieved level ten with a gift, for example voucher for a book. The special importance of incentives and "tangible" rewards for gamification mechanisms is well described in a chapter by Richter et al. [22].

In the context of this work, gamification is primarily seen as a means to achieve "flow"—a mental state in which a person is fully immersed in an activity, experiencing energized focus and believes in the success. It is an area where high skill and adequate challenge converge, first proposed by Csíkszentmihályi in 1975. There are four conditions to achieve this state [23]:

- One must be involved in an activity with a clear set of goals. This adds direction and structure to the task.
- One must have a good balance between the perceived challenges of the task and the own perceived skills. One must have confidence to be capable to do the task.
- The task at hand must have clear and immediate feedback. This helps negotiate changing demands and allows adjusting performance to maintain the flow state.
- The activity is intrinsically rewarding, so there is a perceived effortlessness of action.

Figure 10.3 illustrates the spectrum of mental states and links them to ability and skill. Thus the concept of flow inherently integrates Bailey's human performance model [24]: high skill results in high performance. In addition to Bailey, Csíkszentmihályi highlighted the concept of challenge, the idea that flow results from high skill and high challenge level.

However, for achieving flow, there is no balanced position: the challenge has to be regularly adjusted to meet the user's current skill level. This adaptation requires a high

Fig. 10.3 Mental states resulting from the interaction between level of challenge and skill

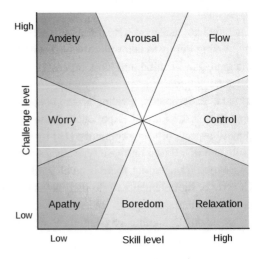

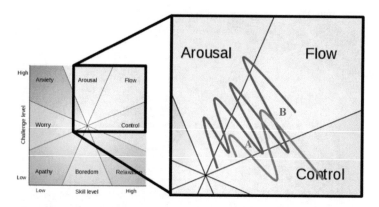

Fig. 10.4 The concept of flow curves expects users to hover between different positions within the areas "Control", "Flow", and "Arousal"

frequency of both input (regarding the user's state) and feedback to the user. To be permanently motivating, an activity must be designed in phases that partly arouse the user and partly give him or her the feeling of control, so that flow comes in waves [25]. The same applies to the gamification of production processes: to maintain flow, the challenge level must hover above and below the "perfect fit" which would be the user's current performance level.

The modeling of the user is simplified in production environments as there is primarily one activity and a worker's skill can be characterized to a high amount by the output. Accordingly, frequency and amplitude of the "flow curve" will be hovering between the two poles of arousal and control. Figure 10.4 illustrates this concept: The two exemplary curves represent two users with different characteristics: User A (blue) needs frequent longer phases of lower challenge level (control or even relaxation) whereas User B needs frequent arousal to maintain flow.

A major challenge of this approach is interpreting the user's actions correctly: a decrease in performance could be the result of either boredom or resignation due to overextension. Thus, an ideal implementation for gamification needs to recognize a user's stress level and emotional state.

10.2.3 Recognizing Emotions to Sustain Flow

We can recognize a users' stress level and emotional state by analyzing physiological features. Examples are facial expressions, the heart rate (which both can be extracted from a high-resolution video), the skin conductance (also called galvanic skin response, GSR, or electrodermal activity, EDA), as well as the eye movements.

A first approach to measure emotions in a production-oriented gamification scenario was described in a 2015 article [26]. However, the only modality used there was facial

Fig. 10.5 Setup for an emotion-aware assistive system in production, using pyramids in various colors with identical geometry to simulate redundant work tasks

recognition, based on SHORE (Sophisticated High-speed Object Recognition Engine) [27]. While facial recognition is helpful to determine the mood (for example happy or angry), the emotional intensity or arousal is best analyzed using heartrate or skin conductance which have not been used in this early and explorative work.

A much more intricate approach on integrating emotion recognition is described in the article "Affective Effects of Gamification: Using Biosignals to Measure the Effects on Working and Learning Users" [28]. The authors describe an experiment where redundant production work is simulated by constructing intricate pyramids out of Lego-bricks—with bricks in varying color patterns but with the same geometry (see Fig. 10.5).

While one group was performing the task with both instructions and gamified interventions, the control group just received regular instructions a traditional assistive system in the production would have provided (see Sect. 10.2.1).

The study confirmed previous findings that gamification highly significantly increases work speed but can also increase in the number of errors. Moreover, the biosignals created a clear picture of the emotional effects of gamification: joy was detected highly significantly more often in the gamification group than in the control group. This finding was reinforced by data on the skin conductivity: while the initial level of arousal was the same, users in the gamification group stayed in a state of consistent arousal, whereas the control group users drifted towards boredom. This difference again was highly significant. An analysis of the emotion anger created additional results: in the gamification group, anger was detected highly significantly more often than in the control group. This suggests that gamification increases not only positive emotions but raises emotionality altogether.

10.3   Designing Gamification in Production from 2012 to 2021

10.3.1 First Steps Towards Gamified Production

Already in early work in 2012, there were specific requirements mentioned for assistive systems in production [17]:

1. Scalability on the competence and process level
2. Small chunks of information at the right time
3. Increasing motivation and fun while working

These requirements were then extended [13], looking especially at how gamification could be applied in the production domain:

1. adaptation to the user's competence level
2. detection of excitement level and type
3. integration of motivating elements
4. protection of the users' personal data

From 2012 to today, these requirements were subsequently addressed, although there are some inherent challenges, like measuring emotions ("excitement") while protecting the users' personal data.

The first design applying gamification to production environments (see Fig. 10.6) was presented in the article "Industrial Playgrounds: How Gamification Helps to Enrich Work for Elderly or Impaired Persons in Production" [29]. Interestingly, this work focuses on persons with impairments who differ from "regular" workers regarding age or health. At that time, gamification was primarily seen as a way to motivate such special groups. This changed in the last years, as games and gamification became a more integral part of society and the lifestyle of both younger and older generations.

Already this first design relies on motion recognition to make work processes transparent, analyzing and visualizing them in real-time (see Sect. 10.2.1). Thus, it is no surprise the concept was presented in 2012, as in 2011 the Microsoft Kinect was launched—the first solution allowing movement analysis with a single affordable device (rather than multi-camera tracking). This shift from a result-oriented perspective (is the assembled part error-free?) to a process-oriented perspective was essential for employing gamification. In the design, each work process is represented by a brick in a puzzle game resembling Alexey Pajitnov's 1984 classic Tetris. After a sequence is completed, the build-up brick rows disintegrate.

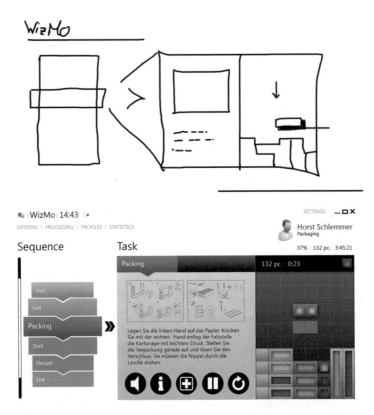

Fig. 10.6 First scribble and a design prototype of a gamification for production environments from [29], inspired by the classic game Tetris

10.3.2 Evaluating Design Variations and Branding

Next to the Tetris design, from 2012 to 2015 two more designs for the gamification of production (and other technology-oriented areas) were proposed (see Fig. 10.7): the circle design [30], and the pyramid design [31]. Regarding the effects of gamification, all studies confirmed the established effects: increased work speed and motivation, while quality decreased if no counter measures were taken: a speed-accuracy tradeoff.

In 2015, a comparative study on designing the gamification of production identified the version with the highest acceptance rate [30]: an improved form of the pyramid design (see Fig. 10.8). In this design, each pyramid step represents a work step, with steps colored according to the required time and error rate. The steps are climbed by a user avatar which can be individualized with a user photo. At the top of the pyramid there is a trophy which dissolves as soon as an error is detected. Thus, the importance of producing high quality with (almost) no errors is highlighted by the design. The users in the comparative study preferred this pyramid design to the alternatives with high significance.

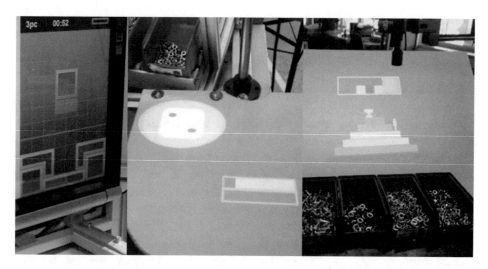

Fig. 10.7 Three gamification designs for production (from left to right): Tetris design (2012), Circle design (2014), and first version of the Pyramid design (2014)

Fig. 10.8 A worker assembling metals shears. The pyramid gamification is projected directly in front of the workspace

These results on gamification design were complemented by a study with instructors of a large automotive manufacturer [32] who also preferred the pyramid design with high significance and recommended it for inclusion in corporate education. It was during the discussions with these instructors that the idea of a gamification design adapted to the automotive brand and thus the concept of "branded gamification" came up.

10.3.3 Exploring Feedback Modalities

The repeated finding that gamification increases both positive and negative emotions required a more thorough investigation. This was pursued in the article with the telling title "Wow, You Are Terrible at This! An Intercultural Study on Virtual Agents Giving Mixed Feedback" [33], where the University of Florida and the German Offenburg University looked at the emotional effects of positive and negative feedback during work tasks. Furthermore, the study introduced a new element: the feedback was not provided as text but contextualized and "embodied" by virtual agents. There already is considerable research regarding agents' likeability, for example regarding their application in augmented reality settings [34]. The process of finding the right agent for a specific work environment is described as a best practice (see Sect. 10.4.2).

Similar to the repetitive construction tasks described in Sect. 10.2.3, the task in this study was deliberately designed to be frustrating. 120 participants from Germany and from the United States had to search for a specific item, which was hidden beneath others. However, some levels hid the item so well that it was almost impossible to find. Thus, negative feedback was inevitable, no matter the user's competence. In the study, the participants were randomly assigned to one of the three agents: a human, a mouse, or a robot. To assess the affective reactions, the study used the International Positive and Negative Affect Schedule Short-Form (I-PANAS-SF).

An analysis of variance revealed that the agents' appearance did neither have a significant effect on the users' perception of the feedback nor on their affective reactions. Also, for the encouraging feedback no significant effect of the agent's appearance on the affective reactions was found. However, criticism from a human agent significantly increased negative affective reactions. This confirms the finding that increased engagement—whether by gamification, branded gamification or human-like agents—increases emotionality altogether. It also indicates that human-like agents can be a wrong design choice (see Sect. 10.4.2).

When investigating cultural differences in agent perception and the affective responses, participants from the United States had higher affective responses for the attentive and determined dimensions and lower responses for the inspired and nervous dimensions when receiving criticism. These participants stated a desire to prove the agent wrong, possibly explaining a heightened sense of attentiveness and determination. These results indicate that we cannot assume an agent's feedback to be appropriate for all cultures. This finding might well extend to over aspect of gamification—like the importance of a real-world connection. However, this is to be determined by future research.

10.4   Best Practices for Designing Gamification in Production

Integrating gamification into a specific work setting works best if it is "tailored" to the specific needs of the workplace and users. This process is well described by Lessel et al. who argue against applying standard gamification in a "top-down" fashion, and towards a "bottom-up" approach [35]. This section provides three examples of how gamification can be tailored to fit the work context.

10.4.1 Designing a Neat Integration into the Workplace

For gamification to be successfully integrated into production work, it needs to be close to the place of action or "in situ". In a factory setting, a remote monitor with a leaderboard is not a good solution, as it does not amplify and enrich the work process itself but rather displays results in a more playful way. Projection or displays integrated directly into the workspace allow the neat process integration required. The projects *motionEAP* and *KoBeLU* both aimed for this close integration (Fig. 10.9).

This neat integration is a requirement for gamification to be an integral and accepted part of the perceived work experience. As pointed out in Sects. 10.2.1 and 10.3.1, production work requires a very high level of focus. Information provided outside of the workspace is either ignored or it forces users to frequently switch the focus, loosing concentration and increasing task completion times and the likeliness of errors.

In production environments, projection has considerable advantages over monitors: oil and dirt which accompany several production activities can be cleaned much easier from a regular worktable than from a monitor. This allows implementing projected feedback even in environments with very high demands for sterility like the cleaning of surgical instruments [36].

A further advantage is the possibility to project information and feedback (Fig. 10.10) directly on work parts and even on the user's hand [37]. Thus, grabbing the wrong of two similar-looking parts in a production sequence can immediately result in a red projection on the user's hand, minimizing distraction from the task at hand while conveying the required information. This way, gamification elements like points can also be integrated without creating distraction.

10.4.2 Designing Branded Gamification for Specific Companies

Existing research has established that brand identification is intertwined with commitment to an enterprise [38]. Accordingly, the identification with a specific brand also has the potential to shape individuals' attitudes, performance and commitment within work contexts and thus increase the benefits from gamification.

Fig. 10.9 Integrating gamified feedback neatly into the workspace is important for acceptance

Fig. 10.10 Projecting information and feedback right into the workspace and even directly on a user's hand reduces the distraction from the work task

These effects were explored in a study with 44 workers performing a soldering task [39]. Elements of branded identification were incorporated in a gamified assistive system. The study investigated task performance and emotions. Instead of the established pyramid-design, the branded version visualized progress through the image of a compressor representative for the company Mahle. This compressor gradually builds while the participants work on the soldering task. With each completed step, a new part of the compressor appears (Fig. 10.11).

The results show that brand identification impacts the users' attitude towards the task at hand: while activating positive emotions, aversion and reactance did also rise. On one hand, this points towards a dual nature of branded gamification: while instigating positive

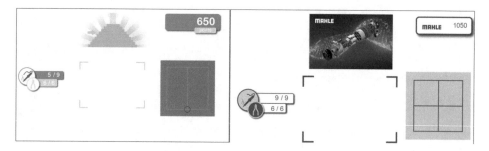

Fig. 10.11 Regular gamification layout with pyramid (left) and branded version with a compressor (right) as projected into the workspace

emotions, aversion and reactance also arise. Such increased emotion al involvement at work is not a bad thing: previous research has highlighted the benefits [40], arguing that emotionality and rationality are an integral part of work life—and organizational effectiveness can be improved by promoting, rather than attempting to suppress emotional variance.

10.4.3 Designing Gamified Agents for Specific User Groups

While the projects *motionEAP* and *KoBeLU* mentioned above allowed optimizing both the design and the integration of gamification in production workflows, the information was communicated in an abstract way, for example by the pyramid. It has already been shown that in production environments, the identification of workers can be raised by choosing a well-known product as a visual representation (see "branded gamification" as described in Sect. 10.3). However, a product is still an abstract entity—adding an agent can further improve the gamified work scenario.

Especially in the education domain, there is considerable evidence that communicating by agents or "companions" yields better results [41–43]. Indeed, the topic of agents or virtual assistants—combined with gamification—has gained importance in the recent years. Young researchers like Stefan Werrlich from the BMW group investigated new playful agents—in this case the little robot "Embly", using an augmented reality scenario with a head-mounted display [44]. However, Werrlich does not elaborate on why "Embly" was designed as a robot rather than a human or an animal. Similarly, in the study on virtual agents discussed in Sect. 10.3, there are three different versions of agents, ranging from a robot to a mouse. However, how was the "right agent" selected or designed? Indeed, the creation of such "individualized gamification" for specific user groups or "bottom-up gamification" has recently gained more attention [35]. In the following, we describe a best practice for this approach.

In the project *incluMOVE*, the suitable agent was realized in a comprehensive participatory design process—in this case with users with impairments are were being prepared for re-entry into the first job market [45]. In focus group discussions these users and their

Fig. 10.12 Design variation of a virtual agent created in the project incluMOVE

caregivers were asked how they would imagine a virtual character, supporting their work tasks. Interestingly, caregivers seemed to have different opinions on adequate character representations than the users—which shows the importance of user-centered design processes.

After defining the underlying traits of the characters, an iterative design process started, where a designer created numerous drafts shown in Fig. 10.12. The resulting character designs were then presented to the users and the most agreeable design was selected.

Clearly, creating many design variations is a laborious and time-consuming process which will not always meet the time and budget restrictions of a project. However, the process of discussing alternatives and selecting the most agreeable one with the users in a participatory process will greatly increase the acceptance of the final gamified solution. Thus, a bottom-up gamification design is preferrable and ideally integrated in the planning phase of a gamification project.

10.5 Design Recommendations

Gamification is only one aspect of assistance in production. An overview paper by Funkt et al. [46] provides eight general recommendations when providing context-aware feedback in production environments:

1. Keep the shown feedback simple
2. Display direct feedback
3. Design for hands-free usage
4. Equip the environment rather than the user
5. Strive for intuitive natural interaction
6. Design for personalized feedback
7. Enable users to control the speed
8. Add motivating quantified-self information

While only the eighth recommendation explicitly addresses gamification, most of the others can be applied to gamified production scenarios. Given this domain's necessary obsession with quality (see Sect. 10.2.1), recommendations 1 to 5 mostly address the need to not distract users by additional information—no matter if it is a visualization like the pyramid (see Sect. 10.3), or information on the work procedure itself. How such a "neat" integration of information can be achieved has been described as a best practice (see Sect. 10.4.1).

A key element surely is the personalization or individualization reflected in the recommendations 6 and 7: as described in Sect. 10.2.2, successful gamification needs to put the user's abilities and mood into the equation to create the right interventions. Otherwise, an ambiguous behavior (for example slower than usual assembly) could trigger the wrong reaction (for example a faster sequence based on the assumption the user was bored while in fact he or she was stressed). Obviously, the tracking of emotions or mood is difficult to achieve in company settings with high demands for personal data protection. On the other hand, a system designed primarily to make work more engaging and avoid boredom, frustration and corresponding illnesses like burnout should be looked at more favorably in the coming years.

Based on the work presented in this chapter and the existing general recommendations described above, I want to conclude with a set of more specific recommendations for the gamification in production environments. These should be designed according to the following guidelines:

1. Keep it simple: Use simple static designs to avoid distraction from the production work task.
2. Keep it close: Use monitors or even better projection integrated in the workspace to avoid distraction and provide information directly on top of the work product.
3. User control: Always allow users to switch off gamification elements to avoid distraction in challenging situations like small lot sizes.
4. Mind the quality paradigm: Ensure that the gamification design does not only incorporate feedback on speed but also on quality (like the dissolving trophy in case of an error).
5. Challenge matching: Integrate basic emotion tracking to avoid wrong adaptations of the challenge level.
6. Challenge variation: Within the user-specifically adapted corridor of performance, provide fluctuations and variety to avoid boredom.
7. Anonymity: Personal data like the history of user emotions and performance must only be used to create suitable gamified interventions and not be stored or communicated to other systems.
8. Adequate gamification design: In a "bottom-up" design process, evaluate with users what kind of gamification elements (specific agents, achievements etc.) work best in a specific production setting. While abstract solutions are feasible, many users prefer gamification designs with a connection to the brand or the product.

References

1. Deterding, S., Sicart, M., Nacke, L.; O'Hara, K., & Dixon, D. (2011). Gamification. Using game-design elements in non-gaming contexts. In *Proceedings of the 2011 annual conference extended abstracts on Human factors in computing systems, CHI EA'11* (Vol. 2, pp. 2425–2428). ACM.
2. Korn, O., & Tietz, S. (2017). Strategies for playful design when gamifying rehabilitation: A study on user experience. In *Proceedings of the 10th International Conference on PErvasive Technologies Related to Assistive Environments, PETRA'17* (pp. 209–214). ACM.
3. Eisner, S. P. (2005). Managing Generation Y. *SAM Advanced Management Journal.*
4. Tulgan, B. (2013). *Meet Generation Z: The second generation within the giant "millennial" cohort.* Rainmaker Thinking.
5. Castellani, S., Hanrahan, B., & Colombino, T. (2013). Game mechanics in support of production environments. In *CHI'13 Proceedings of the ACM SIGCHI Conference on Human Factors in Computing Systems, Workshop on Gamification.* ACM.
6. Heimburger, L., Buchweitz, L., Gouveia, R., & Korn, O. (2020). Gamifying onboarding: How to increase both engagement and integration of new employees. In R. H. M. Goossens & A. Murata (Eds.), *Advances in social and occupational ergonomics* (Vol. 970, pp. 3–14). Springer.
7. Bierkandt, J., Preissner, M., Hermann, F., & Hipp, C. (2011). Usability und human-machine interfaces in der Produktion. In D. Spath & A. Weisbecker (Eds.), *Studie Qualitätsmerkmale für Entwicklungswerkzeuge.* Fraunhofer-Verlag.
8. Ulmer, J., Braun, S., Cheng, C.-T., Dowey, S., & Wollert, J. (2020). Human-centered gamification framework for manufacturing systems. *Procedia CIRP, 93,* 670–675. https://doi.org/10.1016/j.procir.2020.04.076
9. Warmelink, H., Koivisto, J., Mayer, I., Vesa, M., & Hamari, J. (2018). Gamification of the work floor: A literature review of gamifying production and logistics operations. In *Proceedings of the 51st Hawaii International Conference on System Sciences* (pp. 1108–1117). University of Hawaii.
10. Cowie, P. (2013). The phenomena of gamification – The next big thing for employers? Retrieved November 5, 2013, from http://www.enterprise-gamification.com/index.php?option=com_content&view=article&id=167:the-phenomena-of-gamification-the-next-big-thing-for-employers
11. VDI Verein Deutscher Ingenieure. (1990). *VDI 2860 Montage-Und Handhabungstechnik [Technology for Assembly and Handling].*
12. Alm, R., Aehnelt, M., & Urban, B. (2015). Plant@Hand: From activity recognition to situation-based annotation management at mobile assembly workplaces. In *Proceedings of the 2nd International Workshop on Sensor-Based Activity Recognition and Interaction - WOAR'15* (pp. 1–7). ACM Press.
13. Korn, O. (2014). *Context-aware assistive systems for augmented work: A framework using gamification and projection.* University of Stuttgart.
14. Pinhanez, C. S. (2001). The everywhere displays projector: A device to create ubiquitous graphical interfaces. In *Proceedings of the 3rd international conference on Ubiquitous Computing, UbiComp'01* (pp. 315–331). Springer.
15. Korn, O., Schmidt, A., & Hörz, T. (2013). Augmented manufacturing: A study with impaired persons on assistive systems using in-situ projection. In *PETRA'13 Proceedings of the 6th international conference on PErvasive technologies related to assistive environments* (Vol. 21, pp. 1–21). ACM.
16. Shotton, J., Fitzgibbon, A., Cook, M., Sharp, T., Finocchio, M., Moore, R., Kipman, A., & Blake, A. (2011). Real-time human pose recognition in parts from single depth images. In *Proceedings of the 24th IEEE conference on computer vision and pattern recognition* (Vol. 2). Springer.

17. Korn, O., Schmidt, A., & Hörz, T. (2012). Assistive systems in production environments: Exploring motion recognition and gamification. In *PETRA'12 Proceedings of the 5th international conference on PErvasive Technologies Related to Assistive Environments, PETRA'12* (Vol. 9, pp. 1–9). ACM.

18. Reeves, B., & Leighton Read, J. (2009). *Total engagement: Using games and virtual worlds to change the way people work and businesses compete.* Harvard Business Press.

19. Arrasvuori, J., Boberg, M., Holopainen, J., Korhonen, H., Lucero, A., & Montola, M. (2011). Applying the PLEX framework in designing for playfulness. In *Proceedings of the 2011 Conference on Designing Pleasurable Products and Interfaces, DPPI'11* (pp. 1–8). Association for Computing Machinery.

20. Sailer, M., Hense, J., Mandl, H., & Klevers, M. (2013). *Psychological perspectives on motivation through gamification* (Vol. 10). Springer.

21. Korn, O., Rees, A., & Dix, A. (2017). Designing a system for playful coached learning in the STEM curriculum. In *Proceedings of the 2017 ACM Workshop on Intelligent Interfaces for Ubiquitous and Smart Learning, SmartLearn'17* (pp. 31–37). ACM.

22. Richter, G., Raban, D. R., & Rafaeli, S. (2015). Studying gamification: The effect of rewards and incentives on motivation. In T. Reiners & L. C. Wood (Eds.), *Gamification in education and business* (pp. 21–46). Springer.

23. Csíkszentmihályi, M., & Nakamura, J. (2002). The concept of flow. In *The handbook of positive psychology* (pp. 89–92). Oxford University Press.

24. Bailey, R. W. (1989). *Human performance engineering: Using human factors/ergonomics to achieve computer system usability.* Prentice Hall.

25. Korn, O., Funk, M., & Schmidt, A. (2015a). Assistive systems for the workplace: Towards context-aware assistance. In L. B. Theng (Ed.), *Assistive technologies for physical and cognitive disabilities: Advances in medical technologies and clinical practice* (pp. 120–134). IGI Global.

26. Korn, O., Boffo, S., & Schmidt, A. (2015). The effect of gamification on emotions - The potential of facial recognition in work environments. In M. Kurosu (Ed.), *Human-computer interaction: Design and evaluation* (Vol. 9169, pp. 489–499). Springer.

27. Küblbeck, C., & Ernst, A. (2006). Face detection and tracking in video sequences using the modifiedcensus transformation. *Image and Vision Computing, 24*(6), 564–572. https://doi.org/10.1016/j.imavis.2005.08.005

28. Korn, O., & Rees, A. (2019). Affective effects of gamification: Using biosignals to measure the effects on working and learning users. In *Proceedings of the 12th ACM International Conference on PErvasive Technologies Related to Assistive Environments - PETRA'19* (pp. 1–10). ACM Press.

29. Korn, O. (2012). Industrial playgrounds: How gamification helps to enrich work for elderly or impaired persons in production. In *Proceedings of the 4th ACM SIGCHI Symposium on Engineering Interactive Computing Systems, EICS'12* (pp. 313–316). ACM.

30. Korn, O., Funk, M., & Schmidt, A. (2015c). Towards a gamification of industrial production: A comparative study in sheltered work environments. In *Proceedings of the 7th ACM SIGCHI Symposium on Engineering Interactive Computing Systems, EICS'15* (pp. 84–93). ACM.

31. Korn, O., Funk, M., & Schmidt, A. (2015b). Design approaches for the gamification of production environments. A study focusing on acceptance. In *PETRA'15 Proceedings of the 8th International Conference on PErvasive Technologies Related to Assistive Environments.* ACM.

32. Korn, O., Muschick, P., & Schmidt, A. (2016). Gamification of production? A study on the acceptance of gamified work processes in the automotive industry. In W. Chung & C. S. Shin (Eds.), *Advances in affective and pleasurable design. Proceedings of the AHFE 2016 International Conference* (pp. 433–445). Springer.

33. Wang, I., Buchweitz, L., Smith, J., Bornholdt, L.-S., Grund, J., Ruiz, J., & Korn, O. (2020). Wow, you are terrible at this! An intercultural study on virtual agents giving mixed feedback. In *Proceedings of the 20th ACM International Conference on Intelligent Virtual Agents, IVA'20* (pp. 1–8). Association for Computing Machinery.

34. Wang, I., Smith, J., & Ruiz, J. (2019). Exploring virtual agents for augmented reality. In *Proceedings of the 2019 CHI Conference on Human Factors in Computing Systems, CHI'19* (pp. 1–12). Association for Computing Machinery.

35. Lessel, P., Altmeyer, M., Müller, M., Wolff, C., & Krüger, A. (2016). 'Don't whip me with your games': Investigating 'bottom-up' gamification. In *Proceedings of the 2016 CHI Conference on Human Factors in Computing Systems, CHI'16* (pp. 2026–2037). ACM.

36. Rüther, S., Hermann, T., Mracek, M., Kopp, S., & Steil, J. (2013). An assistance system for guiding workers in central sterilization supply departments. In *Proceedings of the 6th International Conference on PErvasive Technologies Related to Assistive Environments, PETRA'13* (Vol. 3, pp. 1–3). ACM.

37. Funk, M., Bächler, A., Bächler, L., Kosch, T., Heidenreich, T., & Schmidt, A. (2017a). Working with augmented reality? A long-term analysis of in-situ instructions at the assembly workplace. In *Proceedings of the 10th International Conference on PErvasive Technologies Related to Assistive Environments, PETRA'17* (pp. 222–229). ACM.

38. Kimpakorn, N., & Tocquer, G. (2010). Service brand equity and employee brand commitment. *Journal of Services Marketing, 24*(5), 378–388. https://doi.org/10.1108/08876041011060486

39. Schulz, A. S., Schulz, F., Gouveia, R., & Korn, O. (2018). Branded gamification in technical education. In *2018 10th international conference on virtual worlds and games for serious applications (VS-Games)* (pp. 1–8). IEEE.

40. Ashforth, B. E., & Humphrey, R. H. (1995). Emotion in the workplace: A reappraisal. *Human Relations, 48*(2), 97–125. https://doi.org/10.1177/001872679504800201

41. Baylor, A. L., & Kim, Y. (2005). Simulating instructional roles through pedagogical agents. *International Journal of Artificial Intelligence in Education, 15*(2), 95–115.

42. Kim, Y. (2007). Desirable characteristics of learning companions. *International Journal of Artificial Intelligence in Education, 17*(4), 371–388.

43. Veletsianos, G. (2010). Contextually relevant pedagogical agents: Visual appearance, stereotypes, and first impressions and their impact on learning. *Computers & Education, 55*(2), 576–585. https://doi.org/10.1016/j.compedu.2010.02.019

44. Werrlich, S., Nguyen, P.-A., Yanez, C. E. F., Lorber, C., Daniel, A.-D., & Notni, G. (2018). *Design recommendations for HMD-based assembly training tasks* (Vol. 12). Springer.

45. Grund, J., Umfahrer, M., Buchweitz, L., Gay, J., Theil, A., & Korn, O. (2020). A gamified and adaptive learning system for neurodivergent workers in electronic assembling tasks. In *Proceedings of the Conference on Mensch und Computer (MuC'20)* (pp. 491–494). ACM Press.

46. Funk, M., Kosch, T., Kettner, R., Korn, O., & Schmidt, A. (2017b). *MotionEAP: An overview of 4 years of combining industrial assembly with augmented reality for Industry 4.0.* Springer.

New Industrial Work: Personalised Job Roles, Smooth Human-Machine Teamwork and Support for Well-Being at Work

Eija Kaasinen, Anu-Hanna Anttila, and Päivi Heikkilä

Abstract

Industry 4.0 is changing factory floor work towards knowledge work and novel digital tools provide new opportunities for factory operators, as illustrated in Operator 4.0 visions. We have studied worker expectations towards the change in large-scale surveys with factory floor workers in Finland. Based on the results of the surveys, Operator 4.0 literature and our earlier studies, we analyse how factory operators' work is changing and how well-being at work can be supported in the Industry 4.0 transformation. We present a vision of future industrial work with personalised job roles, smooth teamwork in human-machine teams, and support for well-being at work. We conclude the results by giving recommendations on how the new industrial work should be considered when designing and introducing Industry 4.0 solutions for the factory floor. The focus of the design should be human-machine teams and collaboration in them. The work community should be engaged in designing the new solutions, as well as related new job roles, training and on-the-job support. Impacts on well-being at work should be considered in the design and monitored during the transformation. Future factories are dynamic and resilient, and that is why new industrial work should be designed for continuous change.

Keywords

Operator 4.0 · Human-machine collaboration · Well-being

E. Kaasinen (✉) · P. Heikkilä
VTT Technical Research Centre of Finland Ltd, Tampere, Finland
e-mail: eija.kaasinen@vtt.fi; paivi.heikkila@vtt.fi

A.-H. Anttila
Finnish Industrial Union, Helsinki, Finland
e-mail: anu-hanna.anttila@teollisuusliitto.fi

11.1   Introduction

Industry 4.0 is changing industrial work, the jobs available and the required skills. People are dealing less and less with actual physical artefacts but increasingly with their digital counterparts in the virtual world. In the Industry 4.0 vision this change has been named 'cyber-physical systems', highlighting the trend of seamless integration of physical and virtual in work environments and in work tools. The impacts of Industry 4.0 on industrial jobs have been described in Operator 4.0 visions [e.g. 1, 2]. Despite the Operator 4.0 research, Neumann et al. [3] claim that human factors are still underrepresented in the Industry 4.0 research stream, resulting in a substantial research and application gap. In this chapter, we will present survey results (2019, 2020) of Finnish blue-collar worker experiences and expectations towards new industrial work. Based on these results, Operator 4.0 visions and our earlier studies, we analyse how factory operators' work is changing and how well-being at work can be supported in the Industry 4.0 transformation. The focus is on factory floor workers in the manufacturing industry.

We present a vision of future industrial work where each worker can utilise and develop their personal skills and where well-being at work is at the core. We also highlight teamwork, where the specific skills of each worker are recognised, and machine intelligence complements practical human knowledge and intuition. The vision includes personalised job roles, smooth teamwork in human-machine teams and support for well-being at work. We start in Sect. 11.2 with an overview of related research regarding Industry 4.0 from the worker's point of view, well-being at work, Operator 4.0 visions and human-centred design of industrial systems. In Sect. 11.3 we present the results of three member surveys conducted by the Finnish Industrial Union. Section 11.4 presents our vision of new industrial work with personalised job roles, human-machine teamwork as well as well-being considerations. We conclude the chapter in Sect. 11.5 by providing recommendations for how the new industrial work should be considered when designing and introducing Industry 4.0 solutions to the factory floor.

11.2   Related Work

In this section we give an overview of related work regarding Industry 4.0 transformation. We start by describing studies on Industry 4.0 from factory workers' point of view. We also discuss well-being at work in industry, and the factors that influence it. The change in factory floor work has been termed 'Operator 4.0' and we describe what kinds of Operator 4.0 visions have been presented in the literature. We also describe approaches to the human-centred design of industrial systems. We conclude this section by summarising the research gap that we have identified and that we have been responding to in our study.

11.2.1 Industry 4.0 from Workers' Point of View

By utilising advanced digitalisation, the industrial internet and smart technologies, Industry 4.0 is expected to result in shorter development periods, individualisation in demand for customers, flexibility, decentralisation and resource efficiency [4, 5]. These changes will also radically influence industrial work. Digitalisation is changing factory floor operators' work increasingly towards knowledge work, blurring the traditional division between blue- and white-collar jobs. There will be significantly greater demands on the workforce in terms of managing complexity, abstraction and problem-solving [6]. Subsequently, factory workers are likely to act much more on their own initiative, to possess excellent communication skills and to organise their personal workflow; i.e. in future industrial environments, they are expected to act as strategic decision-makers and flexible problem-solvers [7, 8]. Taylor et al. [9] emphasise the creative role of future operators. With their knowledge of customer needs, the production process, the machinery and the tooling, they can create product innovations, especially at small-scale manufacturers. Industry 4.0 is expected to provide opportunities for factory workers through the qualitative enrichment of their work: a more interesting working environment, greater autonomy and opportunities for self-development [5].

Asik-Dizdar and Esen [10] describe how sensemaking both at the individual worker and organisational level can create meaningful work experience in a dynamic work environment. On an individual level, sensemaking is supported by job crafting, i.e. empowering workers to influence the contents of their job. On an organisational level, sensemaking is supported by organisational learning that further allows all members of the work community to participate in crafting the dynamic strategy. Involving workers in designing both the digital solutions and the related new work system supports ensuring work well-being and performance during the transformation [11, 12]. Kaasinen et al. [11] propose digital tools, with which workers can participate in designing their work and training and can share their knowledge with each other. These tools support adopting Industry 4.0 solutions smoothly. Kaasinen et al. [11] also suggest that the future factory environment should adapt to the skills, capabilities and needs of the worker and should support the worker to understand and to develop his/her personal skills. Workers can have individually defined work roles, which empowers them and increases well-being at work. Holm [13] carried out a survey with over 200 high school students focused on technical programmes, who are thus seen as future shop floor operators. The respondents' views on future shop floor work were predominantly positive, and they highlighted development and teamwork as the two most popular characteristics of future work. This provides evidence that future factory workers are expecting to take a participatory role in future factories.

Recently, in their policy brief [14], the European Commission has presented a vision for the future of European industry, coined 'Industry 5.0'. Industry 5.0 complements the techno-economic vision of the Industry 4.0 paradigm by emphasising the societal role of industry. The policy brief calls for a transition in European industry to become sustainable, human-centric and resilient. The focus in industry should move from sole shareholder value to stakeholder value for all concerned. Industry 5.0 calls for human-centric solutions that adapt the production process to the needs of the workers and guide and train them. New technologies should not impinge on workers' fundamental rights such as privacy, autonomy and human dignity. Romero and Stahre [15] describe how resilient Industry 5.0 systems will require smart and resilient capabilities both in next-generation manufacturing systems and with human operators.

11.2.2 Well-Being at Work

The disruption of industrial work alters the workers' tools, working environment, work roles and familiar ways of working. As the changes require learning new skills and tolerating insecurity, they have a considerable effect on workers' well-being. Future factory work is expected to become more autonomous and require problem-solving skills and the managing of complexity [8], which may induce stress but also provide opportunities for self-development.

Well-being at work can be defined broadly to concern all aspects of working life, from the quality and safety of the working environment to workers' feelings about their work [16]. Schaufeli and Bakker [17] define the concept more precisely as a psychological state consisting of job satisfaction and work engagement. Job satisfaction refers to a positive emotional state resulting from the appraisal of one's job or job experiences [18]. It incorporates various work-related aspects, such as the nature of work, salary, colleagues and working conditions [19]. Work engagement refers to a more intrinsic construct, a fulfilling work-related state of mind, which makes workers energetic, willing to invest efforts into their work and be persistent when facing challenges [20]. In addition, it has been argued that job motivation, especially high intrinsic motivation, can positively affect well-being at work [21].

As the transition to Industry 4.0 is ongoing, most of the related research has been theoretically driven [e.g. 22, 23] and empirical studies of actual impacts on well-being are still scarce. Kadir and Broberg [12] studied the effects of introducing new Industry 4.0 enabling technologies on workers' perceived well-being through ten industrial case studies. In their study, when the workers were informed about the upcoming implementation of new technologies, they expressed both excitement about learning new skills and competencies as well as concerns about working with the new solutions, learning new skills, demands to work faster, safety of the new tools, and even losing their job. Involving the workers in the design and implementation of the new solutions increased their motivation, while in some cases, lack of participation and information and introducing partially developed solutions resulted in frustration and a reluctance to work with new tools, as well as cautiousness and nervousness towards them. While in general the workers' perceived well-being improved after the solutions had been fully implemented, limited training and lack of standard operation procedures still caused errors and frustration.

The results of Kadir and Broberg [12] emphasise the benefits of informing and involving workers early in the design and implementation process of new technologies. Kaasinen et al. [11] came to the same finding in a study where 44 factory workers were interviewed to explore their expectations and concerns related to proposed Operator 4.0 concepts. The interviewed workers were willing to be involved in the design of manufacturing processes and their workplace, and they thought that involvement would decrease the number of work problems that they currently have. Early involvement and a participatory design approach were also found to be beneficial in a study where an idea about a worker feedback tool that tracks workers' well-being and work performance metrics was introduced to factory workers [24]. The idea initially aroused rather negative feedback, but when the workers were able to see an early prototype and give their ideas for improving it, the attitudes became more positive. Later, when ten factory workers used the application for 2 or 3 months, they perceived several benefits of its use related to their well-being at work, such as recognising their own accomplishments at work, which provided motivation to make efforts to improve one's performance [25].

Worker involvement and more generally a worker-centric design process of new digital technologies can be guided by utilising relevant design frameworks. Kaasinen et al. [26] created a framework for the Operator 4.0 context to guide the design and evaluation of new solutions so that they would have a positive impact on well-being at work. The framework is modified from a more general framework by Danna and Griffin [27] to include five design and evaluation perspectives: usability, user experience, user acceptance, safety and ethics. Focusing on these aspects facilitates creating safe and ethically sound solutions that support smooth work practices, evoke positive experiences, and are accepted by workers. Eventually, the targeted impacts include job satisfaction, work engagement and job motivation as well as company benefits, such as productivity, quality and optimised processes.

Besides changing the nature of industrial work and thus affecting workers' well-being, advances in digital technologies have enabled new ways to measure physiological indicators related to well-being at work. In future factory work, wearable trackers are expected to become more widely used in supporting workers' occupational health, safety and productivity [28]. Despite the potential benefits, tracking well-being also involves ethical issues, such as concerns related to privacy, data security and voluntariness of tracking [29–32], and thus ethics need to be considered both when designing the new solutions as well as when implementing them at workplaces.

11.2.3 Operator 4.0 Visions

Visions of future factory work have been created to analyse and illustrate the Industry 4.0 transition from a factory floor worker's point of view. Romero et al. [2] first coined the term 'Operator 4.0' to describe the industrial transformation of factory floor work. Operator 4.0 is a smart and skilled operator who cooperates with robots and whose work is aided by smart machines and digital tools. The Operator 4.0 concept refers to human-cyber-physical production systems that improve the abilities of the operators [33]. Romero et al. [2] have

Table 11.1 Operator types by Romero et al. [2] and ACE Factories Cluster [1]

Romero et al. [2]	ACE Factories Cluster [1]
Augmented operator	Augmented and virtual operator
Virtual operator	Social and collaborative operator
Social operator	Super-strong operator
Collaborative operator	Healthy and happy operator
Super-strength operator	One-of-a-kind operator
Healthy operator	
Smarter operator	
Analytical operator	

proposed seven operator types, and the ACE Factories Project Cluster [1] has further modified them to five operator types based on the results of the cluster projects (Table 11.1). These operator types illustrate how Industry 4.0 technologies [34] can show in operators' work and support them:

1. The **augmented operator** can utilise augmented reality solutions to get contextual guidance and information, e.g., by overlaying real-time information on their field of view [1, 2].
2. The **virtual operator** can use virtual reality solutions for example to train safety-critical work tasks beforehand or remotely operate the system [1, 2].
3. The **social operator** utilises mobile and social network services to interact and share (tacit) knowledge between other operators [1, 2]. Romero et al. [35] propose a social factory architecture, which is based on an adaptive, collaborative and intelligent multi-agent system for enabling cooperation between social operators, social machines and social software systems.
4. The **collaborative operator** smoothly collaborates with robots and smart machines in co-working spaces without the need for safety barriers [1, 2].
5. The **super-strength/strong** operator can utilise robots and exoskeletons to ease physically demanding tasks [1, 2].
6. The **smart operator** utilises artificial intelligence (AI) to support work tasks and interaction with smart factory machines [2].
7. The **analytical operator** uses large sets of data to find useful information, to make better decisions and to take better actions [2].
8. The **healthy operator** concept addresses concerns regarding increasing workforce stress levels, the state of psycho-social health and the new potential physical risks in emerging cyber-physical production environments being 'disrupted' by the introduction of new Industry 4.0 technologies [1, 28, 36].
9. The **one-of-a-kind operator** reminds us that each operator's individual skills, capabilities and preferences should be considered [1].

Beyond these technology mediated operator types that illustrate how new technologies could support operators in their work, there are also other aspects of how to approach the Operator 4.0 concept via operators' skills, capabilities and preferences [11] or human values [37]. Similar to the one-of-a-kind operator type by the ACE Factories Cluster [1], Kaasinen et al. [11] emphasise that each operator has individual capabilities, skills and preferences to which future smart factories should adapt. Furthermore, they suggest that future operators should be empowered to understand their own skills and to take responsibility for how they want to develop their skills. Gazzaneo et al. [37] propose that human values should be embedded in the Operator 4.0 concept. Their Operator 4.0 Compass describes how the Industry 4.0 key enabling technologies match and extend human cognitive, physical, sensorial and interaction capabilities.

11.2.4 Human-Centred Design of Industrial Systems

Human-centred design is quite an established practice when designing work tools. However, the focus has been mainly on usability in a setting of one user and one tool. Some researchers have highlighted the need to extend the focus of the design to the whole sociotechnical system [3, 20, 22, 38]. Cagliano et al. [22] suggest studying the link between technology and work organisation both at the micro level (work design) and the macro level (organisational structure) for a successful implementation of smart manufacturing systems. Neumann et al. [3] propose a framework to systematically consider human factors in Industry 4.0 designs and implementations, integrating technical and social foci in the multidisciplinary design process. Stern and Becker [38] propose that the design should focus on the cyber-physical production systems. The design of the sociotechnical system should focus at the micro level on sensory, cognitive and physical demands for the workers [3, 22], human needs and abilities [38], change of work [3, 22, 38], number of tasks [22], job autonomy [22], supervising [3], co-worker support [3] and interfaces to semi-automated systems [38]. The design of the sociotechnical system should focus at the macro level on work organisation [22], decision-making [22], social environment at the workplace [3] and division of labour [38].

Pacaux-Lemoine et al. [39] claim that in the techno-centred design of intelligent manufacturing systems, the human operator is often considered a 'magic human', who can handle any unexpected situations efficiently. Sgarbossa et al. [40] describe how design teams determine the perceptual, cognitive, emotional and motoric demands on the user. If these demands exceed an individual's capacity, negative consequences both on system performance and worker well-being can be expected. For better design outcomes, designers should understand employee diversity and human factors demands of new technologies [40]. Pacaux-Lemoine et al. [39] suggest that human-machine collaboration should be designed so that situational view can be shared at a plan level, a plan application level (triggering tasks) and at the level of directly controlling the process.

Kadir and Broberg [12] recommend involving workers early in designing the new solutions as well as (re)designing the work systems. Virtual reality-based environments provide good tools for participatory design [41–43]. Each team member can see the production process entity and their foreseen tasks illustrated in a realistic environment and can even try the proposed new tasks. In the virtual environment it is easy to test alternative options to organise the work.

The necessary competencies and training needs should also be evaluated before introducing the change [12]. When starting to use the new solutions, the workers should be well trained but there should also be room for exploration and adaptation [12].

11.2.5 Research Gap

Despite the Operator 4.0 research, the research into sociotechnical systems and human factors in industry is still underrepresented in Industry 4.0 research, as pointed out by Neumann et al. [3]. Human factors are crucial to industrial transformation, as pointed out in the vision for the future of European industry presented by the European Commission, referred to as 'Industry 5.0' [14].

Operator 4.0 visions illustrate well how Industry 4.0 technologies influence factory floor work. Even if the visions introduce the healthy operator as one dimension of the change, empirical studies of actual impacts of the industrial transformation on well-being are still scarce. More attention should be paid to well-being at work and operators' increasingly active roles in future factories [9–13]. Future operators should be provided with opportunities to participate in designing their own work [10, 11, 13] and use their creative potential [9, 12], and their tacit knowledge should be better utilised [10, 11]. Operators' personal skills, capabilities and preferences should be better considered [10, 11].

In the design of Industry 4.0 solutions, the focus should be extended to the sociotechnical system at the factory, including both the micro level (individual worker) and the macro level (work community) [3, 22, 38]. Designers should better understand and support the diversity of the factory workers as users of technical solutions [39, 40]. Participatory design is a promising approach to involving employees and extend the design focus to new work processes and work organisation. Well before introducing new solutions at the workplace, the necessary competencies and training needs should be evaluated [12]. When introducing new solutions to the factory floor, there should be room for exploration, adaptation and creativeness [9, 12].

Operator 4.0 visions constitute a good starting point for understanding future industrial work. In this study we wanted to focus on worker experiences and expectations towards the Industry 4.0 transformation. We also wanted to understand the new skills required, and how the change should influence the design of Industry 4.0 solutions.

11.3   Surveys of Finnish Industry Workers

11.3.1   Industrial Work in EU and in Finland

In Europe, the polarisation between south and north becomes visible when analysing digitalisation and automation. In southern European countries, production is merely labour intensive: machine investments are low and production processes require large amounts of human work [44]. In northern European countries, on the contrary, digitalisation has been adopted on a large scale [44]. There, industrial production is quite capital intensive. Capital-intensive techniques mean machine-intensive production processes and fewer humans to operate the machines.

It is understandable that highly skilled workers will benefit most from Industry 4.0 or 5.0 solutions. If the investments in digitalised, sophisticated and automated machines are high, the expected skills of blue-collar operators are also high.

According to Fernández-Macías and Hurley [45], routine tasks are concentrated in low-skilled jobs in Europe. As Zehn [46] has argued, there is a growing concern about the extent to which workplace automation affects employment patterns in western European countries. Investments in automation and digital solutions will increase the risk of low-skilled workers of becoming unemployed. The most automation- and digitalisation-vulnerable workers are those who are low-skilled and who work with routine tasks.

According to the European Union Labour Force Survey [47], there were 32.5 million manufacturing workers in EU countries. Every sixth worker (14.2%) was a metal and machinery worker, and approximately every tenth worker was a machinery or plant operator (12.3%) or was in another manufacturing job (9.1%). The share of electro-engineering workers (2.8%) and handicraft and printing workers (2.3%) was low in the EU.

The European Working Condition Survey Eurofound [44] asked about working use of digital devices when analysing labour skills, discretion and other cognitive factors. This survey showed that only a minority of industrial and agricultural workers in southern Europe utilise digital devices in their work. The share of the option 'almost never' was highest in Portugal (59%), Greece (58%), Italy (54%) and Spain (49%). In these southern countries the polarisation between high-skilled digital technology users and non-users was also strongest.

According to Eurofound [44], industrial and agricultural workers in northern Europe said their work involved working with computers, laptops, smartphones etc. 'almost all the time' or '25–75% of the time'. The top five countries were: Denmark (78%), Norway (76%), Sweden (75%), UK (69%) and Finland (66%). When focused on 'almost all the time' results, France came to the fore beside the above-mentioned countries. Finland was one of the countries where the results were sharply divided into three: Finnish workers used digital devices 'almost all the time' (30%), '25–75% of time' (33%), and 'almost never' (35%).

The figures presented above show that Finland represents the average European country quite well in terms of Industry 4.0 transformation. The Ministry of Economic Affairs and

Employment in Finland has published results in their Working Life Barometer [48]. Based on their recent results, 44% of industrial workers used an electronic workspace and new communication technology in their daily work in 2020. The share is growing each year; in 2019 it was less than one third (30%).

11.3.2 Survey Methods

In the following sub section, we will analyse and present results from three specific surveys executed by the Finnish Industrial Union. According to Statistics Finland [49], in 2019 there were 323,000 manufacturing workers employed in the Finnish industry sectors. One third of them were members of the biggest trade union of manufacture workers, the Finnish Industrial Union.

The member panel of the Industrial Union consists of more than 3000 voluntary union members who provide responses to electronic questionnaires concerning, for example, work life, education, work health and digitalisation issues. The response rate has been typically high. We refer to the surveys as MPS2019, MPS2020 and GMS2020. MPS2019 was executed in October 2019 (N = 2565) and MPS2020 in January 2020 (N = 2047). The third survey is the grand member survey GMS2020. This was emailed in February 2020 to all union members with a registered email address. The response rate was 16% (N = 13,500), which is high for electronic questionnaires. GMS2020 was closed just before the COVID-19 pandemic caused the declaration of a state of emergency in Finland entering into force on 18 March 2020. Thus, the possible influence of the pandemic is not shown in the results.

All three survey questionnaires comprise multiple choice, Likert scale and open-ended questions. The research unit of the Industrial Union conducted the surveys taking care of research ethics. All respondents are asked for their consent to use the data as part of research. Beneficence is the focus at every step of analysis and reporting the results. No individual or any of their workplaces could be identified from any of the responses.

Respect for anonymity, confidentiality and privacy has been taken seriously. The qualitative data collected was anonymised and coded by the research unit of Industrial Union, using randomised numbering. Quantitative data analysis was conducted using SPSS software.

The descriptions quoted here have been selected from among the thousands of answers to the open-ended questions. The respondents have written 'in their own words' quite widely and have added detail to their answers. These descriptions were coded with randomised numbers and background information on the respondent's occupation, age, gender and industry was added. Most of the respondents worked in the metal, chemical or wood industries, which are the biggest industries in Finland. Other respondents worked in other manufacturing or service sectors that the Industrial Union represents. These industries include the printing industry, distribution, the mining industry, and servicing and maintenance.

Digi-investments and changes in work are still moderate in Finland. The majority of all industrial companies have invested in new technology to some or a moderate extent, with the remaining one third having not invested at all. Only one in every tent are so-called digital forerunners. Typically, these forerunners are major export companies.

Digitalised manufacturing companies have utilised different innovations—and developed and produced new smart ones. They will also have a broad vision of the future. Investments are necessary when the manufacturing of goods is increasing. Smarter and more effective machines are needed to respond to the growth in demand. Companies also try to respond to the demands of the workforce by reorganising tasks.

In the last decade, investments in robotics and automation have been made in the wood industry. In the metal and chemical industries, smart technology investments, such as new-generation robots, have become reality in recent years.

11.3.3 Many Decades of Experience

The member panel survey (MPS2020) conducted by the Finnish Industrial Union studied the use of digital equipment. The respondents typically worked in the metal, chemical, wood and graphics industries. Some of them worked in services and maintenance and in primary production. At the maximum there were more than a hundred robots in one big workplace, and some of them were smart and mobile with several AI solutions. We can say the experiences of manufacturing workers with Industry 4.0 solutions have been growing for decades.

Almost one half (47%) of the respondents said that robots are used at their workplace. However, only some of them (13%) have worked with or alongside collaborative robots, and every tenth (10%) of them oversaw and controlled robots' work tasks. Most of those who have experience of robots do not believe that their use has freed up their working time for more interesting and varied tasks.

The robots that were used were without exception industrial robots. Industrial robots are used to execute certain types of tasks, especially in mass production. MPS2020 respondents stated how the robots helped the human actors: 'Robots assist in tasks containing tiresome, repetitive and continual stages of work tasks' and 'the automation concerns mainly heavy and monotonous work tasks'. In some factories 'equipment robots will execute product testing and welding' by using artificial intelligence.

Traditional industrial robots are fixed machines with automation, such as sanders and stacking and warehouse robots. Some of them resemble a human torso with robot arms, and they work side by side with humans on a production line. Many of them are just robot hands and arms with turning joints programmed to carry out precise tasks. Some of the robots are mobile, such as transport robots and automatic forklifts, which move and store items according to AI manoeuvring.

Robots can work completely alone according to computer programming. One of the respondents describes the robots in their workplace: 'One that makes the product from the start up to the packing pallet. About fifty that need a human by their side.' They need people for the programming. The simplest robots carry out repetitive tasks while the more complex robots have "machine intelligence"' (MPS2020).

Collaboration and working with robots and automated machines themselves is not common to all. The member survey (MPS2020) shows that respondents have commonly worked with reprogrammed self-acting robots that are controlled by humans (8%) and simple automated machines (5%). The smarter robots are not so familiar in their workplaces and tasks. Blue-collar workers do not often collaborate with remote robots controlled by humans (2%) or self-acting, mobile and reacting robots (2%).

The work use of mobile phones, tablets and computers is familiar. The majority (70%) of all MPS2020 respondents work with those digital work tools daily. This was not a surprising result. According to Statistics Finland (2018), almost every 16–55-year-old Finn owns a smartphone and uses the internet daily. This so-called Nokia effect can be seen in the survey results when asked about mobile technology.

11.3.4 Investments in Human Capital

At best, the investment in new technology and labour education will grow concurrently on the shop floor. One of the respondents (MPS2020) defines his needs: 'Everyday use. Just needed continuous training to keep skills in good shape in the digital evolution'. The change of manufacture work into more computerised work is releasing, but on the other hand, it can be a bit disturbing. Another craftsman described this ambivalent situation interestingly:

'As a machinist, I have worked with multifunction lathe and industrial automation robots for almost 15 years. I have noticed that in problematic situations there is lots of work to do. The need to improve your skills has also grown. When those digital workmates work normally, they really lighten my job. Sometimes even too much, which makes me passive, so, there you sit and wait for the upcoming problems. And when they show up, you just fix them. I think that all this waiting and controlling will make me passive. Moreover, I have done many extra tasks with a computer besides my "normal" work. For example, I write different reports, minutes, starting and ending acknowledgements, and I also do programming, etc. Nowadays, I think that most of my working hours are full of all kinds of keyboarding.' (MPS2020).

Blue-collar workers are producing, analysing, using and sharing data on industrial production every day. For example, one of the metalworkers describes his work quite concisely: 'My job is to send samples to a specific robot via pneumatic mail. The robot handles them completely automatically. As a result, the analyses show up in the system when ready.' (MPS2020) What he is actually doing is more complicated.

In a digitalised workplace, individuals are part of a complex network including human and non-human actors. Digitalised cooperation is combined with interaction between humans; humans and machines; and humans, computers and mobile devices. And as the previous quotation shows, pneumatic mail, analysing robots, computers and data-producing software constitute what is termed an actor network (cf. [50]). There the digital interaction happens between machines and computers.

One respondent's description (MPS2020) of his work is laconic: 'I monitor a chemicals factory, which is operating completely automatically.' Here is a black box to open. When human-non-human collaboration works fine, there are no problems. The reality, however, is more often like another respondent wisely said: 'Humans create machines to help their lives, but they should just serve humans a little bit more reliably. Nowadays many things and situations will become almost impossible to solve if the systems collapse.' (MPS2020).

It is quite common for manufacturing workers to cooperate with their novel digital workmates on the shop floor and in their personal digital workspaces. Quite a lot of them have been trained to operate with production processes with their smartphones, tablets and laptops. Despite this, many industrial workers do not realise that they are actors in digitalised networks when they are utilising digital services like sensor data, clouds and big data. However, some of them are aware and can see the real value of cooperation. One of these workers with forethought says that "collecting data is The Thing in the future." (MPS2020).

If manufacturing workers have a special interest in coding and they understand software as well as their superiors, their job role can transform. An encouraging management will allow those experts to participate more actively in a digitalised co-operative network. For them, the possibilities to exercise with augmented reality, digital twins, and other AI solutions can open up.

On the other hand, digitalisation may bring about a shortage of an expert workforce. This has a lot to do with those who have applied to work in the industry. In other words, the educational level of some workers is low, or their vocational education was in a completely different field. Quite often these workers' language skills are not good enough: 'English operating directions are too difficult for me.' (MPS2020). Some of them have dyslexia or learning difficulties. Moreover, some people underestimate their digital skills at work, even though they happily use smartphones and other home electronics in their leisure time.

The more there is workmanship, information and communication technology (ICT) knowledge and other work-life skills needed in the workplace, the easier it is to receive new information. In any case, without good supervising and continuous training, workers will face real difficulties:

> For me the use of the workplace computer and software is like trying to control a Russian moon rocket. I really do not understand how the software works. I have to say that the computer or mobile phone solutions are really not user-driven. (MPS2020)

11.3.5 Attitudes and Expectations Are Mainly Positive

The grand member survey (N = 13,500) conducted by the Finnish Industrial Union (GMS2020) studied the attitudes of manufacturing workers towards Industry 4.0 solutions. Attitudes especially among male and younger industrial workers towards the technical revolution and digitalisation are more supportive.

In the reference population of the GMS2020 survey, 80% are male. However, the share of female respondents in GMS2020 is a bit higher: they account for one quarter (25%) of all respondents. The average age of union members is 45. One fifth (21%) of all respondents are under 35, and most of these young digital natives are male.

The survey (GMS2020) showed that the attitudes and expectations towards digitalisation and AI solutions are quite positive (Fig. 11.1). In the survey, this question was presented: 'What do you think about new technology? How will robots, automation, AI, and other digital devices affect your work? Choose only those options in which you could answer "Yes, I agree"'. The respondents could choose all options that they agreed with.

More than half the respondents (60%) thought that new technology will decrease the physical strain of work. Quite a lot of these blue-collar workers (44%) believe that digitalisation will improve productivity and almost as many (37%) think that it will improve work safety and well-being.

Divisive attitudes towards digitalisation became visible when the results were compared between gender and age variables. The experiences and expectations on the shop floor were quite divided. Differences between gender are understandable, because in Finland sectoral gender segregation is one of the strongest in Europe. Roughly speaking, blue-collar work in

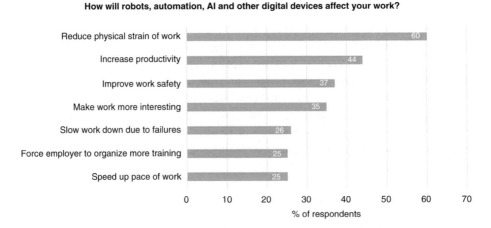

Fig. 11.1 Expectations regarding how digital devices will affect work, % (N = 13,337) Source: GMS2020

Table 11.2 Expectations of how digital devices will affect work, % by gender and age (N = 13,337)

	Age 18–35 years	Age 36–65 years
Female	1. Reduce physical strain of work (66%) 2. Increase productivity (44%) 3. Slow work down due to failures (38%) 4. Make work more interesting (33%) 5. Improves work safety (30%) (n = 767)	1. Reduce physical strain of work (60%) 2. Increase productivity (35%) 3. Slow work down due to failures (31%) 4. Improve work safety (27%) 5. Make work more interesting (25%) (n = 2608)
Male	1. Reduce physical strain of work (63%) 2. Increase productivity (52%) 3. Make work more interesting (45%) 4. Improve work safety (40%) 5. Force employers to organise more training (26%) (n = 2242)	1. Reduce physical strain of work (59%) 2. Increase productivity (45%) 3. Improve work safety (41%) 4. Make work more interesting (35%) 5. Speed up pace of work (26%) (n = 7720)

Source: GMS2020

steel mills, machine shops and shipyards is predominantly male, as well as work in service stations and car repair shops.

Male and younger manual workers are more open to digital solutions in their workplaces (Table 11.2). Especially workers under age 36 (n = 3009) expect digitalisation will make their work more interesting: almost a half (45%) of men under 36 and every third (33%) young woman thought so. However, they also saw that it might increase mental stress and psycho-social strain.

One quarter of all respondents (24%) are quite critical, expecting that the pace of work will become faster, and that failures in production processes will become more common. Older and female workers in particular were suspicious of digitalisation. Here experience speaks rather than expectations. When asked about possible fears in the future, the most worrying for blue-collar workers were lower wages (24%) and finding job in their hometown (17%). The changes to work or adequate competencies did not worry them as much. The share of those concerned about losing jobs because of automation (5%) and digitalisation of work (2%) were quite low (GMS2020).

There are several challenges as well. In real life, digitalised and social networks will become more complex and demanding. Working on the real shop floor and digital workspace can become a big challenge for manufacturing workers. In any case, they must improve their skills and competence. Knowledge-intensive manual work requires improved ICT, communication and professional skills. For older workers, these demands and changes will be more challenging than for younger ones.

In another survey (MPS2019), members of the Finnish Industrial Union were asked about their attitudes towards artificial intelligence (AI). In general, the results are promising, as Fig. 11.2 shows.

Looking more closely by gender and age, the results of this survey (MPS2019) are parallel with the grand member survey (GMS2020) results.

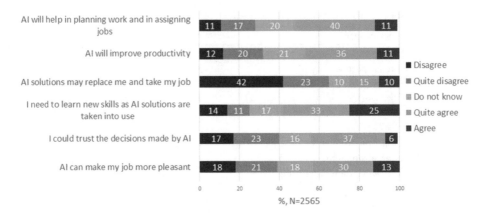

Fig. 11.2 Industrial workers' attitudes towards AI, % (N = 2565) Source: MPS2019

Half of the respondents (51%) agree that AI will help planning work and assigning tasks. Blue-collar workers also mostly (58%) understand that the need to improve their competence will increase. Quite a lot of them trust AI-made decisions (43%) and one third (33%) believe that AI will make their job more pleasant. There is no big fear of losing jobs or lack of trust in AI-based decision-making. Approximately every fifth respondent is more critical of AI.

Male respondents (n = 1922) of the member panel (MPS2019) are more open to AI solutions than females (n = 609). Half of all male respondents (50%) think that AI will improve productivity, but the share of females is much lower (38%). In general, female respondents are more unsure of their opinions. One in every five chose the option 'do not know' in every other proposition except 'AI solutions may replace me and take my job'. To this question, only 12% do not know.

The youngest respondents in the 18–35-year-old age group are the most optimistic towards AI. They see the propositions 'I could trust the decisions made by AI' and 'AI can make my job more pleasant' more agreeable than the older respondents. Older workers have more fears and suspicions. Along with female respondents, workers over 55 are more often afraid that AI solutions will take their jobs.

11.3.6 Work Safety and Well-Being

Positive experiences have an impact on attitudes towards new technological solutions as the results of the analysed surveys—MPS2019, MPS2020 and GMS2020—have shown. In GMS2020 more than half (60%) of all respondents think that new technology will decrease physical strain at work and one in every three (37%) think that work safety will increase. However, a quarter (25%) of blue-collar workers think that the pace of work will increase. In the written comments, the respondents describe their experiences of mental strain, such as boring monitoring tasks or overwhelming problem-solving situations. There are also positive experiences. Below, two manual workers (MPS2020) describe how digitalisation has lightened their work and given them extra skills in production processes:

> All kinds of smart devices (tablet and smartphone) will speed up the actual working processes. For example, it is much easier to find pictures and information with them.
>
> I think that with digitalisation we are moving in the right direction. Workers can avoid unergonomic work tasks, which no one wants to do, and robots will take care of them. With robotics the repeatability is 100%. Also, the accuracy of computer vision is really unbeatable.

The physical stress and the monotonous nature of the work, as well as ailments and diseases caused by them, will decrease. Digitalisation is also a question of health and safety, not just output or effectivity. In many workplaces the health and safety situation has improved because of automation and digitalisation. In smart factories, multiple work tasks have been automated and utilise robotics. This will decrease health problems and diminish risks on the shop floor. For example, before digitalisation, production line workers got Raynaud's disease, carpal tunnel syndrome and other serious health problems in the nape and shoulders.

In metal industry welding and soldering are two of the most dangerous work tasks. For example, welding lead and silver produces dangerous gases. The digitalised solution is clever when the welding robot works in a separate enclosure space. Welding robots engender homogeneous solder results and workers will monitor work tasks and maintain the robots. There are also serious health risks with wood dust, chemical motes and other cancer-causing substances. When human-non-human collaboration is clear, there will no longer be any health risks for humans. That will all reduce work strain and release human work resources to other tasks.

11.3.7 A Range of Individuals

All three surveys (MPS2019, MPS2020 and GMS2020) show that industrial workers are not one homogeneous group. The respondents of the union member surveys differ in terms of age, gender, educational background, ICT and communication skills, occupations, industry, mental and physical strength as well as stress control.

Those blue-collar workers, who are keyboarding and coding operators producing artefacts or services with computers are mainly vocationally trained craftspeople. Working with ICT or alongside robots or AI does not necessarily require highly trained skills or education. Industrial work with digital resources is also varied. While one manual worker is employed in a steel mill, another has their job, for example, in a sawmill, a printing house, an oil refinery, a paint factory, a plywood mill or a car factory. Industrial workers with relatively similar backgrounds may have very different preferences regarding attitudes to new digital tools.

One must also take into consideration that there are quite significant differences when it comes to the starting levels of employees' digital skills. Overall, not everyone wants to or can use digital devices in their work. The GMS2020 results show that approximately one in every four industrial worker has no interest in or they are about to be left out of digital skills. Also, tasks in industrial work have continued to be divided into professional tasks that require a variety of skills and tasks that are less demanding. Industrial work performed by blue-collar workers can include mainly routine tasks such as working on a production line. When the tasks are more demanding, creative digital problem-solving and expertise handcraft skills are needed. The various backgrounds and the multiple new tasks introduce challenges for training and upskilling. Then, individuals' starting levels and goals should be considered on an equal footing as the requirements of the work itself.

11.4 A Vision of New Industrial Work: Personalised Job Roles, Smooth Human-Machine Teamwork and Support for Well-Being at Work

Based on existing Industry 4.0 and Operator 4.0 studies, the results of the worker surveys and our own studies, we have defined a vision of future industrial work on the factory floor. Future factory floor workers will utilise a range of new technologies that increase their capabilities but also require new skills. The possibilities of these technologies and the required new skills are illustrated in Operator 4.0 visions. In our vision we utilise the Operator 4.0 types to illustrate the increased capabilities and related new skills. In our vision, well-being and operator's personal capabilities, skills and preferences are placed at the centre. Future industrial work should support well-being at work and personal compe-tence development opportunities. We also highlight teamwork, where the specific skills of each worker are recognised, and machine intelligence complements practical human knowledge and intuition. As the worker surveys took place in Finland and our previous studies were carried out in Europe, our vision reflects how industrial work will change in the western world. In this section we present this vision, starting with the new capabilities, skills and related new job roles, then we envision teamwork in human-machine teams, and finally we analyse how well-being at work can be supported in the new work environment.

11.4.1 Operator 4.0 Skills Dimensions and Personalised Job Roles

Operator 4.0 visions [1, 2] provide good insights into understanding how Industry 4.0 technologies can be seen in factory floor work by assisting operators and increasing their capabilities. The Operator 4.0 visions highlight the new opportunities, but it is good to remember that there are still many tasks that will remain unaltered and manual work will also be carried out in future factories, even if less commonly than today.

Based on Operator 4.0 visions from the literature [1, 2], we have defined six skills dimensions in Operator 4.0 work: collaborative, super strong, social, extended, smart and analytical. These dimensions are based on key Industry 4.0 technologies: collaborative robots, exoskeletons, virtual and augmented reality, connectivity and artificial intelligence. Worker skills can be extended by utilising Industry 4.0 technologies and by adopting related new ways of working. The Operator 4.0 skills dimensions are described in Table 11.3.

In earlier research, the healthy and happy operator concept [1, 2, 36] and the one-of-a-kind operator concept [1] emphasise the importance of supporting well-being at work and taking into account the individual capabilities and preferences of each operator. These issues should be supported in all skills dimensions, and that is why well-being and individuality are at the core of our skills dimensions framework, as illustrated in Fig. 11.3.

Previous operator 4.0 typologies describe one person interacting with one enabling technology [1, 2]. However, in practice the operator works with many different Industry 4.0 technologies and also with traditional manufacturing machines. In actual job roles, the different operator types are mixed, and the operator can have different skills related to the different operator types. We describe this variety of skills as skills dimensions. The skills dimensions from an individual operator's point of view are illustrated in Fig. 11.3. The skills dimensions framework illustrates the skills that Operator 4.0 can gain by utilising Industry 4.0 technologies and by adopting related new ways of wroking. The core of the change is ensuring well-being at work and considering the personal capabilities, initial skills and preferences of each operator. By utilising Industry 4.0 technologies and by learning new ways of working, the operators can develop their skills. In the following we will focus on supporting the individuality of the operators while we will discuss well-being aspects in detail in Sect. 11.4.3.

The worker surveys presented in Sect. 11.3 highlight the diversity of the factory operators. The different starting levels and different attitudes are shown in the results: some workers see new, interesting career opportunities as experts of different ICT solutions, while some workers doubt and even underestimate their digital skills. As operators are different, there should be different development paths in operator skills. Skills can be developed both by adopting related digital tools and by learning new ways of working. The skills dimensions framework can be used to describe both current skills and skills development plans. The operator should be able to choose the development path and the pace according to his/her individual interests. As an example, Fig. 11.4 illustrates the skills of one imaginary operator using orange dots: the operator has specialised in utilising AI solutions and big data, while also having some experience of utilising social media applications and mixed reality technologies.

Future factories can be connected to each other. In connected factories, everyone has the opportunity to develop even into a global expert. For instance, a novice augmented operator can utilise augmented reality tools to get support in daily tasks, while an expert augmented operator can produce AR-based material to teach good work practices to others, then also proceeding in the social dimension of the skills framework.

Table 11.3 The skills dimensions in Operator 4.0 work, illustrated by worker citations from the surveys

	**Collaborative:** The operator collaborates with robots and smart machines. Humans and machines work together, complementing each other in the different tasks combining workers' dexterity, flexibility and problem-solving skills with robots' repetitiveness, speed and precision. *'The robot will combine a part of the product and I another part. After my check, I will pack the product.'* (MPS2020)
	**Super strong:** Robots support the operator in physically demanding tasks. Workers still have manual tasks that cannot be allocated to robots. Some manual tasks require repetitive movements in non-ergonomic positions. Exoskeletons support workers in such activities. *'My job is to input the place in storage and number of items to computer. A robot will bring the product to the conveyor belt. At the end of the line there is a grapple which lifts the packed product onto a pallet.'* (MPS2020)
	**Social:** Operators possess a lot of tacit knowledge regarding good working practices and problem-solving, for example. This knowledge can be shared with mobile and social media-based tools. *'During my shift, I'm working with conveyor, lift, slitter, bagging machine, staging robot, decking robot, riming machine, wheel machine, pallet wrapping machine, and labelling machine. I'm interacting directly with people at the shipping room, loader and people working with the longitudial cutters.'* (MPS2020)
	**Extended:** In virtual environments, operators can practise working with the manufacturing systems well before those systems have been installed on the factory floor. Augmented reality-based solutions can give contextual guidance and they support remote operation and support. *'I work in quality control. As a part of my all-around education I have practised using and handling robots.'* (MPS2020)
	**Analytical:** The operator can utilise data from different sources in managing the production process and in forecasting changes. Artificial intelligence can support analysing the data. *'An electronic delivery guide shows how to move along the route. It also saves working hours and places automatically.'* (MPS2020)
	**Smart:** Artificial intelligence-based solutions assist operators in complex operations, problem-solving and decision-making. *'We have a controlled production system in the factory. Electronic control systems, accounting and invoicing. These tasks are increasingly executed by bots, specific software applications that humans control. These bots execute simple routine tasks. For example, bots will receive and deal with work tasks.'* (MPS2020)

Fig. 11.3 Operator 4.0 skills dimensions framework

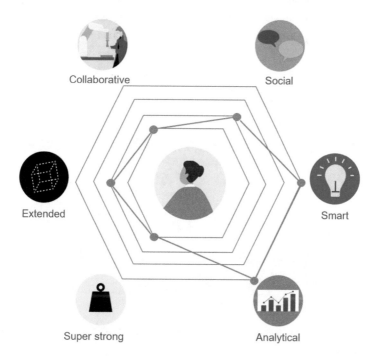

Fig. 11.4 An example of the skills dimensions of an imaginary worker

Within each skills dimension, both off-the-job and on-the-job training are needed. Off-the-job training ensures that the operators are well prepared for the change, but continuous on-the-job learning is also important so that operators can gradually increase their skills levels.

11.4.2 Smooth Collaboration in Human-Machine Teams

Future factories will combine human and non-human actors with multiple variations. In the survey results presented in Sect. 11.3, the respondents describe their experiences of collaboration with automation, robots and smart machines. Quite often the experience is two-fold: on one hand the workers feel that the work has become passive and boring monitoring, while on the other hand they describe overwhelming challenges when they must solve malfunctions of the smart machines. There is clearly a need to develop work allocation and teamwork in human-machine teams so that human workers feel they are in the loop and that human jobs remain meaningful and manageable.

The focus in Operator 4.0 visions has been on individual operators. In practice, the operator is part of one or more work teams that can include both human and machine actors on site and remotely. As illustrated in Fig. 11.5, the teams can include human actors with different skills, collaborative robots and AI-based systems for assistance or supervision. These kinds of human-machine teams should be designed so that they optimally utilise both human and machine skills and provide meaningful jobs for human operators. New collaboration and communication skills are required to interact with smart machines and remote human actors. Human-machine teams also propose changes to labour processes and work organisation.

Fig. 11.5 Human-machine teams optimally utilise the skills of each team member and the capabilities of technology

Smart solutions based on collaborative robots and AI have a lot of potential in easing physically and mentally demanding tasks. Still, there are many tasks where humans are superior. Some manual tasks are far too challenging to automate and human practical knowledge and intuition cannot be replaced with the analytical intelligence of AI. Allocating the tasks that the technology can do to technology and letting human workers take care of the 'leftovers' is a simple and far too often used solution but it does not result in well-functioning human-technology teams. A better approach is to consider how to integrate the complementary skills of humans and smart machines. AI is good at learning from past data while humans can think 'out of the box' and provide creative solutions. Integrating these two complementary viewpoints can bring about an ensemble that is bigger than the sum of the capabilities of either [51]. The work organisation should be considered as a whole, and from the viewpoint of providing meaningful jobs to the human workers. Introducing smart solutions provides an opportunity to consider the roles of all actors. What are the new tasks that each team member is responsible for? How should the team members collaborate and communicate? How can the competence development plans of each team member be supported so that they can utilise and practise their gradually developing new skills?

As suggested by Cagliano et al. [22], Neumann et al. [3], Kadir and Broberg [12] as well as Stern and Becker [38], Industry 4.0 systems should be designed as sociotechnical systems consisting of technical tools and human workers in different roles. Worker roles and skills demands related to the roles should be considered in the design of these systems. This is in line with the initial value system in sociotechnical design [52]: while technology and organisational structures may change in industry, the rights and needs of all employees must always be given a high priority. These rights and needs include varied and challenging work, good working conditions, learning opportunities, scope for making decisions, good training and supervision, and the potential for making progress in the future. Now that industrial work is changing radically, these original values are still valid in guiding the design of human-machine systems.

Moreover, in the dynamic Industry 4.0 environment, workers should be able to change their roles flexibly on the fly, based on learning new skills or just for change. In this kind of environment, work allocation is becoming a complex task but AI-based solutions have good potential for supporting dynamic work allocation based on each employee's personal skills, capabilities and preferences [11].

11.4.3 Well-Being at the Centre

To ensure a motivated workforce with an open-minded attitude towards new technologies and changes due to Industry 4.0, well-being at work needs to be considered as a crucial factor when adopting new tools and work processes in workplaces. Both development towards personalised job roles and smooth collaboration in human-machine teams have the potential to foster well-being at work. Personalised job roles incorporate the pursuit of well-being by enabling ways to develop at work—based on the worker's own interests and learning pace. Developing optimal collaboration in human-machine teams fosters well-being at work by letting machines to perform repetitive, secondary and unergonomic tasks while allowing workers to use their skills for creative, value-adding tasks. Removing or reducing unergonomic tasks is likely to increase job satisfaction, while creative and suitably challenging tasks may make the workers more engaged. New technology will not only eliminate unsatisfactory work, but it also complements and augments the capabilities of the worker in physical, mental and cognitive tasks. To support well-being, the augmentation should not be technology-driven but worker-driven, optimally with the possibility to gradually advance the role of technology in one's work.

The worker surveys presented in Sect. 11.3 highlight that industrial workers have different experiences and expectations of how Industry 4.0 is influencing well-being at work. Most of the respondents agree that physical strain at work will reduce but the opinions vary regarding whether work safety will increase or whether work will become more interesting. The respondents described their experiences of how work has on one hand become a passive monitoring role and on the other hand become a stressful and far too demanding problem-solving job. If the worker does not fully understand the reasoning behind the automated system, it is very difficult to jump in and solve problem situations where the automation does not work as expected. Digitalisation has also added many extra tasks to the work. A minority, but still a large proportion of the respondents foresee that the pace of work will become faster. While Industry 4.0 solutions will decrease physical strain, the solutions may create other kinds of well-being challenges, especially regarding mental strain. Thus, well-being at work should be ensured, particularly during the transformation.

As raised in studies by Kaasinen et al. [11] as well as Kadir and Broberg [12], involving workers in designing their own work environment and work processes is beneficial. The opportunity to participate as such supports well-being at work, and worker involvement in design can lead to changes that make working smoother or easier. Digital solutions and related new work practices should be designed in parallel, in collaboration with workers. Involving workers early on in a design process is important—to understand the context of use and work requirements but also to achieve commitment from workers, management and other key stakeholders. Co-creation has the advantages of bringing together different stakeholder perspectives, adding ways to influence as well as keep the personnel informed, thus reducing uncertainty about the upcoming changes. In addition to adoption of participatory design practices, a holistic human-centeed design approach, as suggested by Kaasinen et al. [26], aims at creating safe and ethical solutions with a positive user experience and acceptance and is thus likely to foster well-being at work.

New technologies do not only change the nature of industrial work, but they also provide new solutions to enhance the well-being of workers. Digital technology may be used to provide the user with information and insights into one's own physiology, to ensure adequate recovery from work during the workday and in their free time. Wearables and well-being applications may increase awareness of the work-related factors affecting one's well-being as well as provide data-driven feedback to enable positive behaviour change (see e.g [24].). However, to support workers' autonomy, adopting such solutions needs to be voluntary.

In addition to supporting workers in considering their well-being, future digital solutions enable monitoring worker well-being in a wider scope by providing anonymous data on the well-being of a work community. Anonymous monitoring may lead to benefits from noticing changes in workers' overall stress, for example, which could facilitate well-being interventions or restructuring of work processes. However, despite the anonymity, the process of monitoring needs to be transparent and the purpose clear for workers.

11.5   Recommendations for the Design of Factory Floor Solutions

In the previous section, we presented our vision of new industrial work, in which factory floor operators can utilise and develop their individual skills, human-machine teams utilise the best characteristics of both human and machine actors, and well-being at work is supported. How should this vision then guide the design and the continuous development of the production environment?

To get full benefit of Industry 4.0, new industrial work should have positive impacts both on productivity and well-being at work [26]. New digital solutions can support employees in the disruption of industry, which alters working tools and requirements, work roles and familiar working methods. As at least part of industrial work is expected to gradually become more and more like knowledge work [2], solutions that support workers' independent decision-making and well-being are needed.

In the worker surveys, almost half of the respondents stated that robotics solutions are used at their workplace. However, only 10% of the respondents had themselves worked with robots. Those who had worked with robots mostly had a dual experience of boredom and excess challenges. They felt that their job was mainly boring monitoring of the robot and occasionally busy and difficult problem-solving when the robot was not working as expected. The respondents did not have much experience of artificial intelligence-based solutions. Similar experiences to those with robots will be probable if AI solutions are similarly taken into use without considering worker roles and how to keep human jobs meaningful.

In the following, based on our vision of future industrial work, we will present six recommendations for the design and deployment of Industry 4.0 systems for the factory floor.

Recommendation 1: When designing Industry 4.0 solutions, the focus of the design should be the human-machine teams and collaboration within them.

Industry 4.0 is equipping the factory floor with cyber-physical systems that involve different human and smart machine actors. These systems have a lot of potential to improve the production process, and to provide more versatile and interesting jobs for employees. Getting the full benefit of this potential requires that the new solutions are designed as sociotechnical systems, as suggested by Cagliano et al. [22], Neumann et al. [3] and Stern and Becker [38]. The design should cover in parallel the technical solutions and work organisation in human-machine teams. Task allocation in the human-machine teams should be designed to utilise the best capabilities of all actors, to ensure fair task allocation and to keep human work meaningful.

Recommendation 2: Identify the new skills needed and define new work roles

As part of the design of Industry 4.0 solutions, the kinds of new skills that are required from workers should be analysed. We are extending the earlier Operator 4.0 visions [1, 2] with our Operator 4.0 skills dimensions framework. The framework supports identifying how Industry 4.0 technologies can extend workers' skills. The framework also supports defining individual skills development plans. The different combinations of skills represent new work roles. The work roles should be personally motivating and meaningful for the workers, and that can be ensured by letting the workers participate in defining those roles.

Recommendation 3: Provide training and on-the-job support to workers before and during the change

According to the worker surveys, the attitudes of workers towards Industry 4.0 are mainly positive. Digital solutions are expected to improve both productivity and well-being at work, but only a few respondents suspected that they would lose their jobs because of digitalisation. Younger workers in particular expected their work to become more interesting. This constitutes a good starting point for development. Special support should be ensured to those worker groups who seem to have concerns about the change and who are unsure whether they will be able to manage with the new solutions. Training should be planned to consider the different initial skills levels of the workers and their personal competence development plans. On-the-job learning and peer support should be planned to support continuous competence development.

Recommendation 4: Engage the work community in designing the work organisation, task allocation and work roles

Work motivation and well-being can be supported by engaging workers in planning their new work roles [11, 12, 27]. Participatory design of the overall production system can be supported with virtual and augmented reality-based tools that illustrate the planned systems and even make it possible to try out proposed new work tasks [41]. Similar tools can be utilised for continuous development by gathering development ideas while the systems are used.

Recommendation 5: Consider impacts on well-being at work in the design, and monitor well-being during the transformation

As related studies have stressed worker involvement and worker-centric design [e.g. 10, 13, 11, 12], we want to put the workers in an even more central position when their work changes. The workers should not only be involved in designing their work, but also to more holistically foster well-being at work, their capabilities, skills and preferences should be taken into account and meaningful work roles in human-machine teams should be key to the design process. Industry 4.0 technologies should be seen as a means to increase well-being at work by providing an opportunity for self-development and extending one's skills—preferably in the desired dimensions. Workers' well-being will be the key to productivity, learning new skills, finding fulfilling job roles and fluent collaboration in human-machine teams. Providing interesting and personalised job roles and supporting smooth collaboration will as such have positive impacts on well-being. Design aspects relevant to well-being, such as user experience, safety and ethics, should be considered in designing new work tools and work environments [26].

Well-being at work should be studied when introducing the systems and during their use. New technical solutions can be utilised for personal well-being monitoring. However, the workers should have the right to decide whether they want to utilise these opportunities.

Recommendation 6: Prepare for continuous change and get the most out of it

The Industry 4.0 environment is dynamic and should be resilient in reacting to changing customer needs and the production environment [6–8]. In this dynamic environment, worker roles may also experience continuous change. Continuous changes are more manageable if worker skills, development targets and preferences are considered. The changes can even make work more versatile and interesting. Continuous development can be supported by providing tools to share development ideas from and on the factory floor. Continuous competence development can be supported by personalised on-the-job learning solutions.

11.6   Conclusions

Building on Operator 4.0 research and the results of large-scale worker surveys, we have presented a vision of new industrial work, where well-being at work is central and workers are seen as individuals. Each worker's individual skills are utilised in personalised job roles and in smooth collaboration in human-machine teams. Based on the vision, we have suggested how new industrial work should be supported when developing and deploying Industry 4.0 solutions on the factory floor. Industry 4.0 solutions provide many opportunities to make factory floor work more versatile and interesting, thus offering new, attractive work roles. However, many current workers doubt whether their skills will be sufficient for future work demands and whether new technologies will just increase the pace of work by forcing human actors to adopt to the speed of machines. In this chapter we have focused on the new opportunities brought about by Industry 4.0 technologies. It should be kept in mind that in the future manual work will still be needed and there will still be a need for skilled craftspeople.

To manage the Industry 4.0 transformation, the personal capabilities, skills, and preferences of each worker should be considered. Not all workers need to be experts in everything. Worker skills can be extended by adopting different Industry 4.0 technologies, and by learning related new ways of working. Each worker should be provided with the possibility to develop their skills in preferred dimensions and at their preferred pace. The skills dimensions approach presented in this paper can support personal skills development planning and monitoring.

The vision of future industrial work we presented here is based on worker surveys in Finland, our earlier studies and related research. The vision needs to be evaluated with relevant experts and revised accordingly. That is our plan for the future.

When introducing the transformation, well-being at work should be at the core. Well-planned Industry 4.0 solutions will increase both well-being at work and productivity. The focus of the design should be extended from mere technology to work organisation, new work tasks and task allocation. The design should target smooth collaboration in human-machine teams, providing meaningful and motivating work roles for human workers.

Acknowledgments We wish to thank the members of Finnish Industrial Union for their insightful comments in the surveys. We also wish to thank our colleagues at VTT for participating in developing the vision of new industrial work, and Hanna Lammi for the graphic illustrations.

References

1. Human-Centred Factories (ACE) Cluster White paper. (2019). Accessed October 14, 2021, from https://www.effra.eu/news/ace-factories-white-paper
2. Romero, D., Stahre, J., Wuest, T., Noran, O., Bernus, P., Fast-Berglund, Å., & Gorecky, D. (2016). Towards an operator 4.0 typology: A human-centric perspective on the fourth industrial revolution technologies. In *Proceedings of the International Conference on Computers and Industrial Engineering (CIE46)* (pp. 29–31). IEEE.
3. Neumann, W. P., Winkelhaus, S., Grosse, E. H., & Glock, C. H. (2021). Industry 4.0 and the human factor – A systems framework and analysis methodology for successful development. *International Journal of Production Economics, 233*, 107992.
4. Lasi, H., Fettke, P., Kemper, H. G., Feld, T., & Hoffmann, M. (2014). Industry 4.0. *Business & Information Systems Engineering, 6*(4), 239–242.
5. MacDougall, W. (2014). Industrie 4.0: Smart manufacturing for the future. *Germany Trade & Invest.*
6. Kagermann, H., Wahlster, W., & Helbig, J. (2013). *Securing the future of German manufacturing industry. Recommendations for implementing the strategic initiative Industrie 4.0, final report of the Industrie 4.0 Working Group.* Forschungsunion.
7. ElMaraghy, H. A. (2005). Flexible and reconfigurable manufacturing systems paradigms. *International Journal of Flexible Manufacturing Systems, 17*(4), 261–276.
8. Gorecky, D., Schmitt, M., Loskyll, M., & Zühlke, D. (2014). Human-machine-interaction in the Industry 4.0 era. In *12th IEEE International Conference on Industrial Informatics (INDIN)* (pp. 289–294). IEEE.
9. Taylor, M. P., Boxall, P., Chen, J. J. J., Xu, X., Liew, A., & Adeniji, A. (2020). Operator 4.0 or Maker 1.0? Exploring the implications of Industrie 4.0 for innovation, safety and quality of work in small economies and enterprises. *Computers and Industrial Engineering, 139*, 105486.
10. Asik-Dizdar, O., & Esen, A. (2016). Sensemaking at work: Meaningful work experience for individuals and organizations. *International Journal of Organizational Analysis, 24*(1), 2–17.
11. Kaasinen, E., Schmalfuß, F., Özturk, C., Aromaa, S., Boubekeur, M., Heilala, J., Heikkilä, P., Kuula, T., Liinasuo, M., Mach, S., Mehta, R., Petäjä, E., & Walter, T. (2020). Empowering and engaging industrial workers with Operator 4.0 solutions. *Computers & Industrial Engineering, 139*, 105678.

12. Kadir, B. A., & Broberg, O. (2020). Human well-being and system performance in the transition to Industry 4.0. *International Journal of Industrial Ergonomics, 76*, 102936.
13. Holm, M. (2018). The future shop-floor operators, demands, requirements and interpretations. *Journal of Manufacturing Systems, 47*, 35–42.
14. Breque, M., De Nul, L., & Petridic, A. (2021). *Industry 5.0. Towards a sustainable, human-centric and resilient European industry. R&I Paper Series, Policy brief*. European Commission.
15. Romero, D., & Stahre, J. (2021). Towards the resilient operator 5.0: The future of work in smart resilient manufacturing systems. In *54th CIRP Conference on Manufacturing Systems*. IEEE.
16. ILO (International Labour Organisation). *Workplace wellbeing*. Accessed May 07, 2021, from https://www.ilo.org/global/topics/safety-and-health-at-work/areasofwork/workplace-health-promotion-and-well-being/WCMS_118396/lang%2D%2Den/index.htm
17. Schaufeli, W., & Bakker, A. (2010). Defining and measuring work engagement: Bringing clarity to the concept. In A. Bakker & M. Leiter (Eds.), *Work engagement. A handbook of essential theory and research* (pp. 10–24). Psychology Prepp.
18. Locke, E. A. (1976). The nature and causes of job satisfaction. In M. D. Dunnette (Ed.), *Handbook of industrial and organizational psychology* (pp. 1297–1349). Rand McNally.
19. Spector, P. E. (1997). *Job satisfaction: Application, assessment, causes, and consequences*. Sage.
20. Schaufeli, W. B., Salanova, M., González-Romá, V., & Bakker, A. B. (2002). The measurement of engagement and burnout: A two sample confirmatory factor analytic approach. *Journal of Happiness Studies, 3*(1), 71–92.
21. Nie, Y., Chua, B. L., Yeung, A. S., Ryan, R. M., & Chan, W. Y. (2015). The importance of autonomy support and the mediating role of work motivation for well-being: Testing self-determination theory in a Chinese work organization. *International Journal of Psychology, 50*(4), 245–255.
22. Cagliano, R., Canterino, F., Longoni, A., & Bartezzaghi, E. (2019). The interplay between smart manufacturing technologies and work organization: The role of technological complexity. *International Journal of Operations and Production Management, 39*, 913–934.
23. Frank, A. C., Dalenogare, L. S., & Ayala, N. F. (2019). Industry 4.0 technologies: Implementation patterns in manufacturing companies. *International Journal of Production Economics, 210*, 15–26.
24. Heikkilä, P., Honka, A., & Kaasinen, E. (2018). Quantified factory worker: Designing a worker feedback dashboard. In *Proceedings of the 10th Nordic Conference on Human-Computer Interaction (NordiCHI'18)* (pp. 515–523). ACM.
25. Heikkilä, P., Honka, A., Kaasinen, E., & Väänänen, K. (2021). Quantified factory worker: Field study of a web application supporting work well-being and productivity. *Cognition, Technology & Work, 2021*, 1–16.
26. Kaasinen, E., Liinasuo, M., Schmalfuß, F., Koskinen, H., Aromaa, S., Heikkilä, P., Honka, A., Mach, S., & Malm, T. (2018). A worker-centric design and evaluation framework for operator 4.0 solutions that support work well-being. In B. Barricelli et al. (Eds.), *Human work interaction design conference proceedings* (pp. 263–282). ACM.
27. Danna, K., & Griffin, R. W. (1999). Health and well-being in the workplace: A review and synthesis of the literature. *Journal of Management, 25*(3), 357–384.
28. Romero, D., Mattsson, S., Fast-Berglund, Å., Wuest, T., Gorecky, D., & Stahre, J. (2018). Digitalizing occupational health, safety and productivity for the operator 4.0. In *IFIP international conference on advances in production management systems* (pp. 473–481). Springer.
29. Heikkilä, P., Honka, A., Mach, S., Schmalfuß, F., Kaasinen, E., & Väänänen, K. (2018). Quantified factory worker – Expert evaluation and ethical considerations of wearable self-tracking devices. In *Proceedings of the 22nd International Academic Mindtrek conference* (pp. 202–211). ACM.

30. Lupton, D. (2016). The diverse domains of quantified selves: Self-tracking modes and dataveillance. *Economy and Society, 45*(1), 101–122.
31. Mattsson, S., Fast-Berglund, Å., & Åkerman, M. (2017). Assessing operator wellbeing through physiological measurements in real-time—Towards industrial application. *Technologies, 5*(4), 61.
32. Moore, P., & Piwek, L. (2017). Regulating wellbeing in the brave new quantified workplace. *Employee Relations, 39*(3), 308–316.
33. Romero, D., Bernus, P., Noran, O., Stahre, J., & Fast-Berglund, Å. (2016). The operator 4.0: Human cyber-physical systems & adaptive automation towards human-automation symbiosis work systems. In *IFIP International Conference on Advances in Production Management Systems* (pp. 677–686). Springer.
34. Ruppert, T., Jaskó, S., Holczinger, T., & Abonyi, J. (2018). Enabling technologies for operator 4.0: A survey. *Applied Sciences, 8*(9), 1650.
35. Romero, D., Wuest, T., Stahre, J., & Gorecky, D. (2017). Social factory architecture: Social networking services and production scenarios through the social internet of things, services and people for the social operator 4.0. In H. Lödding, R. Riedel, K. D. Thoben, G. von Cieminski, & D. Kiritsis (Eds.), *Advances in production management systems. The path to intelligent, collaborative and sustainable manufacturing. APMS 2017. IFIP advances in information and communication technology 513.* Springer.
36. Sun, S., Zheng, X., Gong, B., Garcia Paredes, J., & Ordieres-Meré, J. (2011). Healthy operator 4.0: A human cyber–physical system architecture for smart workplaces. *Sensors, 20*(7), 2011.
37. Gazzaneo, L., Padovano, A., & Umbrello, S. (2020). Designing smart operator 4.0 for human values: A value sensitive design approach. *Procedia Manufacturing, 42*, 219–226.
38. Stern, H., & Becker, T. (2019). Concept and evaluation of a method for the integration of human factors into human-oriented work design in cyber-physical production systems. *Sustainability, 11*(16), 4508.
39. Pacaux-Lemoine, M. P., Trentesaux, D., Zambrano Rey, G., & Millot, P. (2017). Designing intelligent manufacturing systems through human-machine cooperation principles: A human-centered approach. *Computers and Industrial Engineering, 111*, 581–595.
40. Sgarbossa, F., Grosse, E. H., Neumann, W. P., Battini, D., & Glock, C. H. (2020). Human factors in production and logistics systems of the future. *Annual Reviews in Control, 49*, 295–305.
41. Aromaa, S. (2017). Virtual prototyping in design reviews of industrial systems. In *Proceedings of the 21st International Academic Mindtrek Conference* (pp. 110–119). ACM.
42. Bruno, F., & Muzzupappa, M. (2010). Product interface design: A participatory approach based on virtual reality. *International Journal of Human-Computer Studies, 68*(5), 254–269.
43. Davies., R. C. (2004). Adapting virtual reality for the participatory design of work environments. *Computer Supported Cooperative Work, 13*, 1–33.
44. Eurofound: European Working Condition Survey. (2015). *Eurofound, European Foundation for the improvement of living and working conditions.* Accessed October 14, 2021, from https://www.eurofound.europa.eu/surveys/european-working-conditions-surveys/sixth-european-working-conditions-survey-2015
45. Fernández-Macías, E., & Hurley, J. (2017). Routine-biased technical change and job polarization in Europe. *Socio-Economic Review, 15*(3), 563–585.
46. Zhen, J. I. (2021). *Status decline and welfare competition worries from an automating world of work: The implications of automation risk on support for benefit conditionality policies and party choice.* Publications of the Faculty of Social Sciences, University of Helsinki.
47. European Union: Labour Force Survey 2019. Accessed October 14, 2021, from https://skillspanorama.cedefop.europa.eu/en/dashboard/browse-sector?sector=02&country=#2

48. Ministry of Economic Affairs and Employment, Finland. *Työolobarometri 2020 – ennakkotiedot (Working Life Barometer 2020 – preliminary data)*. Accessed October 14, 2021, from http://urn.fi/URN:ISBN:978-952-327-753-3

49. Statistics Finland: Employed persons and employees aged 15–74 by employer sector and industry in Finland. Accessed October 14, 2021, from https://pxnet2.stat.fi/PXWeb/pxweb/en/StatFin/StatFin__tym__tyti__vv/statfin_tyti_pxt_11qj.px/

50. Latour, B. (2007). *Reassembling the social – An introduction to actor-network theory*. Oxford University Press.

51. Chakraborti, T., & Kambhampati, S. (2018). *Algorithms for the greater good!* Yochan.

52. Mumford, E. (2000). *Socio-technical design: An unfulfilled promise or a future opportunity?* Springer.

Which Factors Influence Laboratory Employees' Acceptance of Laboratory 4.0 Systems?

12

Sarah Polzer ⓘ, Milena Frahm ⓘ, Matthias Freundel ⓘ, and Karsten Nebe

Abstract

Laboratory 4.0 systems provide a central ecosystem in the lab that connects people, processes, devices, and environmental data, comparable to the concept of smart home. Laboratory 4.0 enables laboratory employees to organize their lab and allows users to combine products from different vendors to create their personal laboratory infrastructure. Contrary to the field of smart home, to our knowledge, there is no study investigating the laboratory employees' acceptance and intention to use the technology of laboratory 4.0. Therefore, this study aims to examine the factors which influence the acceptance of laboratory 4.0 of potential users by applying the technology acceptance model (TAM) adopted from smart home. Partial least squares—structural equation modeling (PLSSEM) was used to describe the TAM and extended by trust and perceived risk, which pose potentially important factors for users in the sensitive field of laboratory data. The results revealed that users' attitude toward laboratory 4.0 is heavily affected by users' perceived usefulness which, in turn, impacts the intention to use laboratory 4.0. By determining the total effects, perceived usefulness is the most important factor influencing attitude toward and intention to use laboratory 4.0. In comparison to smart home, attitude toward use and perceived usefulness seem especially important in the context of laboratory 4.0 and appear to play a decisive role

Supported by Fraunhofer Institute for Manufacturing Engineering and Automation IPA.

S. Polzer (✉) · K. Nebe
Hochschule Rhein-Waal, University of Applied Sciences, Kamp-Lintfort, Germany
e-mail: sarah.polzer@ipsos.com; karsten.nebe@hochschule-rhein-waal.de

M. Frahm · M. Freundel
Fraunhofer Institute for Manufacturing Engineering and Automation IPA, Stuttgart, Germany
e-mail: milena.frahm@ipa.fraunhofer.de; matthias.freundel@ipa.fraunhofer.de

© The Author(s), under exclusive license to Springer Nature Switzerland AG 2023   303
C. Röcker, S. Büttner (eds.), *Human-Technology Interaction*,
https://doi.org/10.1007/978-3-030-99235-4_12

regarding the establishment of this infrastructure. The current study can serve as a foundation for future research on improving laboratory 4.0 systems by considering the relevance of influencing factors on user acceptance.

Keywords

User acceptance · Laboratory 4.0 · Smart lab · Smart home · Technology acceptance model · Partial least squares structural equation modeling

12.1   Introduction

Smart spaces support services that actively involve digital devices and Internet services in the surroundings [1]. Based on the Internet of Things (IoT), smart spaces enable smart objects to exchange, collect, and share data with other physical or virtual objects through a dynamic network [2]. The third generation of IoT includes embedded sensors, big data, machine learning, cloud computing, artificial intelligence, augmented reality, and nearfield-communication (NFC) [3]. Most known smart spaces are residences equipped with high-tech devices, appliances, sensors, and networks, which allow remote access, control, and monitoring to enhance the convenience of the inhabitants, called smart homes [4]. In smart home environments, smart objects include sensors for motion detection, heating, lighting, sockets, speakers, TVs, fridges, curtains, ovens, microwaves, coffee makers, doors, fire alarm systems, and further intelligent devices and sensors. Furthermore, smart spaces are represented in the industry sector, especially in the field of manufacturing and Industry 4.0 such as life science laboratories.

Laboratory 4.0, smart lab, or laboratory of the future [5] can also be defined as a smart space and has a similar basic concept compared to smart home. The two notions, smart lab and laboratory 4.0 have different word origins. While smart home and other smart spaces shaped the term smart lab, laboratory 4.0 is derived from the term Industry 4.0. Nevertheless, both terms are widely used as synonyms throughout the literature. In a laboratory 4.0, analyzers and measuring devices, sensors, processes, and data are connected through a network. Consisting of hardware, software, and electronics, this network has to be regulated by automation and systems e.g. laboratory information management systems (LIMS), assisted by users [5]. Laboratory 4.0 systems (see Fig. 12.1) enable laboratory employees to connect, manage and monitor instruments, and to examine and control laboratory sensors for temperature, humidity, air pressure, ambient light, UV, oxygen, etc. Furthermore, a laboratory 4.0 infrastructure can allow to visualize the workload of each laboratory technician and to register automatically in which part of the lab the employee works by NFC. The management of the laboratory and inventory can be improved by a direct link of consumables and chemicals to processes and a visualization of the laboratory structure in a digital twin. Automation, digitalization, and miniaturization are the desired outcome of laboratory 4.0, which shall simplify a laboratory employees' routine [5]. Within the framework of "Lab.Vision" in 2019, digitalization was selected among others like

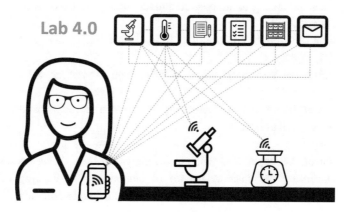

Fig. 12.1 Laboratory 4.0 infrastructure

networking, automation, data security, reliability, and data integrity as one of the most important trends in the laboratory sector for the coming 5 years [6].

Despite its potential, the usage and acceptance of laboratory 4.0 systems are still far from expectations. In every respect, the industrial and scientific laboratories have to overcome a huge backlog to become a laboratory 4.0. Examples of areas with considerable room for improvement include the concept and organization of the laboratory facilities, the laboratory technique and equipment, and the networking, integration, and automation of devices, processes, and data [7]. An important reason for the low adoption rate is the lack of connectivity of outdated laboratory infrastructure and the usage of highly specified equipment of different manufacturers with incompatible interfaces and data formats, which, in turn, complicate the networking between the diverse devices and the integration into the existing IT-structure [8]. Another reason is a lack of understanding of how users accept laboratory 4.0 environments. Shown by results of the workshop "Smart Drug Discovery" at the "Lab.Vision" in 2019, laboratory employees desire simplified equipment instead of over-engineered solutions, which exceed time and financial resources. Furthermore, the consideration of the end-user needs is strongly demanded during and beyond the development process of devices and systems [6]. Thus, understanding the laboratory employees' attitude toward and intention to use laboratory 4.0 systems is important for their success and can increase the rate of users' acceptance. To the best of our knowledge, no previous study has focused on the exploration of user acceptance and intention to use laboratory 4.0.

To address this gap, this study aims to evolve a better understanding of factors, which influence the acceptance of laboratory 4.0 users. Based on recent studies of users' acceptance of smart home [9, 10], a theoretical model was built to analyze these potential factors. To evaluate the empirical strength and the relationship between the latent variables used in this model, partial least squares—structural equation modeling (PLS-SEM) was applied.

The major contributions of this paper are:

- Development and testing of an integrated model to explain users' acceptance of laboratory 4.0 systems.
- Identification of different contributions of the factors influencing users' acceptance of laboratory 4.0 systems.
- Understanding the impact of users' beliefs about laboratory 4.0 systems on their acceptance.
- Comparison of the influencing factors regarding users' acceptance between laboratory 4.0 systems and smart homes.

The remainder of this paper is organized as follows. In Sect. 2, we review the existing literature, concerning the state of the art of laboratory 4.0, existing data to factors influencing users' acceptance in smart home, and the methodological background of exploratory factor analysis, partial least squares—structural equation modeling, and technology acceptance model. Section 3 deals with the research model and hypotheses. The next section presents the methodology used in this research. In Sect. 5, results and findings are addressed, followed by their discussion in Sect. 6. Section 7 concludes the paper and addresses further research topics.

12.2 Literature Review

12.2.1 Laboratory 4.0

Growth from almost USD 5 billion in 2020 to USD 6.9 billion by 2026 is estimated for the global market of laboratory automation, which leads to a compound annual growth rate (CAGR) of 5.6 percent for the named period [11]. Besides, the global laboratory informatics market is expected to rise at a CAGR of 4.7 percent from 2021 to 2028, from USD 3.2 billion in 2021 to USD 4.4 billion by 2024 [12]. In particular, the previously stated growing demand for laboratory automation and the development of integrated lab informatics solutions are responsible for the CAGR. Further reasons are the increasing need for biorepositories/biobanks and the rising demand to fulfill regulatory standards. Despite its economic upswing, factors such as high maintenance and service costs together with a lack of standards for integration are estimated to mitigate the growth of the global laboratory informatics market to a certain degree [13].

   In a McKinsey article, Han et al. [14] claim that "[t]he emerging technologies that characterize Industry 4.0—from connectivity to advanced analytics, robotics, and automation—have the potential to revolutionize every element of pharma-manufacturing labs within the next five to ten years" [14]. First, real-life use cases demonstrate that laboratory 4.0 can increase productivity by 30–40 percent in comparison to conventional laboratory environments. Hence, cost savings for an average microbiological laboratory could estimate 15–35 percent. Additionally, more agile working methods and shortened test phases could decrease the lead times for quality assurance laboratories by 60–70 percent [14]. Three main trends are:

- Efficiency and reliability: the digitalization of working processes will lead to a lower risk of errors compared to analog processes.
- Managing the increase of complexity: growing demand for individual medical care can accelerate the trend toward laboratory 4.0 in the medical and pharmaceutical industry.
- New services and business models: data-based services will guide laboratories to a sustainable and focused use of resources [15, 16].

As Dr. Michael Schreiber, Head of Marketing Central Europe Mettler-Toledo GmbH (2018, p. 4), states:

> [w]ith digital, software-based systems and direct data networking, it is already possible to optimize essential processes in the laboratory [...]. Here the focus is on data integrity, automation of repetitive activities for greater efficiency, economical use of resources, and user safety [5].

Laboratory 4.0 is more than the usage of single instruments, transferring data to software, and the connection between the measuring device and different equipment such as pumps, injection valves, or samplers [17]. As a basis for laboratory 4.0 systems, laboratory network elements such as data analysis, processes, collection of big data, and combining equipment are indispensable. For instance, automated analytical technology will be embedded in the process control [18].

However, most laboratory employees are dealing with manual tasks, overutilized instruments, elaborate ordering processes, and non-value added activities such as lengthy routes, long search times, and waiting periods. In a study by Fraunhofer IPA and PTC, around 26 percent of the daily working time of the laboratory technicians and managers is dedicated to documentation, which is still mostly done analogy. Additionally, reproducibility is another challenge through immature digitalization, networking, and automation of the existing laboratories [19].

Thus, there is still a gap between the "laboratory of the future" and the current situation in most laboratories, which needs to be filled. The trend toward laboratory 4.0 environments, as mentioned above, means an immense change for the entire laboratory staff. One barrier, respondents addressed in the study mentioned above, is gaining the acceptance of the whole laboratory team [19]. Therefore, it is important to focus on user needs from the beginning and to find out which factors are influencing the users' acceptance of laboratory 4.0 systems. These factors will impact laboratory employees' willingness to learn and adapt to laboratory 4.0 environments and could facilitate the transformation to a laboratory 4.0 era [20].

12.2.2 Smart Home

Contrary to the field of laboratory 4.0, in the industry of smart home, several recent studies have focused on factors that influence users' acceptance of smart home, as presented below.

In 2012, one of the first studies regarding this topic examined the adoption of IoT in UK homes. The results revealed that ease of use, perceived usefulness, privacy, and security, along with knowledge of the technology are relevant factors for the adoption of IoT in homes [21].

In 2014, a qualitative study compared consumer perceptions toward smart homes in the UK, Germany, and Italy. As drivers of smart homes, tangible benefits, and increases in quality of life, were reported. Whereas, lack of understanding of smart home technology, concerns on technology failure or difficulties in use, privacy and/or security concern, and loss of consumer freedom, were named as barriers for the smart home expansion [22].

In the USA, a new model for sustainable acceptance of smart household technology was built and tested, based on the unified theory of acceptance and the use of technology model (UTAUT). The results revealed that product attributes such as compatibleness, performance, and hedonic expectancy, together with sustainable innovativeness and consumer characteristics, play a significant role in the intention of adoption. Instead, social pressure, effort expectancy, and environmentalism are not significant regarding the adoption intent [23].

Another study from 2016, conducted six focus group discussions on consumers' needs for smart home in Korea. Tangible benefits and users' lifestyle are influencing factors when it comes to the adoption of smart home. In contrast, concerns about customer care and privacy are critical barriers to its adoption [24].

In 2017, another Korean study examined the adoption of IoT smart home services by combining the technology acceptance model (TAM) and the value-based adoption model (VAM). The new approach was developed and analyzed with factors from the UTAUT and the elaboration likelihood model (ELM). The results showed that perceived benefit and perceived sacrifice influence perceived value. Especially, the perceived benefit has a strong positive effect on perceived value. Contrary, factors, which harm the perceived value are innovation resistance and privacy risk [25].

A new theoretical model, which extends the theory of planned behavior (TPB) was developed and analyzed by partial least squares (PLS) in another Korean study from 2017. Factors, which affect the user acceptance of smart services are mobility, security/privacy risk, and trust [26].

A third Korean study of 2017, examined key factors for user acceptance of IoT technologies in a smart home environment by a new research model including TAM and five other potential factors. By using structural equation modeling (SEM), the analyzed data revealed that compatibility, connectedness, and control have a positive and cost a negative influence on the users' acceptance of IoT technologies. Furthermore, the results indicate that the greatest influential factor of intention to use is the users' attitude toward IoT technologies [9].

By applying the TAM, core motivations for adopting smart home services were investigated in another Korean study from 2018. The data were analyzed using SEM and confirmatory factor analysis (CFA). The results designated that perceived compatibility, connectedness, control, system reliability, and enjoyment of smart home services were positively related to the users' intention to use the services, whereas the perceived cost had a negative influence on the intention [27].

In 2019, a Jordanian study extended the TAM with four factors perceived risk, trust, awareness, and enjoyment in order to study the intention to use and the acceptance of smart home. PLS-SEM was used to analyze the data. Results showed that perceived risk, trust, awareness, enjoyment, perceived usefulness, and perceived ease of use significantly affect the attitude toward smart homes which, on the other hand, impact the intention to use smart homes [10].

In 2020, a Korean study focused on barriers, which keep possible users from applying smart home services. Based on the perceived risk model and the resistance theory, a new model was developed to examine the impact of technology uncertainty, service intangibility, and perceived risk on resistance to smart home services. Perceived risk was divided into privacy risk, performance risk, financial risk, and psychological risk. These four types of perceived risk are affected by technology uncertainty and service intangibility. Except for financial risk, all factors had a positive influence on the resistance to smart home services [28].

12.2.3 Commonalities and Differences Between Laboratory 4.0 and Smart Home

As mentioned before, by their nature, smart home and laboratory 4.0 can both be assigned to smart spaces. Smart home and laboratory 4.0 are based on a dynamic network, which allows smart objects to collect, exchange, and share data with other physical or virtual objects. It can be assumed that relating to user acceptance, laboratory 4.0 exhibits an essential overlap with smart home. Then again, laboratory 4.0 systems certainly have different requirements to be fulfilled when it comes to accepting, integrating, and using new technologies in a professional work context.

Furthermore, enterprise user experience for laboratory 4.0 tries to simplify the work of laboratory employees and thereby increase the work efficiency to contribute toward the company's productivity and revenue, whereas smart home systems are designed to improve the personal home experience [29, 30]. While smart home users decide on their own whether or not to implement smart home systems in their private environment, laboratory employees have limited influence on the used laboratory equipment or systems at their workplace. Nevertheless, in the end, the successful integration of laboratory 4.0 will be dependent on final users' acceptance of laboratory 4.0. To sum up, these differences between smart home and laboratory 4.0 can likely make different contributions of the previously analyzed factors mentioned above to explain the users' acceptance of laboratory 4.0 environments, which will be examined in this study.

12.2.4 Exploratory Factor Analysis

Exploratory factor analysis (EFA), which has been used in the following study, is a multivariable analysis method either for data reduction through assigning items to components or for validating latent variables of questionnaires. In general, the EFA is a hypothesis-generating analysis method. Before conducting an EFA, the extraction method, stop criterion and the method for the factor rotation are determined. The results specify how many components/latent variables are recommended for the used indicator items and which items are assigned to which component [31].

Additionally, the sampling adequacy can be measured by the Kaiser-Meyer-Olkin criterion (KMO), which lies between 0 and 1. The sampling adequacy can be interpreted by the KMO value as follow: in the .90s—marvelous, in the .80s—meritorious, in the .70s—middling, in the .60s—mediocre, in the .50s—miserable and below .50—unacceptable [32].

12.2.5 Structural Equation Modeling

Structural equation modeling (SEM) is an analysis and modeling method, which was used in several studies about users' acceptance of smart home [9, 10, 23, 25–28]. In general, SEM is used to analyze independent and response variables and their influence on each other, based on a developed predictive model. There are different SEM approaches divided into covariance-based SEM and variance-based SEM [33]. Variance-based SEM is preferable for exploratory or prediction modeling, whereas covariance-based SEM is rather used for confirmatory modeling. Hereinafter, we will focus on variance-based SEM, in particular partial least squares—structural equation modeling (PLS-SEM), which was devised in the 1960s by Herman Wold [34].

As Henseler et al. [35, p. 282] stated, "PLS path modeling is recommended in an early stage of theoretical development in order to test and validate exploratory models" [35]. PLS enables relating multiple independent variables to several dependent (response) variables. Furthermore, multicollinearity of predictors and different independent variables are realizable by variance-based SEM. Garson G. D. (2016, p. 8) explains that:

> PLS may be implemented as a regression model, predicting one or more dependents from a set of one or more independents; or it can be implemented as a path model, handling causal paths relating predictors as well as paths relating the predictors to the response variable(s) [36].

Both, reflective and formative models can be built with latent variables (constructs) and their indicator variables/items in PLS-SEM. For a reflective model, at least three items are recommended to describe a latent variable [37]. Nevertheless, PLS-SEM can handle single-item variables without identification problems [38]. As one of the most fundamental and frequently used psychometric methods in educational and social sciences research, the Likert scale is the preferred measurement tool for the indicator items [39].

To estimate the minimum sample size in PLS-SEM, the "10-times rule" implies that the sample size should be equal or greater than the larger of either 10 times the largest number of formative indicators used to measure a single construct, or 10 times the largest number of structural paths directed at a particular construct in the structural model [40].

For validation of the model construct in PLS-SEM, several validity measurements are conducted stepwise to assess the measurement and structural model. Reflective and formative measurement models are analyzed differently. Our model presented in Chap. 3 is a reflective model, because we assume that the causality flows from the construct to the indicators. Therefore, the following description only deals with this model type's validation.

Firstly, the outer measurement model is examined. To prove convergent validity, the outer loadings of the indicators should be reviewed. According to Hair et al. [40], an outer loading above 0.60 means that the indicator can be retained, whereas indicators with an outer loading between 0.40 and 0.60 should be considered for deletion if the removal increases the composite reliability or the average variance extracted (AVE) above the suggested threshold. Outer loadings below 0.40 cause elimination of the indicator. More recent literature of Hair et al., recommends loadings above 0.708, as this indicates acceptable item reliability and that the construct explains more than 50 percent of the indicator's variance [41].

Additionally, the intern consistency reliability should be proven by the composite reliability and Cronbach's alpha, which both should exceed the 0.70 threshold [40]. More-over, the convergent validity is represented by the AVE value, which should be equal to or greater than 0.50 to indicate that the construct explains at least 50 percent of the variance of its items [41].

The empirical distinction from one construct to another in a structural model, namely discriminant validity, can be assessed by the Fornell-Larcker criterion and the examination of cross-loadings. To state the discriminant validity of a structural model, the correlations between two constructs should be lower than the square root degree of the AVE (Fornell-Larcker criterion) and the item loading on an associated construct should be greater than all of its loadings on other constructs (cross-loadings) [40].

Secondly, the inner structural model needs to be verified. To examine the collinearity of the model, the variance inflation factor (VIF) is calculated by using the latent variable values of the independent variables in a partial regression. The VIF scores should fall below the threshold of five [41]. In the next step, the model's explanatory power can be described by R^2, which indicates if a dependent variable's variance is explained by the assigned independent variables. An R^2 value of 0.25 stands for weak, 0.50 moderate, and 0.75 substantial explanatory power [41].

Afterward, the Q^2 is calculated to examine the predictive relevance of the PLS-SEM. Based on the blindfolding procedure, single values in the data matrix are removed and imputed with the mean to estimate the model parameters. The omission distance has to be chosen so that the number of observations in the data set divided by the omission distance is not an integer. To predict the accuracy of the structural model for a specific dependent variable, the Q^2 value of the construct should be above 0. A Q^2 value equal to or higher than 0.50 can be determined as a large, 0.25 as a medium, and 0 as a small predictive relevance of the structural model [41].

Finally, an investigation of the statistical significance and relevance of the path coefficients can be performed. The path coefficients represent the theoretically expected relations between the constructs, which are generally between -1 and 1. Bootstrapping allows the estimation of the significant levels of path coefficients through random resamples from the original samples by the standard error, t-statistics, and confidence interval. The sample size for bootstrapping should be five times the original sample size of the survey [37].

12.2.6 Technology Acceptance Model

Based on well-established psychological theories, such as the theory of reasoned action (TRA) [42] and the theory of planned behavior (TPB) [43], several models have been developed to describe the mechanism behind and factors influencing technology adoption.

Both, the unified theory of acceptance and the use of technology model (UTAUT) and the technology acceptance model (TAM) are well-established instruments in this field. The UTAUT divides user intention and actual use of technology into four factors: performance, effort expectancy, social influence, and facilitating conditions [44]. The hypothesized moderation effects are applied by respondents' gender, age, experience, and the voluntariness of technology use [45]. Therefore, the UTAUT is more difficult to test than the TAM, even though the setup is otherwise comparable [46]. To describe users' acceptance of technology, the TAM dominates the area of investigation and has demonstrated a high level of predictiveness in many contexts [47–49]. Taking this into consideration and the fact that the study of Shuhaiber and Mashal [10] applied this method as well, we decided to choose TAM over UTAUT within this study.

The technology acceptance model exists of four main characteristics: perceived usefulness, ease of use, attitude toward use, and intention to use. As illustrated in Fig. 12.2, perceived usefulness positively influences attitude toward use and intention to use. Perceived ease of use has a positive effect on attitude toward use and perceived usefulness, as well as attitude toward use on intention to use.

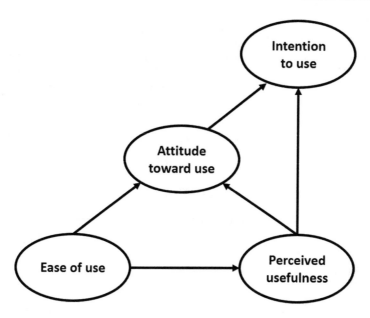

Fig. 12.2 Technology acceptance model

12.3 Research Model and Hypothesis

12.3.1 TAM Factors

Based on the literature review, the research model for examining the users' acceptance of laboratory 4.0 was developed as shown in Fig. 12.3. In alignment to earlier studies [50–54] the technology acceptance model, as explained before, was extended by two user-related factors: perceived risk and trust. Perceived risk negatively affects trust, whereas trust has a positive influence on attitude toward use and intention to use.

In the following, the TAM is set in the context of laboratory 4.0. Firstly, perceived usefulness is described as the degree to which a user believes that using laboratory 4.0 will improve the quality of the laboratory. Secondly, perceived ease of use is defined as the degree to which a user believes that using laboratory 4.0 would be without physical and mental effort. Thirdly, attitude toward laboratory 4.0 is the degree to which users have positive or negative feelings about using laboratory 4.0. Finally, the intention to use laboratory 4.0 is an indicator of the user's willingness to use laboratory 4.0 [10].

Thus, we hypothesize the following:

H1. Perceived usefulness will have a significant positive influence on intention to use laboratory 4.0.

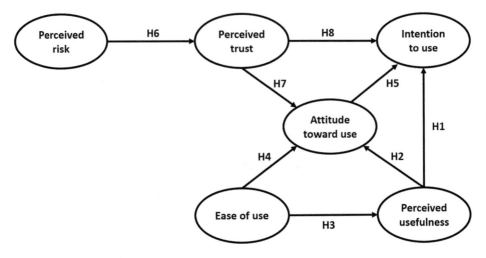

Fig. 12.3 Extended technology acceptance model

H2. Perceived usefulness will have a significant positive influence on attitude toward using laboratory 4.0.

H3. Perceived ease of use will have a significant positive influence on the perceived usefulness of laboratory 4.0.

H4. Perceived ease of use will have a significant positive influence on attitude toward using laboratory 4.0.

H5. Attitude toward using laboratory 4.0 will have a significant positive influence on intention to use laboratory 4.0.

12.3.2 Personal Factors

Trust In general, if users trust a technology, it will have a highly positive influence on attitude toward using this technology. Therefore, it is assumed, that users need trust in laboratory 4.0 services to overcome doubts about using them, such as concerns about risks as mentioned below. In the context of smart homes, several studies have proven that users' perceived trust in smart home environments positively influences their attitude toward and their intention to use smart homes [10, 55]. Based on DIN EN ISO 92419-11:2020 (2020), which explains the principles of ergonomic design of assistive systems, the operating principles of a system should lead to trust in the system by the user [56]. Thus, we define users' trust in laboratory 4.0 as the degree to which users confidently rely on laboratory 4.0 and anticipate the service to meet their expectations. Moreover, we assume that users' trust in laboratory 4.0 services increases their attitude toward and intention to use them, which leads to the following hypotheses:

H7. Trust will have a significant positive influence on attitude toward using laboratory 4.0.
H8. Trust will have a significant positive influence on intention to use laboratory 4.0.

Perceived Risk Most important trends in the laboratory sector for the coming 5 years include data security and data integrity. In response to the question "Regarding digital transformation, which homework does the laboratory industry still need to do?", the balance between flexibility/comfort together with data security, and the question "To whom do the data belong?" were named among other buzzwords at the "Lab.Vision" (p. 17) [6]. For laboratory 4.0 environments, a ramified, network is needed to access, collect, and control data, whereas protection against cyber criminality is more complicated. Thus, comprehensive surveillance and protection of the whole laboratory network, along with every device and terminal is of high priority. If there is any atypical finding, the IT-security shall intervene and take countermeasures to prevent sabotage and espionage [20].

According to DIN EN ISO 92419-11:2020 (2020), the usage of assistive systems to which laboratory 4.0 systems belong to, shall not lead to negative consequences regarding security, protection and privacy [56].

As described in DIN EN ISO/IEC 17025:2017 (2018), laboratories are responsible for the management of all information, which has been created or collected during laboratory activities. The control of data and information management is described as the duty to check all laboratory information management systems for functionality, before introducing them. Three steps are compulsory: authorization, documentation, and verification, before implementing any changes regarding data infrastructure and information management [57].

Perceived risk can include four variables such as security, privacy, financial and technical risks [58]. However, in a recent study about barriers to consumers' adoption of smart home services [28], mentioned before, perceived risk was divided into privacy risk, performance risk, financial risk, and psychological risk, and all variables, except financial risk, have significantly increased the resistance against smart home services.

Laboratory 4.0 can lead to severe risks regarding data confidentiality, information privacy, and data authentication. In our study, perceived risk is defined as the degree to which users believe that using laboratory 4.0 is safe, secure, and will protect their personal and their laboratory data.

Therefore, we assume that:

H6. Perceived risk will have a significant negative influence on trust toward laboratory 4.0.

12.4 Methodology

The questionnaire and model design by Shuhaiber and Mashal [10] was used as a basis for item determination in this quantitative study to allow for a later comparison between factor contributions for users' acceptance in a TAM of smart home and laboratory 4.0. The study by Shuhaiber and Mashal [10] was selected as a reference study due to its hypothetical design. It investigated intended users instead of users already familiar with smart home systems comparable to this study on intended user's acceptance of laboratory 4.0 systems. Additionally, the model design is based on PLS-SEM, which is preferable for exploratory modeling, as explained above. A hypothetical study is important to make an investigation on user's acceptance in an early stage of development to facilitate the market lounge of laboratory 4.0 systems. Participants of the following study were laboratory employees including CEOs, managers, project leaders, scientists, and technicians, older than 18 years and with a minimum work experience of 1 year.

Due to restrictions in the number of participants, as the field of intended users of laboratory 4.0 was expected to be smaller than for smart home users, a simplification of the above-mentioned model was required, excluding enjoyment and awareness.

A minimum sample size estimation with the "10-times rule" showed that a minimum sample size of 30 participants is necessary to validate the PLS-SEM model for laboratory 4.0.

As we asked laboratory employees to take part in our study in Germany, the survey was available in German only. Regarding the six remaining categories, all 26 items have been translated into German and modified to the context of laboratory 4.0. The latent variables trust, perceived risk, perceived ease of use, and intention to use laboratory 4.0 were defined through four measuring items, while perceived usefulness and attitude toward using laboratory 4.0 were measured through five items each. Items PU2, IU4, and TR4 needed to be changed to match the field of laboratory 4.0, whereas all other items remained equal to the previous questions to achieve the semantic equivalent. Afterward, the measuring items were revised by three professional experts of laboratory 4.0, computer science, and user experience.

Eleven variables about demographics, previous experiences, and professional background were recorded to allow a description of the study population in regard to: gender, age range, work experience, position, involvement in technical/ appliance check/ maintenance tasks, previous experiences with laboratory 4.0, previous experiences with smart home, company sector, company size, type of laboratory, and laboratory size.

To evaluate the items of the survey, a seven-point Likert scale was used (1 = "extremely disagree," 4 = "neutral", 7 "extremely agree"), because it offers a wider range of agreement levels to a statement than the traditional five-point Likert scale [39].

As the technology of laboratory 4.0 is still not widely established and the questionnaire was likely to address laboratory employees that were new to the concept of laboratory 4.0, a two-minute explanatory film was produced to explain the basics through icons, text, and audio to achieve a common understanding of the topic. The study survey included four parts in the following order: clarification and consent page, explanatory film, the questionnaire about laboratory 4.0, and the specifications about personal criteria. After completion of the input, the survey was created in Questback. Data aggregation and management were separated from analysis and conducted by the Information management department (Fraunhofer Cooperation).

A pilot test was performed with four laboratory employees of Fraunhofer IPA to clarify whether the explanatory film and all items were clear and usable. As recommended and observed in the pilot study, the explanatory film was shortened (from 2:20 to 2:08 min) and four items were rephrased and again revised by all three professional experts. The final 26 questionnaire items, shown in Appendix 1, were used for the survey.

Over 5 weeks (November 16th until December 21st, 2020), the online survey was spread over business-oriented social media and via e-mail to laboratory-related companies and research institutes. From 109 responses, both validation and data filtering methods were examined. Exclusion criteria were less than 1 year of work experience in the laboratory (n = 3), an incomplete survey (n = 1), or only extreme responses (only one or seven on the seven-point Likert scale throughout the whole questionnaire, n = 2). 103 valid responses were retained for analysis.

By using IBM SPSS Statistics 27.0.1.0 (IBM, New York, 2020), an exploratory factor analysis (EFA) was conducted to prove the one-dimensionality of the constructs and to optimize them, if needed.

If the one-dimensionality cannot be secured, further inspection of the factor quality is conducted to detect possible additional latent constructs. A new extended model can be reviewed in comparison with the initial model for further explorative studies if applicable. For the EFA, the principal component analysis was used as the extraction method, varimax with Kaiser Normalization as rotation method and stop criterion, along with no cutting value of extracting factors. To analyze the data, PLS-SEM regression analysis was performed through SmartPLS 3.3.2 (SmartPLS GmbH, Bönningstedt, 2020). The following settings were used: path as the weighting scheme, 300 as maximum iterations, and no weighting vector. The PLS algorithm stops when the change in the outer weights between two consecutive iterations is smaller than a stop criterion value for which 7 as the default value was set. A p value of 0.05 was defined as the level of significance.

As a supplemental analysis, we decided to investigate a further look into the distribution of the two extending latent variables perceived risk and perceived trust regarding the age and work experience of the respondents. We only selected the most descriptive item of each latent variable, namely TR1 and PR2 for this analysis instead of every individual item of the factors to minimalize the risk for type 1 error through multiple comparisons. Two independent samples t-tests were performed, the first with two subgroups differing in age (age range 18–35 and older than 35) and the second with two subgroups differing in work experience (1–10 years and more than 10 years). The boundaries of the characteristic age were chosen carefully to the fact that the respondents of 35 years old and younger are grown up with digital devices and are counted among the millennial and post-millennial generation as digital natives, compared to the older generation (36+ years old), which rather belongs to the generation x and baby boomer as digital immigrants [59, 60]. Due to the low number of participants, the respondents were separated in two groups 1–10 years and more than 10 years of work experience. Laboratory staff with more than 10 years expertise has well-established routines and processes and is less motivated to change them compared with less experienced participants.

12.5   Results

12.5.1 Descriptive Analysis

The 103 valid responses showed an equal distribution of gender (55 males, 46 females, 2 diverse). Most participants were in the age range between 26 and 35 years (46%), followed by over 50 years (26%), 36–50 years (15%), and 18–25 years (13%).

Nearly half of the participants (48%) have been working in a laboratory over 10 years, while the others have seven to 10 years (11%), four to 6 years (25%), or 1–3 years (16%) of work experience. 43 respondents are managers (CEOs, project managers, and laboratory managers), while 60 are laboratory employees without management responsibility.

60 percent of the participants are employed in the sector "research/university", 34 percent in the field of "pharmaceutics and health", 9 percent in "chemistry and resources", 9 percent in "energy and environment", and 9 percent "others" (multiple choice was allowed).

The types of laboratories participants are working in are "research & development" (68%), followed by "analytics" (29%), "quality assurance" (18%), "diagnosis" (11%), and "others" (10%) (multiple choice was allowed). Around half of the respondents (52%) is employed in a company/cooperation/ university with more than 500 employees, 11 percent with 250–500 employees, 16 percent with 50–249 employees, 17 percent with 10–49 employees, and 4 percent with 1–9 employees.

The laboratory of the respondents employs 1–5 (21%), 6–10 (26%), 11–15 (17%), 16–20 (10%), 21–30 (9%) and above 30 staff members (17%).

57 percent of the participants are involved in technical, appliance check, or maintenance tasks. 83 percent of the respondents have no previous experience with laboratory 4.0 and 72 percent no previous experience with smart home.

12.5.2 Exploratory Factor Analysis

To validate the constructs and the items of the questionnaire, exploratory factor analysis was applied. The sampling adequacy was marvelous (0.914), measured by Kaiser-Meyer-Olkin criterion (KMO). The results designated that all construct items could be assigned to components, except PU2 and TR4, which led to an elimination of these two measurement items. Consequently, perceived usefulness was explained through four (PU1, PU3, PU4, and PU5) and perceived trust through three remaining items (TR1, TR2, and TR3) [37].

Table 12.1 Descriptive statistics

Construct	Mean (Standard deviation)
Attitude	4.93 (1.51)
Ease of use	4.88 (1.55)
Intention	4.85 (1.56)
Risk	4.15 (1.77)
Trust	4.70 (1.49)
Usefulness	4.73 (1.53)

12.5.3 Reflective Measurement Model

Table 12.1 summarizes the descriptive results of the presented research model. To analyze our reflective model, the reliability of the items was verified by examining the outer loadings/indicator loadings. As recommended, all 24 items had indicator loadings above 0.708 and could remain in the model (see Appendix 2) [41].

As shown in Table 12.2, the internal consistency reliability was tested by composite reliability and Cronbach's alpha measurements. All values were between satisfactory and good, exceeding the 0.70 thresholds [40].

In the next step, the convergent validity of each construct was assessed by the value of the average variance extracted (AVE) for all items on each construct. The threshold of 0.50 was exceeded by all AVE measurements (see Table 12.2).

The Fornell-Larcker criterion and the examination of the cross-loadings were reviewed to verify the discriminant validity of the PLS-model (see Appendices 3 and 4).

Measurement items ATT3, ATT5, and PU5 were eliminated to fulfill that, firstly, the correlations between the constructs were lower than the square root degree of the AVE (Fornell-Larcker criterion), and secondly, the item loading on an associated variable was greater than all of its loadings on other variables (cross-loadings) [40]. Thus, perceived usefulness and attitude toward using laboratory 4.0 were explained by three items. In total, 21 items remained in the model. After these adjustments, the measurement model could be distinguished as satisfactory.

12.5.4 Structural Model

After verifying the reflective measurement model, the internal structural model was analyzed in the next step.

As represented in Appendix 5, all VIF values stayed below the threshold of five, verifying the collinearity of the model [41].

The R^2 score (see Table 12.3), explaining the model's explanatory power, was weak for the latent variables trust (0.202) and usefulness (0.240), whereas substantial for attitude toward using laboratory 4.0 (0.792) and intention to use laboratory 4.0 (0.876) [41].

Table 12.2 Internal consistency and convergent reliability

Construct	Composite reliability (>0.7)	Cronbach's alpha (>0.7)	AVE (>0.5)
Attitude	0.955	0.929	0.929
Ease of use	0.895	0.843	0.682
Intention	0.938	0.912	0.792
Risk	0.918	0.882	0.737
Trust	0.856	0.751	0.665
Usefulness	0.912	0.855	0.775

Table 12.3 Model's explanatory power and predictive relevance

Construct	R^2	Q^2
Attitude	0.792	0.681
Intention	0.876	0.683
Trust	0.202	0.118
Usefulness	0.240	0.180

The predictive accuracy of the model was accessed through the Q^2 values (see Table 12.3), which were higher than 0. The model shows small predictive relevance for trust (0.118) and usefulness (0.180), whereas for attitude toward using laboratory 4.0 (0.681) and intention to use laboratory 4.0 (0.683) the model has large predictive relevance [41].

Finally, the significance and relevance of the path coefficients (see Appendix 6) were examined through standard error, t-statistics, and confidence intervals by bootstrapping with the recommended sample size of 515.

Six out of eight hypotheses were supported with a p value <0.05 for the path coefficients. The intention to use laboratory 4.0 was shown to be determined by attitude toward using laboratory 4.0 (H5, $\beta = 0.661$, $p = < 0.001$) and usefulness (H1, $\beta = 0.227$, $p = 0.009$). However, no significant connection was indicated between trust and intention to use laboratory 4.0 (H8, $= 0.105$, $p = 0.066$). Two factors, usefulness (H2, $\beta = 0.767$, $p = < 0.001$) and trust (H7, $\beta = 0.170$, $p = 0.034$), were designated to determine attitude toward using laboratory 4.0. Nevertheless, the hypothesized relation between ease of use and attitude toward laboratory 4.0 (H4, $\beta = 0.027$, $p = .652$) remained unsupported. Perceived usefulness was specified to be influenced by ease of use (H3, $\beta = 0.490$, $p = < 0.001$) and trust by perceived risk (H6, $\beta = -0.449$, $p = < 0.001$).

Examining the total effects (see Appendix 7) of the constructs, usefulness showed the strongest effect on intention to use laboratory 4.0 (0.734), followed by attitude toward laboratory 4.0 (0.661). Moreover, ease of use (0.378) and trust (0.217) were specified to have a notable effect on intention to use. The lowest impact on intention was measured by the factor perceived risk (−0.098).

The total effects of the different factors on attitude toward laboratory 4.0, indicate the same ranking. Usefulness (0.767) has the highest influence on attitude. Secondly, the total effects showed that ease of use has an indirect effect on attitude toward using laboratory 4.0 (0.403). Finally, trust (0.170) and perceived risk (−0.077) impact attitude the least.

12.5.5 Supplemental Analysis

Alternative Model The R^2 value for the latent variable trust was weak (0.202) and the EFA showed that TR2 "Ich denke laboratory 4.0 wäre zuverlässig." (in English "I think laboratory 4.0 would be reliable.") had the highest loading (0.726) on the trust component. It can be assumed that reliability has a high influence on trust.

Therefore, an alternative model was built, where perceived risk and a single-item construct reliability were designated to determine trust. Trust was explained through TR1 and TR3, reliability through TR2. The results indicate an improvement of the R^2 value of trust (0.493) and a significantly positive influence of reliability on trust ($\beta = 0.533$, p = < 0.001).

Subgroup Analysis Two-sample t-tests were assessed regarding the demographic and work-related data of the respondents to investigate whether the users' perceived trust and risk patterns of laboratory 4.0 were consistent, similar, or different across homogenous subgroups.

The 60 participants who were in the age range of 18–35 years (M = 4.85, SD = 1.219) compared to the 43 participants in the age group above 35 years (M = 3.95, SD = 1.690) demonstrated a significantly higher level of trust, t(101) = 3.130, p = 0.002. In comparison, younger laboratory employees answered with 0.90 (SE = .286) Points higher on the Likert-Scale on Average

A close to significant decrease in the value of perceived risk was measured for the group with the younger respondents (M = 3.98, SD = 1.662) compared to the older ones (M = 4.60, SD = 1.734), t(101) = −1.837, p = .069. The results from 54 employees with 1–10 years (M = 4.81, SD = 1.361) and 49 employees over 10 years (M = 4.10, SD = 1.558) of work experience indicate a higher level of trust, t(101) = 2.478, p = .015. There was no significant effect for perceived risk, t(101) = −1.284, p = .202, between less (M = 4.04, SD = 1.659) and more experienced employees (M = 4.47, SD = 1.757).

12.6 Discussion

12.6.1 Hypotheses

The current study introduced the technology acceptance model with the integration of perceived risk and trust from smart home to the context of laboratory 4.0. The findings of this study demonstrate how users' intention to use and attitude toward laboratory 4.0 are affected by usefulness, ease of use, trust, and perceived risk.

Six of eight hypotheses were supported, which was shown by a significant connection between these constructs (p = < 0.05), while the relationship between ease of use and attitude toward using lab 4.0 (H4) and between trust and intention to use lab 4.0 (H8) were not validated. H8 only just missed the level of significance likely because of the relatively low sample size (n = 103). Nevertheless, the connection between the constructs can be assumed and could be validated in further studies. H4 had an unexpectedly low path coefficient, which will be further discussed in the following.

12.6.2 Intention to Use

In agreement with TAM postulates, attitude toward using laboratory 4.0 indeed has a significant positive effect on intention to use laboratory 4.0. This path was found to have the highest direct influence on intention to use laboratory 4.0, in accordance with a previous study about IoT technologies in smart home [9].

Perceived usefulness shows a lower, but still significant positive effect on intention to use. Examining the R^2 value indicates that intention to use is substantially explained by the factors: attitude, usefulness, and trust. The total effects revealed that usefulness has the highest impact when its indirect effect via attitude is taken into consideration.

Consequently, both, attitude and usefulness play a key role when it comes to the users' intention to use and their acceptance of laboratory 4.0, while trust only seems to play a minor role.

12.6.3 Attitude Toward Use

The highest positive effect on attitude toward using laboratory 4.0 was shown by usefulness, followed by trust, which has a low, but significant impact on attitude. These factors in combination are substantially explaining the construct attitude, shown by a high R^2 value.

In the smart home-related technology acceptance model by Shuhaiber and Mashal [10], ease of use together with usefulness played a significant role, influencing attitude toward using smart home. Furthermore, ease of use seems to be important for laboratory employees as they prefer simplified equipment instead of over-engineered solutions regarding a workshop at the "Lab.Vision" [6].

Surprisingly, the results in this study only show a minimal influence of ease of use on attitude, which failed to prove statistically significant in this study. This finding is possibly the result of the strong influence of usefulness on attitude in the context of laboratory 4.0 and as a consequence narrows the relative effect of ease of use on attitude toward laboratory 4.0.

A reasonable explanation for this pattern could be that laboratory employees have to complete extensive training regarding the complex laboratory devices and technicians are responsible for the implementation, repair, and maintenance of the laboratory equipment

[19]. Therefore, intended users may have mixed expectations about the difficulties that they may face with the implementation of laboratory 4.0, but this does not necessarily influence their attitude toward the technology, whereas usefulness is the much more dominant factor when it comes to accepting the technology. However, the total effects revealed that ease of use indirectly influences attitude toward and intention to use laboratory 4.0 by usefulness. This indirect strong relation enhances the value of ease of use and determines its permission in the laboratory 4.0 model in common with a high relation to usefulness.

In total, perceived usefulness is still the most important factor influencing attitude toward laboratory 4.0, while ease of use seems to be no key factor by itself, but can be seen as a necessary supportive factor. To summarize the findings above, our data suggests that user experience professionals, designers, developers, manufacturers, and researchers should lay special focus on laboratory employees' perceived usefulness in order to increase the attitude toward and the intention to use laboratory 4.0 systems.

12.6.4 Usefulness and Ease of Use

In addition to the constructs usefulness and ease of use, a low R^2 value of usefulness indicates that ease of use is not sufficient for describing usefulness and this gives a hint that further independent variables could be missing in the context of laboratory 4.0.

Perceived connectedness and perceived compatability are possible factors that could have a positive impact on usefulness in the laboratory sector, as shown before in a smart home study by Park et al. [9].

An extension of the laboratory 4.0 TAM could lead to a better explanation of the construct usefulness. The high relevance of usefulness for the acceptance of laboratory 4.0 warrants further studies to explore its influencing factors and if those factors lead to an improved explanation of the construct.

12.6.5 Laboratory 4.0 and Smart Home

Overall, the PLS-structural model of laboratory 4.0 specified similar results to the smart home-related model by Shuhaiber and Mashal [10], but as well two noticeable differences.

The influence of attitudes on intention to use (laboratory 4.0: $\beta = 0.661$, smart home: $\beta = 0.297$) and perceived usefulness on attitude (laboratory 4.0: $\beta = 0.767$, smart home: $\beta = 0.411$) showed a remarkably higher positive tendency compared to the context of smart home, whereas the impact of ease of use on usefulness (laboratory 4.0: $\beta = 0.490$, smart home: $\beta = 0.795$) was considerably lower in the context of laboratory 4.0, but with a common tendency.

Further studies could clarify the similarities and differences between laboratory 4.0 and smart home regarding users' acceptance and the influencing factors. Furthermore, a within-subjects designed study could show if the measured results differ in the same group of participants by asking laboratory 4.0 users to respond to both questionnaires regarding laboratory 4.0 and smart home.

12.6.6 Trust and Perceived Risk

The results yielded a high and significant negative impact of perceived risk on trust, confirming Shuhaiber and Mashal results in the context of laboratory 4.0 [10]. Thus, manufacturer of laboratory 4.0 systems should inform their intended users about their security regulations to reduce uncertainty and doubts in order to enhance the trust level of intended laboratory 4.0 users.

The influence of trust on attitude (H7) and intention to use (H8) was low with a path coefficient of $\beta = 0.170$ for trust on attitude and $\beta = 0.105$ for trust on intention to use. While H7 could be determined as supported, the p value of H8 only reached a level close to significance but remained statistically unsupported.

This is contrary to our earlier expectation that trust might play a key role in the sensitive field of research data that a laboratory 4.0 would have to deal with. A possible reason for the limited impact of trust and risk on attitude and intention to use laboratory 4.0 might be that intended users may take usefulness as the primary factor into account when they think about integrating a completely new technology with which they have no experience yet. Data and security considerations and building of trust, as described in DIN EN ISO 92419-11:2020 (2020) [56] might be secondary in this regard. As laboratory 4.0 is still in an early stage and there is only a small number of laboratory technicians actually using these systems by the time this study was conducted, the establishment and distribution of the technology could lead to a shift in the factor loadings.

12.6.7 Supplemental Analysis

Alternative Model The R^2 value indicates that trust is weakly explained by perceived risk. Furthermore, the results of the EFA provided a basis for a new model, including the TR2 item as a single-item variable, called reliability. Another TAM model regarding smart home services added system reliability combined with perceived security as independent variables to the field of security value [27].

The results of the alternative model revealed that reliability could have a significant positive effect on trust and that trust is better explained through two independent variables, namely perceived risk and reliability. In 2019, reliability was ranked as one of the most important trends in the laboratory industry for the coming 5 years [6]. A related idea that might explain why reliability seems to be more important in the laboratory sector compared to smart home, is that quality assurance and management, as well as regular calibration of the laboratory equipment, is required and reviewed in order to prevent incorrect measurements [57], which could lead to potential risks or even endanger human health or life [61]. Therefore, reliability could play a major role, when it comes to inspiring confidence in laboratory 4.0.

To enhance the model's explanatory power, further studies may choose to include reliability as a positive influencing factor on trust, and examine if this factor plays a key role in users' acceptance in the field of laboratory 4.0. By a qualitative study and verification through an EFA, suitable questions should be found and validated in order to add more indicator items to the single-item construct reliability before testing a further extended TAM.

Subgroup Analysis The supplemental analysis revealed interesting additional findings regarding four investigated subgroups divided by age and work experience regarding the factors of trust and risk.

Two subgroups (age: below 36 years old, work experience: less than 10 years) showed a significantly higher value of trust, compared to the contrary subgroups (age: 36 years and older, work experience: 10 years and more). On average, younger laboratory workers and employees with less work experience answered with 0.90 points (age) and 0.71 points (work experience) more on the seven-point Likert scale for trust. The values for perceived risk showed the same tendencies but did not reach the level of significance.

The variations between age groups could be attributed to the fact that the group of 35 years old and younger is grown up with digital devices and are counted among the millennial and post-millennial generation as digital natives, compared to the older generation (36+ years old), which rather belongs to the generation x and baby boomer as digital immigrants [59, 60]. This difference could lead to taking risks more into account (older generations) or less (younger generations) when it comes to trust. Moreover, the effect of the respondents' age could be relevant regarding differences in the working experience subgroups as well, as an overlap of the age distribution is given. Besides, laboratory employees with more working experience could have a higher effect of perceived risk on trust due to greater knowledge and experiences regarding data security and safety, such as the DIN EN ISO/IEC 17025:2017 standard [57], compared to less experienced employees.

This additional analysis revealed that it may be necessary to especially concentrate on the older generation regarding building trust in laboratory 4.0. To lower the perceived level of risk, for instance more information about provisions on the protection of personal and laboratory data could be provided for this target group.

Due to the low number of participants, a subgroup analysis regarding different types of laboratories could not be conducted but should be taken into account for further studies. Varieties in the factor distributions are possible as laboratory technicians could have various demands and requirements for different laboratory categories, such as diagnostic test, research and development, and production labs.

For diagnostic test labs, the most relevant requirement is complete tracking and processing of the sample analysis in accordance with certified quality criteria. For the highly standardized sample processing, every step from the arrival of the sample to the report and certificate of the sample analysis can be digitalized and fully automated in a laboratory 4.0 environment. Then again, the key priority of laboratory 4.0 systems in the field of research and development is flexibility. Researchers have different workflows depending on the current study, need to shift between different equipment and various locations, and regularly exchange data with colleagues. Cloud storage and a more flexible laboratory environment allow for rearranging the workplace for different workflows and processes.

Production labs should focus on efficiency, process optimization, security, and flexibility. Production labs need to constantly adapt to changing market needs. Therefore, automation and digitalization of established processes with a high security standard, as well as the customizable design of the laboratory and the entire instrumentation matter [15].

12.6.8 Limitations

This study has some notable limitations. Covid-19 restrictions and an in general very specific industry field, led to a challenging acquisition, resulting in an acceptable, but still, low number of participants compared to previous studies of TAM analysis in smart home. As a consequence, the study can only be seen as a pilot study to gain a first insight about factors influencing users' attitude toward and intention to use laboratory 4.0.

In this study, the results of both lab staff that worked in a laboratory 4.0 before and lab staff completely new to the topic were clustered. The two groups are likely to differ as empirical knowledge can have an influence on the assessment of the factors. Sample sizes for both groups were too low to allow for separate modeling. When laboratory 4.0 is widespread, further studies should examine if laboratory employees and managers, who are already using laboratory 4.0 systems, show the same acceptance patterns as intended users.

Due to the low number of respondents, economic constructs could not be implemented into the model. High maintenance and service costs have been estimated to mitigate the growth of the global laboratory informatics market [13]. It is likely that economic factors play a role in the acceptance of laboratory 4.0 as well and should be implemented in further research.

The results of the structural model of laboratory 4.0 cannot directly be compared with the findings of previous studies like Shuhaiber and Mashal [10] but can only show tendencies between factor loadings and the different relevance of factors for the constructs because obviously, the used items could not be identical and the model excluded two of the original factors.

Additionally, the different nationalities of the users of our study and the study of Shuhaiber and Mashal [10] could have an impact on the results as well. Since our study was conducted in Germany, the outcome cannot be directly applied to other nations. User adoption patterns can significantly although variate due to cultural or national differences, as previous studies demonstrated [62].

Within this study the latent variable 'perceived risk' was described with items that concentrated on the aspect of data security and privacy concerns. To give a broader view, in following studies further items could explore softer factors, e.g. the perceived risk of loss of entitlement due to new technology.

By the introduction of laboratory 4.0 systems, organizational and management changes could influence user acceptance of laboratory 4.0 systems as well. For instance, as activities such as documentation in laboratories are eliminated and new ones are added, the job descriptions in laboratories might change. Additionally, in the laboratory sector new jobs could develop e.g. for data management. In the future, the whole user acceptance including institutional framework conditions, the acceptance of new work routines, and changes in work culture could be within the scope of further studies.

The explanatory film about smart home and laboratory 4.0 has been considered necessary to convey the meaning and a common understanding of both types of smart spaces. Even if the explanatory character was the intention of the film, however, it cannot be excluded that the explanatory film might have influenced their attitude positively through explaining areas of application of laboratory 4.0.

12.7 Conclusion

Overall, this pilot study about laboratory 4.0 indicates that users' intention to use laboratory 4.0 is heavily affected by users' attitude, while trust seems to have a lower impact on intention to use. Attitude toward laboratory 4.0 is mainly affected by perceived usefulness, whereas ease of use plays a minor role in this respect. In comparison to smart home, attitude toward use and perceived usefulness seem to be most important in the context of laboratory 4.0 and appear to play a decisive role regarding the establishment of this infrastructure. As a consequence, the user-centered design process should be more established in the laboratory field and we should concentrate on questions such as "Which user needs can be solved through laboratory 4.0 systems?"

As smart spaces are represented in many industry sectors, this study was an exemplary implementation of the TAM and PLS-SEM in the field of laboratory 4.0 and could be applied to other smart spaces.

Further research should focus on laboratory 4.0 users, who respond to both questionnaires regarding smart home and laboratory 4.0 to examine differences in the same group of participants. User experience professionals, designers, developers, manufacturers, and researchers of laboratory 4.0 environments should focus on users' perceived usefulness and attitude in order to increase users' acceptance and intention to use laboratory 4.0. However, as laboratory 4.0 is still in an early stage and there is only a small number of laboratory technicians using these systems to date, a sustainable establishment and distribution could lead to a change of the factor loadings and potentially other key factors. Future research in modeling TAM for laboratory 4.0 should consider including reliability as a positive influencing factor on trust, to examine if this factor plays a key role in users' acceptance in this context. The degree of laboratory 4.0 acceptance and its key influencing factors should be compared individually between users and intended users of laboratory 4.0 systems to gain additional information in this growing field of technology.

Acknowledgments Thank you to Andrea Siegberg and her C9 Information management team (Fraunhofer Cooperation) for assisting with the questionnaire. Special thanks to Ms. Sabine Lauderbach for her knowledge of statistics.

Appendices

Appendix 1: Questionnaire Items Used in the Survey

Construct	Description
Perceived usefulness (PU)	
PU1	Using laboratory 4.0 would enable me to accomplish laboratory tasks more quickly.
PU2	Laboratory 4.0 would be useful for me to keep track of the laboratory tasks.
PU3	Using laboratory 4.0 would enhance the quality of my laboratory work.
PU4	Using laboratory 4.0 would enable me to accomplish laboratory tasks more easily.
PU5	Overall, I would find using laboratory 4.0 to be advantageous.
Perceived ease of use (PEOU)	
PEOU1	I feel using laboratory 4.0 would be easy.
PEOU2	I feel learning to use laboratory 4.0 would be easy for me.
PEOU3	I feel my interaction with laboratory 4.0 would be clear and understandable.
PEOU4	I feel I would find it easy to get laboratory 4.0 applications to do what I want them to do.
Attitude toward lab 4.0 (ATT)	
ATT1	In my opinion, it is desirable to use laboratory 4.0.
ATT2	I feel using laboratory 4.0 would be a good idea.
ATT3	I feel I would have a generally favorable attitude toward using laboratory 4.0.
ATT4	I feel using laboratory 4.0 would be beneficial for me.
ATT5	I like the idea of using laboratory 4.0.
Intention to use (IU)	
IU1	By offering, I intend to use laboratory 4.0.
IU2	I am willing to use laboratory 4.0 in the near future.
IU3	I would recommend laboratory 4.0 to others.
IU4	If I have laboratory 4.0 applications that could network with laboratory 4.0 I would do so.
Trust (TR)	
TR1	I feel laboratory 4.0 to be trustworthy.
TR2	I feel laboratory 4.0 to be reliable.
TR3	I feel laboratory 4.0 to be controllable.
TR4	I feel laboratory 4.0 to be applicable in a professional work context.
Perceived risk (PR)	
PR1	I have privacy concerns associated with laboratory 4.0.
PR2	I would be anxious about my personal and/or laboratory data by using laboratory 4.0.
PR3	I have security concerns associated with laboratory 4.0.
PR4	I would be anxious about the data security by using laboratory 4.0.

Appendix 2: Laboratory 4.0 Model

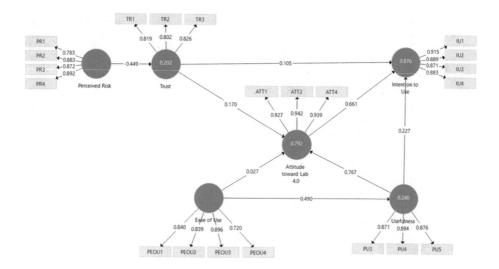

Appendix 3: Discriminant Validity: Fornell-Larcker Criterion

	Attitude	Ease of use	Intention	Risk	Trust	Usefulness
Attitude	**0.936**					
Ease of use	0.503	**0.826**				
Intention	0.925	0.505	**0.890**			
Risk	−0.282	−0.154	−0.312	**0.859**		
Trust	0.621	0.583	0.644	−0.449	**0.816**	
Usefulness	0.877	0.490	0.866	−0.180	0.567	**0.880**

Appendix 4: Discriminant Validity: Outer Loadings/Cross-Loadings

	Attitude	Ease of use	Intention	Risk	Trust	Usefulness
ATT1	**0.927**	0.452	0.847	−0.201	0.549	0.801
ATT2	**0.942**	0.487	0.879	−0.326	0.574	0.818
ATT4	**0.939**	0.472	0.871	−0.264	0.620	0.843
PEOU1	0.464	**0.840**	0.466	−0.136	0.546	0.424
PEOU2	0.349	**0.839**	0.322	−0.078	0.374	0.377
PEOU3	0.462	**0.896**	0.466	−0.181	0.660	0.476
PEOU4	0.369	**0.720**	0.398	−0.099	0.280	0.324
IU1	0.861	0.526	**0.915**	−0.341	0.648	0.804
IU2	0.859	0.470	**0.889**	−0.231	0.521	0.769
IU3	0.770	0.384	**0.871**	−0.314	0.559	0.724
IU4	0.798	0.409	**0.883**	−0.225	0.561	0.783
PR1	−0.239	−0.121	−0.283	**0.783**	−0.269	−0.153
PR2	−0.223	−0.111	−0.271	**0.883**	−0.428	−0.110
PR3	−0.282	−0.112	−0.291	**0.872**	−0.380	−0.215
PR4	−0.235	−0.183	−0.243	**0.892**	−0.431	−0.150
TR1	0.557	0.305	0.546	−0.495	**0.819**	0.431
TR2	0.421	0.555	0.462	−0.267	**0.802**	0.432
TR3	0.522	0.602	0.556	−0.304	**0.826**	0.523
PU3	0.730	0.438	0.705	−0.161	0.473	**0.871**
PU4	0.764	0.471	0.757	−0.163	0.538	**0.894**
PU5	0.817	0.389	0.819	−0.152	0.484	**0.876**

Appendix 5: Collinearity Statistics (VIF)

	Attitude	Ease of use	Intention	Risk	Trust	Usefulness
Attitude			4.799			
Ease of use	1.606					1.000
Intention						
Risk					1.000	
Trust	1.797		1.633			
Usefulness	1.562		4.344			

Appendix 6: Influence Paths and Hypotheses Results

Hypothesis/Path	Original sample (β) (>0.1)	T.statistics (|O/STERR|) (>1.96)	P-value (<0.05)	Result
H1: Usefulness → Intention	0.227	2.619	0.009	Supported
H2: Usefulness → Attitude	0.767	10.799	0.000	Supported
H3: Ease of use → Usefulness	0.490	6.194	0.000	Supported
H4: Ease of use → Attitude	0.027	0.451	0.652	Unsupported
H5: Attitude → Intention	0.661	7.903	0.000	Supported
H6: Risk → Trust	−0.449	4.977	0.000	Supported
H7: Trust → Attitude	0.170	2.125	0.034	Supported
H8: Trust → Intention	0.105	1.840	0.066	Unsupported

Appendix 7: Total Effects

	Attitude	Ease of use	Intention	Risk	Trust	Usefulness
Attitude			0.661			
Ease of use	0.403		0.378			0.490
Intention						
Risk	−0.077		−0.098		−0.449	
Trust	0.170		0.217			
Usefulness	0.767		0.734			

References

1. Korzun, D. G., Kashevnik, A. M., Balandin, S. I., & Smirnov, A. V. (2015). The smart-m3 platform: Experience of smart space application development for internet of things. In *Internet of things, smart spaces, and next generation networks and systems* (pp. 56–67). Springer.
2. Vermesan, O., Friess, P., et al. (2014). *Internet of things-from research and innovation to market deployment* (Vol. 29). River Publishers.
3. Atzori, L., Iera, A., & Morabito, G. (2017). Understanding the internet of things: Definition, potentials, and societal role of a fast-evolving paradigm. *Ad Hoc Networks, 56*, 122–140.
4. Balta-Ozkan, N., Boteler, B., & Amerighi, O. (2014). European smart home market development: Public views on technical and economic aspects across the United Kingdom, Germany and Italy. *Energy Research & Social Science, 3*, 65–77.
5. Analytica, Messe München. (2018). *Towards the laboratory of the future.* Retrieved from https://www.analytica.de/en/press/trend-reports/laboratory-of-the-future/
6. Trendreport 2019 Analysen-, Bio- und Labortechnik: Märkte, Entwicklungen, Potenziale. SPECTARIS. (2019). Retrieved from https://www.spectaris.de/fileadmin/Content/Analysen-Bio-und-Labortechnik/Zahlen-Fakten-Publikationen/Trendreport ABL 2019.pdf
7. NEMO-Projekt smartLab Innovationsnetzwerk. EurA AG. (2016). Retrieved from http://www.smartlab-netzwerk.de/netzwerk.html
8. SPECTARIS. (2021). Lads – Laboratory Agnostic Device Standard. Ein neuer Standard für das smarte Labor. Retrieved from https://www.spectaris.de/analysen-bio-undlabortechnik/vernetzte-laborgeraete/
9. Park, E., Cho, Y., Han, J., & Kwon, S. J. (2017). Comprehensive approaches to user acceptance of internet of things in a smart home environment. *IEEE Internet of Things Journal, 4*(6), 2342–2350.
10. Shuhaiber, A., & Mashal, I. (2019). Understanding users' acceptance of smart homes. *Technology in Society, 58*, 101110.
11. Global Laboratory Automation Market Report and Forecast 2021-2026. Retrieved from https://www.expertmarketresearch.com/reports/laboratory-automation-market, publisher=Expert Market Research, year=2021
12. Grand View Research. (2021, April). *Laboratory informatics market size, share trends analysis report by product (LIMS, ELN, SDMS, LES, EDC CDMS, CDS, ECM), by delivery mode (cloud-based, on-premise), by component, by end-use, by region, and segment forecasts, 2021–2028.* Retrieved from https://www.grandviewresearch.com/industry-analysis/laboratoryinformatics-market

13. Markets and Markets Research. (2019). *Laboratory informatics market*. Retrieved from https://www.marketsandmarkets.com/Market-Reports/lab-informatic-market203037633.html
14. Han, Y., Makarova, E., Ringel, M., & Telpis, V. (2019, January). *Digitization, automation, and online testing: The future of pharma quality control.* McKinsey & Company. Retrieved from https://www.mckinsey.com/industries/pharmaceuticals-and-medicalproducts/our-insights/digitization-automation-and-online-testing-the-future-ofpharma-quality-control
15. Laboratory 4.0: Who needs it, and to what extent? - Smart Lab. LABVOLUTION (2017, January). Retrieved from https://www.labvolution.de/en/news/article/news-details 3712.xhtml
16. Leitfaden für das Labor der Zukunft. PTC (2020). Retrieved from https://www.ptc.com//media/Files/PDFs/IoT/wp-leitfaden-fur-das-labor-der-zukunft.pdf
17. Gauglitz, G. (2018). *Lab 4.0: SiLA or OPC UA.* Springer.
18. Mayer, M., & Baeumner, A. J. (2018). *ABC spotlight on analytics 4.0.* Springer.
19. Frahm, M., Freundel, M., & Zölfl, R. (2021, February). *Qualitative Studie über Digitalisierungsstand von Life-Science Laboren und AR Potenziale - Fraunhofer IPA.* Fraunhofer. Retrieved from https://www.ipa.fraunhofer.de/de/Publikationen/studien/digitalisierungsstandvon-life-science-laboren.html
20. Labor 4.0 - smart in die Zukunft. Bimos. (2017). Retrieved from https://www.bimos.com/B/dede/news2/2885/labor-40—smart-in-die-zukunft
21. Coughlan, T., Brown, M., Mortier, R., Houghton, R. J., Goulden, M., & Lawson, G. (2012). Exploring acceptance and consequences of the internet of things in the home. In *2012 IEEE international conference on green computing and communications* (pp. 148–155). IEEE.
22. Balta-Ozkan, N., Amerighi, O., & Boteler, B. (2014). A comparison of consumer perceptions towards smart homes in the UK, Germany and Italy: Reflections for policy and future research. *Technology Analysis & Strategic Management, 26*(10), 1176–1195.
23. Ahn, M., Kang, J., & Hustvedt, G. (2016). A model of sustainable household technology acceptance. *International Journal of Consumer Studies, 40*(1), 83–91.
24. Kim, S., & Yoon, J. (2016). An exploratory study on consumer's needs on smart home in Korea. In *International conference of design, user experience, and usability* (pp. 337–345). Springer.
25. Kim, Y., Park, Y., & Choi, J. (2017). A study on the adoption of IOT smart home service: Using value-based adoption model. *Total Quality Management & Business Excellence, 28*(9–10), 1149–1165.
26. Yang, H., Lee, H., & Zo, H. (2017). User acceptance of smart home services: An extension of the theory of planned behavior. *Industrial Management & Data Systems, 117*, 68–89.
27. Park, E., Kim, S., Kim, Y., & Kwon, S. J. (2018). Smart home services as the next mainstream of the ICT industry: Determinants of the adoption of smart home services. *Universal Access in the Information Society, 17*(1), 175–190.
28. Hong, A., Nam, C., & Kim, S. (2020). What will be the possible barriers to consumers' adoption of smart home services? *Telecommunications Policy, 44*(2), 101867.
29. Jung, T. W., Yoon, S. Y., Nam, Y. S., Seong, D. I., Yoon, Y. J., Lee, M. H., Song, S. K., & Ha, K. S. (2020). Developing evaluation criteria for enterprise UX. *The Journal of the Korea Contents Association, 20*(4), 99–110.
30. Six, J. M. (2017, Jan). *The differences between Enterprise and consumer UX design.* UXmatters. Retrieved from https://www.uxmatters.com/mt/archives/2017/01/thedifferences-between-enterprise-and-consumer-ux-design.php
31. Brandt, H. (2020). Exploratorische Faktorenanalyse. In *Testtheorie und Fragebogenkonstruktion* (pp. 575–614). Springer.
32. Dziuban, C. D., & Shirkey, E. C. (1974). When is a correlation matrix appropriate for factor analysis? Some decision rules. *Psychological Bulletin, 81*(6), 358–361.

33. Weiber, R., & Mühlhaus, D. (2014). *Strukturgleichungsmodellierung: Eine anwendungsorientierte Einführung in die Kausalanalyse mit Hilfe von AMOS, SmartPLS und SPSS*. Springer.
34. Tobias, R. D., et al. (1995). An introduction to partial least squares regression. In *Proceedings of the twentieth annual SAS users group international conference* (Vol. 20). SAS Institute Inc Cary.
35. Henseler, J., Ringle, C. M., & Sinkovics, R. R. (2009). The use of partial least squares path modeling in international marketing. In *New challenges to international marketing* (p. 282). Emerald Group.
36. Garson, G. D. (2016). *Partial least squares. Regression and structural equation models*. Statistical Publishing Associates.
37. Secka, M. (2015). *Einfluss von Kommunikationsmaßnahmen mit CSR-Bezug auf die Einstellung zur Marke: Entwicklung und Überprüfung eines konzeptionellen Modells*. Peter Lang International Academic Publishers.
38. Hair, J. F., Jr., Hult, G. T. M., Ringle, C., & Sarstedt, M. (2016). *A primer on partial least squares structural equation modeling (PLS-SEM)*. Sage.
39. Joshi, A., Kale, S., Chandel, S., & Pal, D. K. (2015). Likert scale: Explored and explained. *Current Journal of Applied Science and Technology, 2015*, 396–403.
40. Hair, J. F., Hult, G. T. M., Ringle, C. M., Sarstedt, M., Richter, N. F., & Hauff, S. (2017). *Partial Least Squares Strukturgleichungsmodellierung: Eine anwendungsorientierte Einführung*. Vahlen.
41. Hair, J. F., Risher, J. J., Sarstedt, M., & Ringle, C. M. (2019). When to use and how to report the results of PLS-SEM. *European Business Review, 31*(1), 2–24.
42. Fishbein, M. (1979). A theory of reasoned action: Some applications and implications. *Nebraska Symposium on Motivation, 27*, 65–116.
43. Ajzen, I. (1985). From intentions to actions: A theory of planned behavior. In J. Kuhl & J. Beckmann (Eds.), *Action control from cognition to behavior* (Vol. 50). Springer.
44. Venkatesh, V., Morris, M. G., Davis, G. B., & Davis, F. D. (2003). User acceptance of information technology: Toward a unified view. *MIS Quarterly, 27*, 425–478.
45. Dwivedi, M. (2015). The unified theory of acceptance and use of technology (UTAUT). *Journal of Enterprise Information Management, 28*(3), 443–488.
46. Nistor, N., & Heymann, J. O. (2010). Reconsidering the role of attitude in the TAM: An answer to Teo (2009a). *British Journal of Educational Technology, 41*(6), E142–E145.
47. Hsiao, C. H., & Yang, C. (2011). The intellectual development of the technology acceptance model: A co-citation analysis. *International Journal of Information Management, 31*(2), 128–136.
48. King, W. R., & He, J. (2006). A meta-analysis of the technology acceptance model. *Information & management, 43*(6), 740–755.
49. Marangunc, N., & Granic, A. (2015). Technology acceptance model: A literature review from 1986 to 2013. *Universal Access in the Information Society, 14*(1), 81–95.
50. Egea, J. M. O., & Gonza'lez, M.V.R. (2011). Explaining physicians' acceptance of EHCR systems: An extension of tam with trust and risk factors. *Computers in Human Behavior, 27*(1), 319–332.
51. Gao, L., & Bai, X. (2014). A unified perspective on the factors influencing consumer acceptance of internet of things technology. *Asia Pacific Journal of Marketing and Logistics*.
52. Gefen, D., Karahanna, E., & Straub, D. W. (2003). Trust and tam in online shopping: An integrated model. *MIS Quarterly, 27*, 51–90.
53. Pavlou, P. A. (2003). Consumer acceptance of electronic commerce: Integrating trust and risk with the technology acceptance model. *International Journal of Electronic Commerce, 7*(3), 101–134.

54. Xie, Q., Song, W., Peng, X., & Shabbir, M. (2017). Predictors for e-government adoption: Integrating TAM, TPB, trust and perceived risk. *The Electronic Library, 35*(1), 2–20.
55. Luor, T. T., Lu, H. P., Yu, H., & Lu, Y. (2015). Exploring the critical quality attributes and models of smart homes. *Maturitas, 82*(4), 377–386.
56. DIN EN ISO/IEC 92419:2020-01. (2020). *Grundsätze der ergonomischen Gestaltung assistiver Systeme*. Beuth.
57. DIN EN ISO/IEC 17025:2017. (2018). *Allgemeine Anforderungen an die Kompetenz von Prüf- und Kalibrierlaboratorien*. Beuth.
58. Shuhaiber, A. (2016). *Factors influencing consumer trust in mobile payments in the United Arab Emirates*. Springer.
59. Ng, E. S., & Johnson, J. M. (2015). Millennials: Who are they, how are they different, and why should we care? In *The multi-generational and aging workforce*. Edward Elgar.
60. Prensky, M. (2001). Digital natives, digital immigrants part 2: Do they really think differently? *On the Horizon, 9*, 1–6.
61. Lippi, G. (2009). Governance of preanalytical variability: Travelling the right path to the bright side of the moon? *Clinica Chimica Acta, 404*(1), 32–36.
62. Carter, L., & Weerakkody, V. (2008). E-government adoption: A cultural comparison. *Information systems Frontiers, 10*(4), 473–482.

Determinants of Trust in Smart Technologies

13

Jörg Papenkordt and Kirsten Thommes

Abstract

Smart technologies are ubiquitous. Yet, although investments are rising, their positive economic effects are empirically questionable. One main reason for this lack of efficiency improvement are human factors: Humans need to cooperate with technology and, therefore, trust it. In this paper, we analyze particular antecedents of trust in technology both theoretically and empirically. Our results reveal that immaterial, psychological benefits affect trust stronger than material benefits. Thus, addressing advantageous aspects of new technology that benefit users is essential. However, breaking the promise of immaterial benefits may easily lead to distrust.

Keywords

Human-machine interaction · Trust · Benefits

This research and development project is funded by the German Federal Ministry of Education and Research (BMBF) within the "The Future of Value Creation—Research on Production, Services and Work" program (02L19C115) and managed by the Project Management Agency Karlsruhe (PTKA). The authors are responsible for the content of this publication.

J. Papenkordt (✉) · K. Thommes
Organizational Behavior, University Paderborn, Paderborn, Germany
e-mail: joerg.papenkordt@uni-paderborn.de; kirsten.thommes@uni-paderborn.de

13.1   Introduction

Smart technologies are ubiquitous. Yet, research on human relations with smart technologies is still in its infancy. Many different factors contribute to this lack, ranging from diversity in academic disciplines and academic fragmentation (e.g. psychology, engineering, ergonomics) to the fact that research on human-technology interaction is frequently assumed to be artefact-dependent [1–6]. However, numerous research questions concerning the interaction between humans and smart technologies are pressing, e.g. how smart technologies alter human perceptions of technology, how humans can be enabled to use smart technology efficiently, and how smart technologies affect human outcomes. The fundamental research question to be addressed as the groundwork is how human trust in smart technology can be established and enhanced.

Addressing this issue is of utmost importance, as investment volumes in smart technologies are rising but efficiency improvements are still not fully exploited. Already in 2006, Lippert and Davis stated that the investments in new technologies exceeded values over trillions of dollars worldwide, although on average only every second investment achieved the desired return [7]. Since then, the success rate of the implementation process has not noticeably improved. Instead, severe rebound effects were observed [8]. For example, seven out of ten artificial intelligence (AI) initiatives fail or only have a minimal impact on the company [9]. Furthermore, current developments, such as the recent COVID-19 pandemic, serve as another catalyst for digitization [10, 11]. The increasing pace and pressure to introduce new technologies might result in even more erroneous investment decisions. Additionally, individuals and firms are facing significant levels of uncertainty when adopting new technologies, as investments involve high sunk costs and irreversible investments [12, 13] while at the same time promising great opportunities and triggering fear of missing out [14–16]. This uncertainty makes them vulnerable to malinvestments in new technologies. Hence, trust is crucial for the successful implementation of unfamiliar technology [17].

Siau and Wang [18] regard trust as a primary reason for acceptance. However, this begs the question of what kind of trust is needed to guarantee a successful human-machine interaction and how trust can be established if the technology is unknown. Past research on trust has generated a broad notion of trust as a concept with complementary but also conflicting elements. For instance, there is some ambiguity about whether trust is a behavior, attitude, or intervention [19]. Muir [20] stresses that trust between humans and machines is special because possible benefits of the machine are completely irrelevant if the user does not trust the machine. Ejdys adds to this view with her claim: "The total lack of human confidence in technology would make it impossible to use it in everyday life, and thus would hinder the development of humans and entire civilisations." [21, p. 981].

In the dynamic process of trust-building, there is a myriad of multiple factors and attributes that influence the relationship between a user and a certain type of technology [17–19, 21]. For example, just the design of technology affects the emotional response of a user and can, therefore, lead to a feeling of pleasure even when the tasks are disliked

[22]. These specifics of technology, the different circumstances, as well as the multidimensional and contextual nature of trust, render it almost impossible to create a general model of trust in technology [21].

To advance our understanding of how trust between technical devices and humans evolves, we analyze the introduction of a technological device. For this purpose, we closely examine a study conducted on the implementation and usage of smart home technologies in United Kingdom (UK) households. The study aimed to explore how households use and interact with smart home technologies in order to save energy and reduce carbon emissions [23]. In our research, we analyze how the participants weighted benefits and costs of new technologies and how this relates to trust. We are aware that the selection of a dataset derived from the private sector raises some questions regarding the transferability of the results to the industrial context. However, firstly, the interpersonal or human-machine level of trust must be considered for this study anyway, which is why the context is negligible for the time being. Literature from psychology [24] and technology acceptance [25] assumes that trust between humans and machines relies on person-specific elements, such as traits, characteristics, motivation, and emotional aspects, but also heavily draws from individual perceptions. Accordingly, the generic building blocks of trust are equal in private and professional usage settings. Secondly, we selected a study in a private setting so that the answers to benefits and risks would not be influenced by professional pressure or any form of social desirability [26]. Thirdly, we also assume that the responses regarding benefits and risks will be more honest because participants are not forced to use the technology. In a professional context, employees often are not given a choice of whether to adopt new technologies. Therefore, in the study selected, we can assume that respondents do not tend to adjust their answers according to the interviewer's expectations so as to avoid conflict. Finally, it is important to mention that the sample is remarkably heterogeneous and large, and, in our study, we only consider the working-age population.

The remainder of the study is structured as follows. First of all, we describe the complex construct of trust, especially pertaining to new technologies. Secondly, we explicate the study "UK Homeowner Survey: Perceptions of Smart Home Benefits and Risks" and give reasons why we consider it suitable for providing relevant findings to our research. Thirdly, we establish our structural equation model (SEM), the key variables for our analysis, and the main results based on our model. Finally, we discuss the limitations and future implications of our research and place our study in the scientific context.

13.2 Theoretical Background

13.2.1 Trust

"Trust has not only been described as an "elusive" concept, but the state of trust definitions has been called a "conceptual confusion", a "confusing potpourri", and even a "conceptual morass" [27, p. 28]. McKnight and Chervany [27] state two reasons for this confusion: For one, it is difficult to limit the term trust to a specific definition, as meanings vary widely in everyday usage. Secondly, trust is treated differently in a range of scientific communities. "In general, perhaps the common summary of the various views of trust is that it expresses a willingness to take risk under uncertainty." [28, p. 77]. Following Deutsch [29], this definition entails that risk-taking behavior and trust are two sides of the same coin. This is derived from the fact that most people are unwilling to trust in case of a low probability that the given trust is returned—regardless of whether the benefits are exceptionally high. This assumes that trustees weigh up the benefits if the expected trust is fulfilled and the risks if the trust is abused. It also indicates that, at first, it does not matter what context we consider when building trust because the major factors remain the same: the uncertain situation and the weighing of benefits and risks.

13.2.2 Trust in Smart Technologies

Smart technologies offer tremendous advantages in terms of performance, efficiency, cost of ownership, and more because they cannot only detect changes in their environment but also automatically modify and adjust their behavior to the new conditions to improve their functionality [30]. Akhilesh and Möller [31] also identify smart technologies, such as artificial intelligence, internet of things, big data analytics, machine learning, or cloud computing, as the main driving force in the modern world since they are exhibiting a lasting impact on people's lives, business, enterprises, mobility, and more. In business, smart technologies will take the industry to a new level—Industry 4.0—by significantly improving productivity, flexibility, and business efficiency [32]. Thus, smart technologies are impacting a wide variety of environments. "So smart environments, by definition, are designed to exploit rich combinations of small distributed sensing/computational nodes to identify and deliver personalized services to the user when they are interacting and exchanging information with the environment." [33, p. 249]. These so-called smart environments, such as smart factories, smart cities, or smart homes, are, therefore, based on the same smart technologies and usually pursue the same overarching goal, namely user support, which allows for comparability between the fields, at least to a certain extent [31, 34]. To master the advancing future use of smart technologies in firms, the workforce must first be prepared for such an enormous change in the industrial context.

Smart technologies are becoming more and more part of our daily lives, and trust is a crucial factor when it comes to interaction with a machine because it affects the decision, performance, experience, and overall capability of humans [35]. "Automated decision aids are increasingly being modeled as "partners" that support or assist the human in performing functions that may be either difficult or even impossible for the operator to perform without the assistance of a "knowledge teammate." [36, p. 581]. Nevertheless, most of us feel uneasy when asked to team up with new technology. Why do we still prefer to work together with a human being instead of a highly developed machine? What are the key drivers of the trust-building process in these similar but also different kinds of trust? To answer these questions, it is essential to clarify what interpersonal and technological trust have in common and where exactly they differ. Additionally, we need to identify to what extent interpersonal trust serves as a basic element of human-machine trust.

Research on trust in technologies is mostly rooted in the Computer as Social Actors paradigm (CASA) introduced by Nass and colleagues (e.g. [37]). They assume that, on a basic level, people treat machines as they would other people. Even minimal cues as "talking" via text or speech, "looking", or seemingly autonomous actions result in people responding to the technology as if it were human [38]. Very similar to interpersonal relations, trust in machines is crucial as people will refuse to cooperate if trust is lacking.

The explanations and elucidations of interpersonal and technological trust indicate some similarities between the two. "At the most fundamental level, the two types of trust are similar in that they represent situation specific attitudes that are relevant only when something is exchanged in a cooperative relationship characterized by uncertainty." [39, p. 410]. But there are also some differences between them: Atkinson et al. [40] outline that reciprocity is the fundamental principle of interpersonal trust. They also regard this as the most serious counterargument against using interpersonal trust as an analogy for human-technology trust. Accordingly, it is important to mention that trust in machines differs from interpersonal trust despite the features both concepts have in common [36, 39, 41]. For example, Araujo et al. [42] emphasize that people generally build trust in machines at a much slower pace than in humans—even if the technology outperforms humans. They also lose their trust much faster if the machine makes a mistake. Thus, trust is more vulnerable compared to human-human interactions as even small mistakes or problems in usability can destroy trust [43, p. 111] further assume that trust is "notoriously easy to break down." This might be even more likely in human-machine interactions compared to human-human interactions.

Similarly, Hoffmann, Bradshaw, and Johnson [41] stress that people are also less forgiving towards machines, even so, it is always difficult to establish trust once it has been lost. Trust in human-machine interaction also plays a decisive role in effective cooperation. Muir [20] proposes a thin line between too much or too little trust. If an individual does not trust the machine, he may use any available force—even extremely cost- or time-intensive ones—to influence or control the decisions and performance of the machine. On the other hand, too much trust can also impair collaboration negatively since the machine is prone to making mistakes, as well. Lee and See [44] identify the same characteristics and define these two possibilities as "misuse" and "disuse" of the technology. To find the right balance between unjustified trust and unjustified mistrust, Hoffman et al. [41] point out that expertise and experience when using technology are crucial factors in the equilibrium of trust. They also state that people will never achieve the same level of absolute trust in technologies as they have in close relatives because, under these circumstances, trusting is a dynamic process and always depends on the amount of exploration possible and necessary. Several studies also determine the dynamic process between the time domain and distinguish between initial and continuous trust [2, 18, 45, 46]. Initial trust is based on the trustee's first impressions, expectations, and judgments and is required to overcome perceptions of risk and uncertainty before using new technology [46]. Continuous or experimental trust, on the other hand, presupposes familiarity, experience, and knowledge about the new technology and is built on the initial trust [2, 18].

Thus, to avoid distrust and foster trust, we need to understand the initial trust level in interaction. Borum [47] describes interpersonal trust as a willingness to accept vulnerability or risk based on the expectations of another person's behavior in a dynamic environment that can be influenced by individual, organizational, economic, political, social, and cultural factors. He emphasizes that interpersonal trust is a very complex construct when one party relies on the other agent's willingness to cooperate in order to achieve its goal. Based on a literature review of McKnight and Chervany [27], interpersonal trust differs from dispositional or institutional trust since the former is related to a specific individual instead of people in general or environmental structures or situations. As visualized in Table 13.1, interpersonal trust can be divided into three different types that all relate to each other.

Conclusively, McKnight and Chervany [27] summarize that people either trust other people personally—equaling trusting behavior or trusting intentions—or people trust other people because of their attributes, namely trusting beliefs.

When comparing human-human and human-machine trust, some distinct similarities and differences can be observed. For one, it might be easier to establish trust as it is easier to convince people on the basis of the attributes of the technology instead of their personality. At the same time, the trust might be particularly vulnerable in this scenario.

Like McKnight and Chervany [27], Carter et al. [2] distinguish between different types of trust. They differentiate between "Trust in General Technology", "Institutional-based Trust in Technology", and "Trusting Beliefs in a Specific Technology". In contrast to "Trust in General Technology", which deals with the general willingness of a person to depend on a technology, and "Institutional-based Trust in Technology", which denotes that using a specific class of technology in a new way is comfortable in a well-known situation (situational normality) and/or with technical support like guarantees, contracts, etc. (structural assurance), "Trusting Beliefs in Specific Technology" is based on a user's beliefs that a specific technology has the necessary attributes to perform as expected in an uncertain situation. Carter et al. [2] also pursue this approach by equating the attributes competence with functionality, benevolence with helpfulness, and predictability/integrity with reliability in this context. Table 13.2 depicts the definitions of the three characteristics by Carter et al. [2]:

Recent research tends to differentiate even further. In human-human interaction, trust processes depend on individual cues and might also be remarkably diverse. The same might hold for human-technology relations, and, consequently, trust research began to investigate artifact-dependent trust. Reviewing literature by Siau and Wang [18] illustrates some distinct concepts of trust and its antecedents depending on the context, e.g. "interpersonal trust in organizations", "trust in virtual teams", "trust in e-commerce", "trust in information systems", "trust in human-automation interaction", "trust in human-robot interaction", and "trust in applied artificial intelligence". Summing up, they

Table 13.1 Interpersonal trust construct definitions [27]

Interpersonal trust

We trust other people, either personally, as in trusting behavior and trusting intentions, or their attributes, as in trusting beliefs.

Trusting beliefs	Trusting intentions	Trust-related behavior
The extent to which one believes, with feelings of relative security, that the other person has characteristics beneficial to one.	One is willing/intends to depend on the other party with a feeling of relative security, in spite of lack of control over that party, and even though negative consequences are possible.	A person voluntarily depends on another person with a feeling of relative security, even though negative consequences are possible.
• **Competence**: One securely believes the other person has the ability or power to do for one what one needs done. • **Benevolence**: One securely believes the other person cares about one and is motivated to act in one's interest. • **Integrity**: One securely believes the other person makes agreements in good faith, tells the truth, and fulfills promises. • **Predictability**: One securely believes the other person's actions are consistent enough that one can forecast them.	• **Willingness to depend**: One is volitionally prepared to make oneself vulnerable to the other person by relying on them, with a feeling of relative security. • **Subjective probability to depending**: The extent to which one forecasts or predicts that one will depend on the other person, with a feeling of relative security.	• Cooperation • Information sharing • Informal agreements • Decreasing control • Accepting influence • Granting autonomy • Transacting business
The combination of the four trusting beliefs provides a firm foundation for trusting intentions and trust-related behavior. What characteristic is most important depends on the context.	In situations of uncertainty and risks a feeling of security makes the prospect of depending on another more comfortable and thus reflects the affective side of trusting intentions.	Every trust-related behavior occurs under circumstances of risk and either the inability, or lack of desire, to control the trustee.

distinguish, like Carter et al. [2], between three attributes of technology that determine the trust-bonding relationship.

In contrast to Carter et al. [2], Siau and Wang [18] take on a more distant perspective. They consider human and environmental factors instead of only focusing on the technological aspect. According to them, human characteristics refer to the personality (e.g., risk disposition, cultural background, past experiences) and ability (e.g., competence, skills) of the trustee. They also list environmental characteristics as elements that influence the relationship between a human being and technology. Institutional factors like situational

Table 13.2 Trust in technologies [2]

Trusting beliefs	Definitions
Functionality	The belief that the specific technology has the capability, functionality, or features to do for one what one needs to be done.
Helpfulness	The belief that the specific technology provides adequate and responsive help for users.
Reliability	The belief that the specific technology will consistently operate properly.

normality and structural assurance only represent one part of environmental circumstances. The environmental characteristics also depend on the nature of the task, the level of accessibility to new technologies, and cultural backgrounds.

However, all technological artifacts have in common that performance, process, and purpose are key characteristics of the technology and the trust assessment by the human interaction partner: Performance is similar to a mix between functionality and reliability and denotes that the expected results of the technology are accurate and efficient. Process describes that the outcome and the working process of technology are transparent and intelligible for the trustee. Lastly, purpose illustrates that the user should be familiar with the role and the boundaries of the technology in the interaction to set their expectations accurately.

Based on the literature above, we assume that first impressions, expectations, and judgments—in this context the perceived benefits—influence the perceptions of risks and the importance of the arrangements for trust-building. Since first impressions are based on the key characteristics of the technology, and the emphasis lies on the trust-building process, we neglect the contextual factors here. The factors 'human and environment characteristics' will differ in every organization anyway and can, therefore, barely be taken into account in a generally valid way. Accordingly, we examine how perceived benefits can compensate for expected risks in initial trust-building. In addition, we also assume that the risks will enhance the importance of trustworthy properties. The research model (Fig. 13.1) underlying the present study is based on the following assumptions:

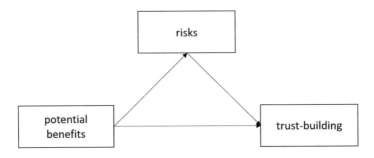

Fig. 13.1 Research model

13.3   Research Design

13.3.1 Data Access and Sample

We use the "UK Homeowner Survey: Perceptions of Smart Home Benefits and Risks" to analyze the introduction of a technical device into a new environment. The project took place between May 2012 and October 2015. The program was initiated to understand the implementation and use of smart home technologies in households of the UK. In the course, it explored how households' usage of and interaction with smart home technologies can be facilitated to save energy and reduce carbon emission [23]. At that time, smart home technologies ranged among the most advanced and innovative technologies. For example, Suresh and Sruthi [48] deemed smart home technology one of the most popular and promising technologies on the rise in the field of automation and management systems. The indisputable relevance of smart home technology is also reflected in their review of 150 publications, considering only conferences, books, and journals with more than 50 citations. Furthermore, smart home technology represents one of ten actions to acceler-ate the energy transition and create jobs in the European (EU) Commission's "Strategic Energy Technology Plan" of 2015 [49]. Even today smart home technology is still listed as a crucial element of the transition towards a climate-neutral energy system [50]. Another aspect that highlights the continuing relevance of this topic is the fact that, since 2015, Google Scholar registered 622,000 with the keyword "Smart Home Technology."

To clarify, a smart home is smart not because of its advanced construction, efficient energy use, or eco-friendly recycling system but due to the interactive technology bundling all these features [51]. Accordingly, smart home technologies are considered smart technologies because they combine a variety of innovative technologies, such as the Internet of Things [52], artificial intelligence [53], machine learning [54] or cloud comput-ing [55]. According to the literature review by Solaimani et al. [56], smart home technologies have been increasingly extended to other subject areas, as well, such as smart cities, smart communities, and smart factories, creating a wave of interest in the concept of smart living overall. Therefore, it is possible to gain important insights from the dataset at hand for human-machine interaction today.

The project was subdivided into four separate datasets and combined various empirical methods. The data were collected at several points in time during the project (Fig. 13.2). The first data collection encompasses detailed qualitative data using semi-structured interviews and a survey at four points in time during a two-year field trial of the installation, implementation, and usage of smart home technologies, which involved 20 recruited households. The second collection includes the measurement of the electrical consumption of the 20 households during the field trial. The third data source entails gas consumption, internal air temperature, local climate data, and data of other sensors in the 20 households (https://www.refitsmarthomes.org/datasets/).

For this analysis, we utilize the fourth dataset: Here, the attitudes of 1054 UK homeowners towards new smart home technologies are surveyed. We assume that

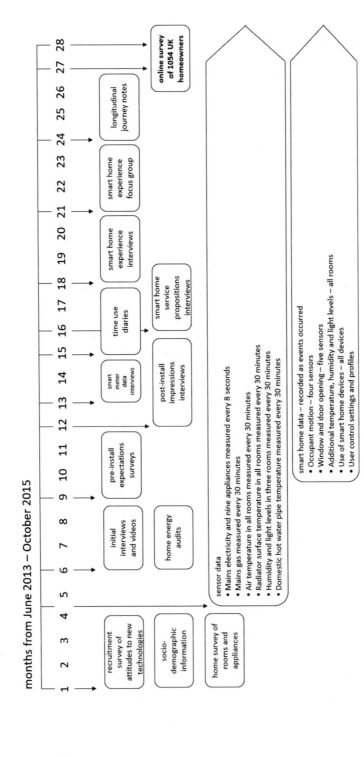

Fig. 13.2 Overview of the REFIT dataset

employees and homeowners share important attributes when making decisions about new technologies as they seek to avoid disinvestments. However, unlike employees, homeowners are more likely to give an honest, unbiased answer since they decide of their own volition if and when to deploy and use smart technology. Employees in firms usually give biased answers due to issues with social desirability. Homeowners, for example, do not have to fear any sort of negative consequences from a third party (e.g. employers). This allows for more accurate measurement of their opinion and their trust.

The survey was conducted online between September and October 2015 and organized by a market research company utilizing a representative sample of UK homeowners. The questionnaire started with socioeconomic and demographic questions, further containing questions about the purposes, benefits, functionalities, design characteristics, consumer confidence, and risks of smart home technologies, and ended with a free-text field for comments.

Trying to apply the results to the professional context, we excluded all individuals older than 64 from further study. As a result, we only consider working-age participants, limiting the number of participants considered to 780. Among the respondents, 5.7 percent were between 18 and 24, 18.85 percent between 25 and 34, 23.59 percent between 35 and 44, 26.41 between 45 and 54, and 25.38 percent between 55 and 64 years old. We can speak of a representative cross-section of the working-age population in the UK, as the age distribution of the working-age population has not changed remarkably in percentage terms between 2015 and 2019, according to data provided by the Office for National Statistics [57]. Similarly, the sample is composed almost equally of male (46.15 percent) and female (53.85 percent) individuals. Concerning the knowledge about smart home technology, 22.05 percent have a vague idea of the technology, while 36.28 percent have a general idea, 37.05 percent have a good idea, and 4.62 percent have already installed some type of smart home technology.

Our research approach aims to point out how the expected benefits and risks influence the relevant key characteristics of trust-building. Therefore, we focus on questions about the potential benefits, as well as perceived risks, of smart home technologies and relevant arrangements to build consumer confidence in smart home technologies. The expected benefits and risks were each queried using 11 items and a free-text field for any benefits or risks not listed. Similarly, the key characteristics for consumer confidence were collected using six items and a free-text field. The "Personalised Retrofit Decision Support Tools for UK Home Using Smart Home Technology" (REFIT) project team used a five-point Likert scale (1 = strongly disagree, 5 = strongly agree) with the additional option 'don't know' to answer the items. The detailed dataset and the used questionnaire are available at www.ukdataservice.ac.uk.

13.3.2 Estimation Strategy

To explore how the potential benefits and risks of smart home technologies influence the trust in technology, we apply SEM. "Structural equation modeling (SEM) is a collection of statistical techniques that allow a set of relationships between one or more independent variables, either continuous or discrete, and one or more dependent variables, either continuous or discrete, to be examined." [58, p. 661]. Weston and Gore [59] regard SEM as a combination of factor analysis and path analysis, with factor analysis describing and validating the relationship of observable and latent constructs and the structural model addressing the interrelationships of these latent constructs.

Since the SEMs focus on the inclusion of latent variables, factor analysis will be conducted first. In this context, an exploratory factor analysis serves to specify the model and to uncover dependencies between the observable variables and the non-observable variables [60]. Thus, we use exploratory factor analysis to identify how well the queried items depict each construct and to disclose whether dividing a construct into different factors for the further specification of our structural equation model is reasonable. First, we delete all missing cases from the dataset. Subsequently, we examine the used items of the REFIT project and sharpen our model for the professional context. The following Table 13.3 lists relevant items for further analysis:

Here, we consider the correlation coefficients since the factor analysis is based on correlations and covariances [61]. The correlation matrices are analyzed, and the unidimensionality of the latent variables is examined. For this purpose, the internal consistency of each latent variable is evaluated with Cronbach's Alpha. A Cronbach's alpha of

Table 13.3 Items of the three latent variables

Number of items	The potential benefits of smart home technologies are to...	There is a risk that smart home technologies...	For there to be consumer confidence in smart home technologies, it is important that they...
1	save time.	increase dependence on technology.	are reliable and easy to use.
2	save money.	increase dependence on outside experts.	can be controlled and over-ridden.
3	save energy.	result in a loss of control.	securely hold all data collected.
4	make things less effort.	are intrusive.	guarantee privacy and confidentiality.
5	provide peace in mind.	monitor private activities.	come with performance warranties.
6	provide comfort.		are made by credible manufacturers.
7	improve security.		

Table 13.4 Factor structure and loadings

Factor	Item	Factor loadings
Potential material benefits	Save money	0.75
	Save energy	0.76
Potential immaterial benefits	Save time	0.60
	Make things less effort	0.54
	Provide peace in mind	0.78
	Provide comfort	0.65
	Improve security	0.64
Risks	Increase dependence on technology	0.52
	Increase dependence on outside experts	0.60
	Result in a loss of control	0.73
	Are intrusive	0.80
	Monitor private activities	0.75
Trust-building	Are reliable and easy to use	0.72
	Can be controlled and over-ridden	0.62
	Securely hold all data collected	0.80
	Guarantee privacy and confidentiality	0.83
	Come with performance warranties	0.78
	Are made by credible manufacturers	0.63

0.7–0.95 is considered an acceptable value for the internal consistency of a factor and serves as a suitable reliability test of a latent variable [62, 63].

In the beginning, we consider the correlation matrices of the items to avoid multicollinearity. However, since no correlations are equal to or greater than 0.85, we do not need to exclude any item and can continue with the study [59].

Next, we execute an exploratory factor analysis with oblique rotation to uncover dependencies between the measured items and the latent constructs [64]. The exploratory factor analysis reveals that, contrary to our previous assumption, the seven items of the potential benefits are divided into two factors. Hence, we split up the potential benefits into the factor's potential material benefits and potential immaterial benefits. The items save money and save energy load on the factor potential material benefits with values of 0.75 and 0.76 respectively. The factor loadings of the other five items range from 0.54 to 0.78 on the factor potential immaterial benefits. The five items of risks provide one factor, with factor loadings ranging from 0.52 to 0.8. Finally, the six items of trust-building load on one factor, as well. The detailed results of the factor analysis are reported in Table 13.4.

The means, standard deviations, alpha reliabilities, and correlations of the four constructs in this study are listed in Table 13.5.

The factor potential material benefits with the items save money and save energy has a Cronbach's alpha of 0.797. The factor potential immaterial benefits with the remaining associate items also reaches an acceptable Cronbach's alpha of 0.836. The other two latent constructs show excellent values of internal consistency with their items, as well.

Table 13.5 Descriptive statistics, reliability estimates, and correlations

Factor	Mean	Std. Dev.	Cronbach's Alpha	1	2	3	4
Potential material benefits	3.978	0.925	0.797	1.000			
Potential immaterial benefits	3.789	0.822	0.836	0.612*	1.000		
Risks	3.232	1.015	0.836	0.036	0.151*	1.000	
Trust-building	4.032	0.981	0.891	0.349*	0.445*	0.305*	1.000

*$N = 780$. Correlations are significant at $p < .05$

13.4 Results

After the specification of our model, we need to control its fit. SEM allows us to validate the measurement of our stated model and review our model fitting. To check the validation of our model, we perform a confirmatory factor analysis (CFA) [65]. "In contrast to its analytic cousin, exploratory factor analysis, CFA explicitly tests a priori hypotheses about relations between observed variables (e.g., test scores or ratings) and latent variables or factors." [66, p. 6].

Thus, to test the fit of our model, the CFA provides numerous criteria of quality. Among them, one of the most widely used is the χ^2-statistic. However, we disregard it in our study since it is sensitive to large samples [67]. According to the review of Schreiber et al. [68], most authors prefer the "Tucker-Lewis index" (TLI), "Comparative fit index" (CFI), and the "Root mean square error of approximation" (RMSEA) to verify their model. "RMSEA is an absolute fit index, in that it assesses how far a hypothesized model is from a perfect model. On the contrary, CFI and TLI are incremental fit indices that compare the fit of a hypothesized model with that of a baseline model (i.e., a model with the worst fit)" [69, p. 409]. In addition, the "standardized root mean square residual" (SRMR) is consulted to detect the differences between the observed data and the model [59]. The cutoff criteria of the indices are based on the renowned research by Hu and Bentler [70].

Once the confirmatory factor analysis confirms the reliability of our hypothetic model and examines covariations among latent constructs, we focus on the structural model [59]. "The structural model displays the interrelations among latent constructs and observable variables in the proposed model as a succession of structural equations—akin to running several regression equations." [68, p. 325]. In our simple mediation model, we examine functional and causal relationships and predictions, noting that a statistical procedure only suggests a causal relationship but is insufficient to determine definite causality by itself [67]. Therefore, we prefer the terminology of total, direct, and indirect effects to avoid misunderstandings and misinterpretations [68, 71]. A direct effect describes the influence of one variable on another, whereas an indirect effect denominates

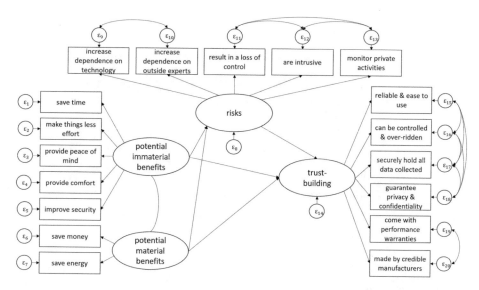

Fig. 13.3 Structural equation model on trust-building in smart technologies (Note: latent variable° = °circle, observed variable° = °boxes; independent variables: potential (im)material benefits; dependent variable: trust-building; mediator variable: risks; ε° = °measurement error of manifest variables. We assume that the variable's 'potential immaterial' and 'potential material benefits' directly affect the variable 'risks' and 'trust-building' while the variable 'risks' affects the variable 'trust-building' itself)

the impact of one variable on another variable through a mediated or intervening variable. Finally, the total effect of one variable on another is the sum of both effects [72].

Based on the correlations, factor loadings, and content factors, we adjust our research model (Fig. 13.3).

Next, we perform a CFA of our specified model to validate the measurement. We apply common fit indices to evaluate the fit of the SEM, namely TLI, CFI, RMSEA, and SRMR. Hu and Bentler [70] recommend the following standards for assessing models: TLI $\geq$ 0.95, CFI $\geq$ 0.95, RMSEA $\leq$ 0.06, and SRMR $\leq$ 0.08. The model achieves an acceptable fit of the data with the TLI = 0.966, the CFI = 0.973, the RSMEA = 0.043, and the SRMR = 0.033. Thus, no further modifications are indicated by the confirmatory factor analysis, and the hypothesized model (Fig. 13.3) appears to be a good fit for the data.

Up to this point, we focused solely on the specification and validation of our model. However, SEM allows us to also answer questions about the interrelationships between the latent constructs. In our case, we assume a mediation effect through the construct risks between the construct's potential immaterial and material benefits on trust-building. Therefore, we consider the different types of effects between the latent variables. Our estimated SEM is described graphically in Fig. 13.4. We choose the maximum likelihood parameter estimation for our model [73].

First of all, we have to distinguish between total, direct, and indirect effects. The total effect of an independent variable on a dependent variable equals the sum of the direct and

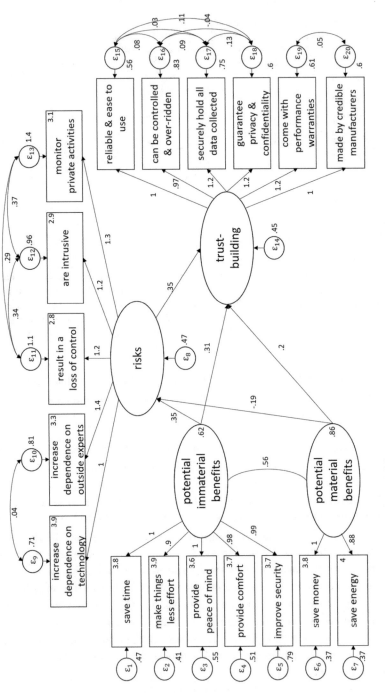

Fig. 13.4 Results of the structural equation model

Table 13.6 Total effects among the variables

Dependent variable	Independent variable	Coef.	Std. Err.	Z	P > |z|	95% Conf. Interval	
Trust-building	Potential immaterial benefits	0.520	0.082	6.34	0.000	0.359	0.681
	Potential material benefits	0.055	0.068	0.82	0.413	−0.077	0.188

indirect effects [71]. Table 13.6 shows that the total effect of the variable potential immaterial benefits on trust-building is significant ($p < 0.05$) with a coefficient of 0.52. The table also illustrates that the total effect of the variable potential material benefits on trust-building is not significant ($p > 0.05$). In this case, Baron and Kenny [74] suggest that, since there is no significant relationship between the variable's potential material benefits and trust-building, there can be no mediating effects between the variables either. At first, this statement feels intuitive because, if no effect exists, it begs the question of which effect should be mediated after all. However, many authors are questioning the requirement of a total effect before assessing mediation [75–78]. Zhao et al. [78] and Hayes [75] consider a total effect the sum of all direct and indirect paths, and not all of these paths may be part of the research model. In addition, they mention the possibility that these paths have a complementary effect on each other, and, therefore, no significant total effect occurs. Thus, authors can miss some potentially interesting, important, and useful mechanisms by which the independent variable exerts some kind of effect on the dependent variable if they do not examine the direct and indirect effects [75, 77]

In a SEM, the direct effects among the variables are depicted by single paths [72]. The direct effects reflect the influence of one variable on another, unmediated by any other variable in the model [71]. In our model, solely the direct effect between the potential material benefits and trust-building is non-significant. This is not surprising if we consider that the total effect between these two variables was not significant either. Yet, the direct effect of the variable material benefits on the risks is significant with a negative coefficient of 0.17 ($p < 0.05$). Table 13.7 also indicates that the variable potential immaterial benefits has the strongest significant direct effect on the mediator variable risks ($p < 0.05$, $\beta = .343$). Furthermore, it is noteworthy that the mediator variable risks has a significant influence on the variable trust-building, as well as the variable potential immaterial benefits.

Due to our selected method, we are also able to not only test the direct effects but also the indirect effects among the variables [58]. The indirect effect describes the effect the two independent variables (potential immaterial and material benefits) have on the dependent variable (trust-building) through the intervening variable (risks) [75]. In short, indirect effects occur when the influence of an independent variable is mediated by an intervening variable [59]. In our simple mediation model, we assume that the two types of benefits have an indirect effect on the variable trust-building through the intervening variable

Table 13.7 Direct effects among the variables

Dependent variable	Independent variable	Coef.	Std. Err.	Z	P > |z|	95% Conf. Interval	
Risks	Potential immaterial benefits	0.343	0.085	4.02	0.000	0.176	0.511
	Potential material benefits	−0.169	0.071	−2.37	0.018	−0.308	−0.029
Trust-building	Risks	0.331	0.053	6.30	0.000	0.228	0.434
	Potential immaterial benefits	0.406	0.080	5.11	0.000	0.250	0.562
	Potential material benefits	0.111	0.066	1.69	0.091	−0.018	0.240

Table 13.8 Indirect effects among the variables

Dependent variable	Mediator variable	Independent variable	Coef.	Std. Err.	Z	P > |z|	[95% Conf. Interval]	
Trust-building	Risks	Potential immaterial benefits	0.114	0.031	3.64	0.000	0.053	0.175
		Potential material benefits	−0.056	0.024	−2.25	0.025	−0.104	−0.007

risks (Table 13.8). Our hypothesis is supported by the fact that the two indirect effects of the variables potential immaterial and material benefits on the variable trust-building are significant (p < 0.05). The variable potential immaterial benefits has a positive indirect effect ($\beta = .114$) on the trust-building components, and the variable potential material benefits has a negative indirect effect ($\beta = -.056$). Thus, due to the significant direct and indirect effects of the variable potential immaterial benefits on the variable trust-building, risks partially mediate the impact of immaterial benefits on trust-building [59]. Contrastingly, in our model, the variable potential material benefits is fully mediated by the variable risks since it has only significant direct effects on the mediator variable and significant indirect effects on the variable trust-building through the intervening variable risks [77].

13.5 Discussion

This study represents an effort to examine how potential benefits and risks influence the components of trust-building. In order to obtain representative and unbiased statements about trust in smart technologies, we examined an online survey of 1054 UK homeowners on their attitudes and perceptions of smart home technologies. After adding some specifications based on exploratory factor analysis, our mediation model achieved a good fit. We tested the fit based on the literature hypothesized simple mediation model to the data through CFA. Our fitted model, therefore, allows us to make the following assumptions:

First, we found that the relationship between potential immaterial benefits and trust-building is partially mediated, so the sum of the two paths in Fig. 13.4 must be considered (potential immaterial benefits → trust-building, and potential immaterial benefits → risks→ trust-building). The potential immaterial benefits have a significant direct and indirect effect on the trust-building components. Therefore, we can conclude that, in total, an increase of the potential immaterial benefits by one unit entails an increase of 0.52 units of the demand for trust-building components. Consequently, it can be assumed that higher potential immaterial benefits, like save time, make things less effort, improve security, and provide peace in mind, as well as the comfort of new technology, will increase the desire for trust-building components to create user confidence.

One might argue that the trust mechanism is different in private technology use compared to professional technology use. However, psychological literature assumes that technology trust is subject to stable individual preferences [24] and situational components, such as perceived costs and benefits. Given that we examine a specified population in our sample, we assume that the effect is even more pronounced in employees: While they do not have to bear the cost of an investment and, therefore, attach less value to the material benefits, they have to cooperate with the machine. Thus, immaterial benefits might be even more important when material benefits are irrelevant. For example, Cobb and Macoubrie [79] identify similar findings in their work on perceived risks, benefits, and trust to nanotechnology. In their survey of 1536 United States citizens, the expected material savings (e.g. cheaper, better consumer products) based on nanotechnology are also of significantly less relevance than the immaterial benefits (e.g. mental improvements for humans, progress in the fight against diseases, cleaner environment, etc.). Similarly, Nguyen and Khoa [80] found a strong positive influence of perceived mental benefits on online trust in their study on the then-new technology of e-commerce in Vietnam.

Moreover, an increase in the variable potential immaterial benefits of technology is also associated with increased perceived risks of said technology, which can be derived from the fact that the variable potential immaterial benefits also has a positive significant influence on the perceived risks, as well as through them on trust-building. An increase in the risks of technology, then, in turn, inevitably leads to increasing demand for trust-building determinants.

Second, our findings indicate that material benefits have a significant mediated effect on the latent variable trust-building, although the total and direct effects on this dependent variable were not significant. This finding is not in line with the popular assumption by Baron and Kenny [74] but rather supports the reasoning of Rucker et al. [77], Hayes [75], MacKinnon et al. [76], and Zhao et al. [78] and allows us to draw further inferences from our model. When examining the direct effects of the potential material benefits, we notice that material benefits significantly influenced the risks of new technology for users. While the immaterial benefits increased the perceived risks of new technology, the material benefits negatively affected them. This suggests that perceived material benefits tend to have a calming effect on the users of new technology, whereas potential immaterial benefits tend to increase the perceived risks and enhance the fears of being dependent on or controlled by a new technology. If we transfer these results to the industrial context, we can assume that the material benefits will not exhibit a major impact on the perceived fears and risks pertaining to smart technology. In the private sector, it is probably more important for people to save money and energy than in the professional context. Nonetheless, this statement can by no means be generalized for all employees.

Besides, the results indicate that, although the variable potential material benefits neither had a total nor a direct effect on the latent variable trust-building, it still had a significant effect on the components of trust-building through the intervening variable risks. We detected that if the potential material benefits increase, then the importance of six items for user confidence decreases. However, it is important to add that this indirect effect on the trust-building variable is very small and, therefore, the relevance of the trust-building components cannot be compensated by material benefits, such as money or energy savings. This finding once again highlights the importance of immaterial benefits in the process of initial trust-building in smart technologies and emphasizes the notion that companies cannot convince their employees of the technology by means of material benefits.

Combining both parts, we find that even in a sample of decision-makers who have to pay the price for the technology, immaterial benefits outweigh material benefits concerning trust. Thus, attractive pricing may be a necessary condition for the initial buy but is not a sufficient condition for trust-building. The general divergence of immaterial and material benefits again gives rise to the CASA framework: Like social actors, technology and trust establishment is rather based on immaterial reliability than on cost aspects.

The interplay between immaterial and material benefits is also interesting for distrust, a similarly important aspect of human-machine collaboration. Past research has shown that trust is fragile—even more so in machine-human contexts compared to a human-human relation [37]. Thus, immaterial benefits might not only be crucial when establishing trust but might also jeopardize trust and result in distrust when missing (e.g. [38]). Therefore, future research should analyze how trust and distrust evolve and depend on perceived immaterial benefits, not only short-term but also in the long run.

13.6 Conclusion

The purpose of this study was to investigate how the initial trust between new smart technology and its user is influenced by the perceived benefits and risks of said technology. Consequently, this study aimed to clarify how the potential benefits, perceived risks, and trust-building influence each other and to disclose effect relationships among these factors. In the course of this research, we emphasized the comparison of human-human and human-machine trust to identify alleged similarities and differences for further investigation. The comparison demonstrated that, on the one hand, building trust in technology might be easier since people are easier convinced on the basis of attributes rather than personalities. On the other hand, in this context, trust might be particularly vulnerable. Based on the literature, we, therefore, assumed that first impressions, expectations, and judgments—in this context the perceived benefits—influence the perception of risks and the importance of agreements to build trust. One of our initial findings was that the factor benefits should be further divided into material and immaterial benefits in order to obtain detailed results of the effect relationships. Hence, we formulated a mediation model that distinguished between these two types of benefits. Subsequently, we empirically tested to what extent the material and immaterial benefits have an impact on trust-building and to what extent these effects are mediated by risks. The results support the proposed model and highlight the strengths of the influences of potential material and immaterial benefits on the trust-building components through the mediator risks. Our results provide particular insight into the trust-building process to new smart technology by illustrating that the two different kinds of benefits have contrary effects on the perceived risks and trust-building. The benefits do not necessarily reduce or compensate for the perceived risks posed by new technology or even reduce the demand for trust-building components as one might expect. Although the material benefits reduce the demand for trust-building components due to the mediator variable, the immaterial benefits increase the expected risks and fears concerning new smart technology and, in turn, enhance the demand for confidence-building arrangements. The results additionally suggest that the immaterial benefits, e.g. provide peace of mind, provide comfort, or make things less effort, have a stronger influence on trust-building than the measurable benefits, like save money or save energy. Therefore, we can assume that trust in smart technology can by no means be generated on the basis of material features. Instead, the user is rather guided by sensations and feelings during the development of trust and, thus, the so-called "Trusting Beliefs in Specific Technology" by Carter et al. [2] have a significant influence.

Finally, the results of this study may be relevant for further trust research. In particular, the results provide important insights into the initial trust formation between humans and technology but, at the same time, also raise several interesting questions. For one, it appears that higher expectations regarding new technology among users lead to increasing concerns, which must be compensated by usage and performance guarantees. Here, factors of transparency, as well as performance warranties, seem to have a decisive influence on initial trust-building. Another point is that, from the user's perspective, the material savings provided by technology barely have an impact on initial trust-building. While this aspect is crucial for the firms, here it recedes into the background. This raises the questions of whether the effect depends on the nature of the material benefits, and whether the strengths of effects of material and immaterial benefits shift or remain constant during the usage of technology. For example, one could assume that material rewards have a stronger effect on trust compared to material savings.

References

1. Ashleigh, M. J., & Nandhakumar, J. (2007). Trust in technologies: Implications for organizational work practices. *Decision Support Systems, 42*(2), 607–617.
2. Carter, M., Thatcher, J. B., Clay, P. F., & Mc Knight, D. H. (2011). Trust in specific technology: An investigation of its components and measures. ACM Transactions on. *Management Information Systems, 2*(2), 1–25.
3. Susi, T., & Ziemke, T. (2001). Social cognition, artefacts, and stigmergy: A comparative analysis of theoretical frameworks for the understanding of artefact-mediated collaborative activity. *Journal of Cognitive Systems Research, 2,* 273–290.
4. Taddeo, M. (2010). Trust in technology: A distinctive and a problematic relation. *Knowledge, Technology & Policy, 23*(3–4), 283–286.
5. Wang, F., & Hannafin, M. J. (2005). Design-based research and technology-enhanced learning environments. *Educational Technology Research and Development, 53*(4), 5–23.
6. Xu, J., Le, K., Deitermann, A., & Montague, E. (2014). How different types of users develop trust in technology: A qualitative analysis of the antecedents of active and passive user trust in a shared technology. *Applied Ergonomics, 45*(6), 1495–1503.
7. Lippert, S. K., & Davis, M. (2006). A conceptual model integrating trust into planned change activities to enhance technology adoption behavior. *Journal of Science, 32*(5), 434–448.
8. Hilty, L. M., Köhler, A., Von Scheele, F., Zah, R., & Rudy, T. (2006). Rebound effects of progress in information technology. *Poiesis & Praxis, 4*(1), 19–38.
9. Boston Consulting Group. (2019). Winning with AI. Pioneers combine strategy, organizational behavior and technology. *MIT Sloan Management Review.*
10. Capgemini. (2020). *Digital mastery. How organizations have progressed in their digital transformation over the past two years.* Capgemini Research Institute. Accessed January 21, 2021, from https://www.capgemini.com/wp-content/uploads/2021/01/Digital-Mastery-Report-1.pdf
11. PricewaterhouseCoopers. (2020). *COVID-19. A digital technology agenda driving an accelerated transition to the new normal.* Accessed January 21, 2021, from https://www.pwc.de/de/deals/covid-19-a-digital-technology-agenda-driving-an-accelerated-transition-to-the-new-normal.pdf
12. Astebro, T. (2004). Sunk costs and the depth and probability of technology adoption. *The Journal of Industrial Economics, 52*(3), 381–399.
13. Keil, M., Turex, D. P., & Mixon, R. (1995). The effects of sunk cost and project completion on information technology project escalation. *IEEE Transactions of Engineering Management, 42*(4), 372–381.
14. Budnick, C. J., Rogers, A. P., & Barber, L. K. (2020). The fear of missing out at work: Examining costs and benefits to employee health and motivation. *Computers in Human Behavior, 104,* 106–161.
15. Cave, S., & Dihal, K. (2019). Hopes and fears for intelligent machines in fiction and reality. *Nature Machine Intelligence, 1,* 74–78.
16. Lucas, H. C., & Goh, J. M. (2009). Disruptive technology. How Kodak missed the digital photography revolution. *The Journal of Strategic Information Systems, 18*(1), 46–55.
17. Bahmanziari, T., Pearson, J. M., & Crosby, L. (2003). Is trust important in technology adoption? A policy capturing approach. *Journal of Computer Information System, 43*(4), 46–54.
18. Siau, K., & Wang, W. (2018). Building trust in Artificial Intelligence, machine learning, and robotics. *Cutter Business Journal, 31*(1), 47–53.
19. Ashoori, M. & Weisz, J. D. (2019). *In AI we trust? Factors that influence trustworthiness of AI-infused decision-making processes.* Accessed January 21, 2021, from https://arxiv.org/abs/1912.02675

20. Muir, B. M. (1987). Trust between humans and machines, and the design of decision aids. *International Journal of Man-Machine Studies, 27*(5–6), 527–539.
21. Ejdys, J. (2018). Building trust in ICT application at a university. *International Journal of Emerging Markets, 13*(5), 980–996.
22. Jeon, M. (2017). *Emotions and affect in human factors and human-computer interaction: Taxonomy, theories, approaches and methods* (pp. 3–26). Academic Press.
23. Palmer, J. & Terry, N. (2016). *Smart homes and saving energy. The REFIT project final report for industry and government.* Accessed January 18, 2021, from https://www.refitsmarthomes.org/publications/
24. Jacques, P. H., Garger, J., Brown, C. A., & Deale, C. S. (2009). Personality and virtual reality team candidates: The roles of personality traits, technology anxiety and trust as predictors of perceptions of virtual reality teams. *Journal of Business and Management, 15*(2), 143–158.
25. Venkatesh, V. (2000). Determinants of perceived ease of use: Integrating control, intrinsic motivation, and emotion into the technology acceptance model. *Information System Research, 11*(4), 342–365.
26. Lee, Z., & Sargeant, A. (2011). Dealing with social desirability bias: An application to charitable giving. *European Journal of Marketing, 45*(5), 703–719.
27. McKnight, D. H., & Chervany, N. L. (2001). Trust and distrust definitions: One bite at a time. In R. Falcone, M. Singh, & Y.-H. Tan (Eds.), *Trust in cyber-societies. Integrating the human and artificial perspectives* (pp. 27–54). Springer.
28. Andras, P., Esterle, L., Guckert, M., Han, T. A., Lewis, P. R., Milanovic, K., Payne, T., Perret, C., Pitt, J., Powers, S. T., Urquhart, N., & Wells, S. (2018). Trusting intelligent machines. Deepening trust within socio-technical systems. *IEEE Technology and Society Magazine, 37*(4), 76–83.
29. Deutsch, M. (1958). Trust and suspicion. *The Journal of Conflict Resolution, 2*(4), 265–279.
30. Worden, K., Bullough, W. A., & Haywood, J. (2003). *Smart technologies.* World Scientific Publishing.
31. Akhilesh, K. B., & Möller, D. P. F. (2020). *Smart technologies.* Scope and Applications.
32. Hernandez-de-Menendez, M., Diaz, C. A. E., & Morales-Menendez, R. (2020). Engineering education for smart 4.0 technology: A review. *International Journal in Interactive Design and Manufacturing, 14*(3), 789–803.
33. Cook, D. J., & Das, S. K. (2005). *Smart environments. Technologies, protocols, and applications.* Wiley Interscience.
34. Preuveneers, D., Tsingenopoulos, I., & Joosen, W. (2020). Resource usage and performance trade-offs for machine learning models in smart environments. *Sensors, 20*(4), 1–27.
35. Yu, K., Berkovsky, S., Taib, R., Zhou, J., & Chen, F. (2019). Do i trust my machine teammate? An investigation from perception to decision. Intelligent User Interfaces 2019: Proceedings of the 24th international conference on Intelligent User Interfaces.
36. Madhavan, P., & Wiegmann, D. A. (2004). A new look at the dynamic of human-automation trust: Is trust in humans compareable to trust in machines? *Human Factors and Ergonomics Society Annual Meeting, 48*(3), 581–585.
37. Lee, J. E. R., & Nass, C. I. (2010). Trust in computers: The computers-are-social-actors (CASA) paradigm and trustworthiness perception in human-computer communication. In *Trust and technology in a ubiquitous modern environment: Theoretical and methodological perspectives* (pp. 1–15). IGI Global.
38. Nass, C., Takayama, L., & Brave, S. (2006). Social consistency: From technical homogeneity to human epitome. In P. Zhang & D. Galletta (Eds.), *Human-computer interaction in management information systems: Foundations* (pp. 373–391). M. E. Sharpe.
39. Hoff, K. A., & Bashir, M. (2015). Trust in automation. Integrating empirical evidence on factors that influence trust. *Human Factors and Ergonomics Society, 57*(3), 407–434.

40. Atkinson, D., Hancock, P., Hoffman, R. R., Lee, J. D., Rovira, E., Stokes, C., & Wagner, A. R. (2012). Trust in computers and robots: The use and boundaries of the analogy of interpersonal trust. *Human Factors and Ergonomics Society 56th Annual Meeting, 56*(1), 303–307.

41. Hoffmann, R. R., Bradshaw, J. M., & Johnson, M. (2013). Trust in automation. *IEEE Intelligent Systems, 28*(1), 84–88.

42. Araujo, T., Helberger, N., Kruikemeier, S., & de Vreese, C. H. (2020). In AI we trust? Perceptions about automated decision-making by artificial intelligence. *AI & Society, 35*(3), 611–623.

43. Rempel, J. K., Holmes, J. G., & Zanna, M. P. (1985). Trust in close relationships. *Journal of Personality and Social Psychology, 49*, 95–112.

44. Lee, J. D., & See, K. A. (2004). Trust in automation: Designing for appropriate reliance. *The Journal of Human Factors and Ergonomics Society, 46*(1), 50–80.

45. Hoehle, H., Huff, S., & Godde, S. (2012). The role of continuous trust in information systems continuance. *Journal of Computer Information Systems, 52*(4), 1–9.

46. Li, X., Hess, T. J., & Valacich, J. S. (2008). Why do we trust new technology? A study of initial trust formation with organizational information systems. *The Journal of Strategic Information Systems, 17*(1), 39–71.

47. Borum, R. (2010). The science of interpersonal trust. *Mental Health Law & Policy, 574*, 1–80.

48. Suresh, S., & Sruthi, P. V. (2015). A review on smart home technology. In *Online International Conference on Green Engineering and Technologies (IC-GET)* (pp. 1–3). IEEE.

49. European Commission. (2015). *Towards an integrated strategic energy technology (SET) plan: Accelerating the European Energy System Transformation.* Accessed June 16, 2021, from https://ec.europa.eu/energy/sites/default/files/documents/1_EN_ACT_part1_v8_0.pdf

50. European Commission. (2019). *The strategic energy technology (SET) plan: At the heart of energy research and innovation in Europe.* Accessed June 16, 2021, from https://op.europa.eu/en/publication-detail/-/publication/064a025d-0703-11e8-b8f5-01aa75ed71a1

51. Harper, R. (2003). *Inside the smart home.* Springer.

52. Möller, D. P. F., & Vakilzadian, H. (2014). Ubiquitous networks: Power line communication and Internet of things in smart home environments. In *IEEE International Conference on Electro/Information Technology* (pp. 596–601). IEEE.

53. Bregman, D. (2010). Smart home intelligence – The eHome that learns. *International Journal of Smart Home, 4*(4), 35–46.

54. Kabir, M. H., Hoque, M. R., Seo, H., & Yang, S.-H. (2015). Machine learning based adaptive context-aware system for smart home environment. *International Journal of Smart Home, 9*(11), 55–62.

55. Lin, Y. (2015). Study of smart home system based on cloud computing and the key technologies. In *International conference on computational intelligence and communication networks* (pp. 968–972). IEEE.

56. Solaimani, S., Keijzer-Broers, W., & Bouwman, H. (2015). What we do—and don't—know about the smart home: An analysis of the smart home literature. *Indoor and Built Environment, 24*(3), 370–383.

57. Office for National Statistics. (2020). *Population estimates for the UK, England and Wales, Scotland and Northern Ireland: Mid-2019.* Accessed June 17, 2021, from https://www.ons.gov.uk/peoplepopulationandcommunity/populationandmigration/populationestimates/bulletins/annualmidyearpopulationestimates/mid2019estimates

58. Ullman, J. B., & Bentler, P. M. (2003). Structural equation modeling. In J. A. Schinka, W. F. Velicer, & I. B. J. Weiner (Eds.), *Handbook of psychology* (Vol. 2, pp. 607–634). Wiley.

59. Weston, R., & Gore, P. A. (2006). A brief guide to structural equation modeling. *The Counseling Psychologist, 34*(5), 719–751.

60. Pohlmann, J. T. (2004). Use and interpretation of factor analysis in The Journal of Educational Research: 1992-2002. *The Journal of Educational Research, 98*(1), 14–23.
61. Mair, P. (2018). *Modern psychometrics with R.* Springer.
62. Taber, K. S. (2018). The use of Cronbach's Alpha when developing and reporting research instruments in science education. *Research in Science Education, 48,* 1273–1296.
63. Tavakol, M., & Dennick, R. (2011). Making sense of Cronbach's Alpha. *International Journal of Medical Education, 2,* 53–55.
64. Osborne, J. W. (2015). What is rotating in exploratory factor analysis. *Practical Assessment, Research, and Evaluation, 20*(2), 1–8.
65. Bowen, N. K., & Guo, S. (2012). *Structural equation modeling.* Oxford University Press.
66. Jackson, D. L., & Gillaspy, J. A. (2009). Reporting practices in confirmatory factor analysis: An overview and some recommendations. *Psychological Methods, 14*(1), 6–23.
67. Bagozzi, R. P., & Yi, Y. (2012). Specification, evaluation, and interpretation of structural equation models. *Journal of the Academy of Marketing Science, 40*(1), 8–34.
68. Schreiber, J. B., Nora, A., Stage, F. K., Barlow, E. A., & King, J. (2010). Reporting structural equation modeling and confirmatory factor analysis results: A review. *The Journal of Educational Research, 99*(6), 323–338.
69. Xia, Y., & Yang, Y. (2019). RMSEA, CFI, and TLI in structural equation modeling with ordered categorical data: The story they tell depends on the estimation methods. *Behavior Research Methods, 51,* 409–428.
70. Hu, L., & Bentler, P. M. (1999). Cutoff criteria for fit indexes in covariance structure analysis: Conventional criteria versus new alternatives. *Structural Equation Modeling: A Multidisciplinary Journal, 6*(1), 1–55.
71. Bollen, K. A. (1987). Total, direct and indirect effects in structural equation models. *Sociological Methodology, 17,* 37–69.
72. Holbert, R. L., & Stephenson, M. T. (2003). The importance of indirect effects in media effects research: Testing for mediation in structural equation modeling. *Journal of Broadcasting and Electronic Media, 47*(4), 556–572.
73. Hox, J., & Bechger, T. (1998). An introduction to structural equation modeling. *Family Science Review, 11,* 354–373.
74. Baron, R. M., & Kenny, D. A. (1986). The moderator – mediator variable distinction in social psychological research: Conceptual, strategic, and statistical considerations. *Journal of Personality and Social Psychology, 51*(6), 1173–1182.
75. Hayes, A. F. (2009). Beyond Baron and Kenny: Statistical mediation analysis in the new millennium. *Communication Monographs, 76*(4), 408–420.
76. MacKinnon, D. P., Lockwood, C. M., Hoffman, J. M., West, S. G., & Sheets, V. (2002). A comparison of methods to test mediation and other intervening variable effects. *Psychological Methods, 7*(1), 83–104.
77. Rucker, D. D., Preacher, K. J., Tormala, Z. L., & Petty, R. E. (2011). Mediation analysis in social psychology: Current practices and new recommendations. *Social and Personality Psychology Compass, 5*(4), 359–371.
78. Zhao, X., Lynch, J. G., Jr., & Chen, Q. (2010). Reconsidering Baron and Kenny: Myths and truths about mediation analysis. *Journal of Consumer Research, 37*(1), 197–206.
79. Cobb, M. D., & Macoubrie, J. (2004). Public perceptions about nanotechnology: Risks, benefits and trust. *Journal of Nanoparticle Research, 6,* 395–405.
80. Nguyen, H. M., & Khoa, B. T. (2019). The relationship between the perceived mental benefits, online trust, and personal information disclosure in online shopping. *Journal of Asian Finance, Economics and Business, 6*(4), 261–270.

Interfaces, Interactions, and Industry 4.0: A Framework for the User-Centered Design of Industrial User Interfaces in the Internet of Production

14

Philipp Brauner ⓘ, Anne Kathrin Schaar ⓘ, and Martina Ziefle ⓘ

Abstract

The Digital Transformation is changing production and creates both new opportunities and requirements to support human operators in their work. To fully exploit the benefits, the integration of digital technology in the work processes needs to be balanced with the human factor by understanding users' requirements and integrating these in the work processes. This article presents a broad series of prototypical Industrial User Interfaces as well as approaches and methods to investigate user interaction and workers' requirements. The presented work is based on research activities from RWTH Aachen University's Cluster of Excellence "Internet of Production" which combines multidisciplinary expertise to work on the future of digital production technology. Along different use cases we present the usage context, specific research question, and the methodological approaches as well as advantages and challenges for evaluating the interface. On the one hand, the article provides an overview of Industrial User Interfaces—from shop floor to strategic dimensions of production—and examples of the breadth of future Industrial User Interfaces (e.g., computer, VR, or Human-Robot-Interaction). On the other hand, it gives insights into current research challenges and their application in industry. We conclude with a research framework building on factors from the underlying production system, the interface, and the users that can inform future research on Industrial User Interfaces.

P. Brauner (✉) · A. K. Schaar · M. Ziefle
Human-Computer Interaction Center, RWTH Aachen University, Aachen, Germany
e-mail: brauner@comm.rwth-aachen.de; schaar@comm.rwth-aachen.de;
ziefle@comm.rwth-aachen.de

Keywords

Interface design · Internet of production · Human factors · Trust in automation · Industry 4.0 · Industrial internet · Smart factory · Ergonomics

14.1 Introduction

The Digital Transformation (DX) is fundamentally changing production as we know it. Regardless of whether we call these changes Industry 4.0 [1], Industrial Internet [2, 3], Industrial Internet of Things [4], or Internet of Production [5], they have in common that machines are increasingly interconnected, automated, and generating masses of data. The consequent capturing and analysis of data will optimize and change both short-term and long-term production processes. Concepts of holistic digital images of the production such as digital shadows [6] or digital twins [7] are current promises of production research to companies to harness the potential of the digitized production. Application areas are as diverse as the production itself: Ranging from new materials, smarter machine tools or additive manufacturing to topics such as predictive or prescriptive maintenance, production management or new forms of human-machine interaction. Moreover these "new" cyber-physical production systems are designed for various production domains and ranging from shopfloor and supply chain operations to factory planning and strategic management. Even if these smart factories and Industry 4.0 is still partly a vision, there are more areas in production where these approaches are taking hold.

A still unsolved question is the role of human workers and decision makers in future socio-technical production systems (STPS) [8–10] and how to design human-machine interfaces to support people working in future production scenarios [11]. On the one hand, more production processes are being automated by smarter and often non-transparent algorithms, raising questions of responsibility and control [12]. On the other hand, smarter systems create new opportunities for collaboration between people and machines, such as worker assistance systems or human-robot collaboration [8]. It is important to note that—even in the era of automation and digitalization—decisions always depend on people in the loop. Meaning that the integration of algorithms and novel human-robot-collaboration does not supersede the human worker but that the tasks and role of the human workers will change fundamentally. Thus, understanding the impact of workers' requirements and integrating human factors' knowledge in the technology development is a chance for the successful design of socio-technical systems but it is also a structural and methodological challenge for research, industry, and education. With this article we aim to give a comprehensive overview of approaches to integrate and support operators in Industry 4.0 and present the specific research questions and methodological approaches.

The article is structured as follows: First, we give a brief outline on current research challenges and present the context of our work in Sect. 14.2. Next, we present specific requirements for different Industrial User Interfaces (IUIs), approaches to development and evaluation, and examples form results of our research in Sect. 14.3. We then summarize the main findings in terms of content and methodology and outline a research framework for future IUI evaluations in Sect. 14.4. Finally, we conclude this article and give a brief overview on open research questions and upcoming topics in Sect. 14.5.

14.2 Industrial User Interfaces in an Internet of Production

The future of industrial work will be characterized by massive changes in the way people work [13] and how work will be organized [14]. Employees will experience the comprehensive digitalization primarily by new types of Industrial User Interfaces [15]. Thus, the adequate design of these interfaces will be an important design parameter of socio-technological production systems and an important necessity for the success of Industry 4.0 [11].

14.2.1 Challenges

To design suitable Industrial User Interfaces, it is essential to consider all areas, levels, and branches of production and their specific challenges from the perspective of human workers and decision makers. Especially in high-wage countries, it is important to integrate human factors knowledge as the interaction between people and production machinery must be particularly efficient and effective to reduce the higher labor costs.

The challenges which digitalization in modern production systems in Industry 4.0 brings for the human workers can be depicted on different dimensions: The *cognitive perspective*, the *affective perspective*, the perspective of *adapting existing knowledge and methods* to the design of IUIs, and *the socio-technical system design* perspective.

From the *cognitive perspective*, the increasing task difficulty due to the handling and understanding of data from different sources is cognitively demanding, in line with the demand to cope with task complexity and information ambiguity on the one hand and the workers' tolerance towards the high volatility of decision taking under time pressure on the other hand. Here, perceptual and memory limitations of the human information processing system need to be taken into account [16] considered by understandable UI, following visual and cognitive ergonomic principles of information representation (see below). Furthermore, the ability and competence to deal with digital media is a critical issue which cannot be taken for granted for most workers [17].

From the *affective perspective*, the ability and willingness to take responsible decisions is diverging across the workforce, especially as workers might bring very different skills and experience to make autonomous decisions in production at different levels, impacted by information processing abilities, emotional stability, but also age, generation factors and performance cultures. Likewise, trust in and acceptance of digital media, decision support systems, and recommender systems are critical and shaped by privacy perception and privacy concerns [18, 19] or prior experience [20]. Here, mismatches between the affective evaluation and the actual system provide ground for misconceptions, thus inadequate user behaviors in forms of over or under trust [20–23].

Adapting existing knowledge and methods poses a challenge as for the design of IUIs, as existing strategies for coping with the cognitive and affective perspectives cannot easily be transferred due to specific requirements of the context. Although the widespread adoption of IUIs is still in its infancy, there is a solid base of guidelines, concepts, and methods from Human-Technology Interaction (HCI), human factors, ergonomics, and other related disciplines to build on for the design and evaluation of IUIs. For decades we know that user-centered and ideally participatory design is essential for designing user interfaces to ensure their usefulness, usability, user satisfaction, and eventually use. This is reflected on the one hand in various textbooks on the topic [24–26], but on the other hand also in relevant ISO standards on usable software [27, 28] and the respective design processes [29]. However, the transfer and application of findings and methods from the HCI community to the broad field of STPS is a big task [30]. Major challenges are to transfer and apply the numerous findings and methods to the field of production and to evolve the methods and findings in a way that they are applicable to the specific industrial use cases.

Lastly, the *socio-technical system* design perspective also provides a challenge when designing IUIs. Socio-technical systems consist of people interacting with technical components [31]. This can be conceptualized with the two layers of a technical and social subsystem, with the task and the used technology belonging to the former layer and people or users as social actors and their roles to the later. However it is important to consider that singular gains in one domain may yield lower overall performance [31, 32]. Consequently, the goal is a joint optimization of the whole system. Hereto, one of the issues is to identify and structure the vast amount of potentially relevant and influencing factors that relate to the evaluation and performance of the created socio-technical systems. Due to the vastness of the possible aspects, we can only touch on this area and more concrete examples will be given in the subsequent sections. Nevertheless, for evaluating IUIs we need to consider factors from the technical and the social subsystems jointly; to ensure that the improvements lay not only in one dimension (such as the IUI) but improves the overall system performance.

We frequently observe that novel designs of IUIs and research approaches on them often address single issues for specific production domains. As a result, the findings are often bound to single domains, which makes the transfer to other production areas difficult. However, the diversity of production with regard its different branches and its multiple tiers ranging from shop floor to production management poses is a further challenge to achieve specific fit to the respective system and at the same time to enable synergies and transfer. To take this aspect into account, interdisciplinary large-scale projects facilitate the interdisciplinary exchange between the various production technology domains and to validate findings from other areas (see Sect. 14.2.2).

14.2.2 Context of the Research

The demonstrators we present in the following stem from the large-scale interdisciplinary research cluster *"Internet of Production"*[1] and its predecessor *"Integrative Production Technologies for High-wage Countries"* at RWTH Aachen University. The goal of the projects funded by the German Research Foundation (DFG) is the Digital Transformation (DX) of production by using and extending the capabilities of the Internet of Things (IoT) towards an Internet of Production (IoP) [5].

With researchers from mechanical engineering, computer science, material science, mathematics, philosophy, and communication science we work on concepts for a highly digitized production and evaluate its implications on products, processes, and the work-force. Hereto, we address a broad range of topics from the technical infrastructure [6], how to make data accessible by means of digital twins and digital shadows [33, 34], over new materials and machines, to production management [35], agile product development, and the requirements of the various stakeholders [36]. Both, the consortium, and the structure of the IoP focus on disciplinary fundamental research as well as the interdisciplinary transfer between production domains.

The following chapter is a journey through the different IUIs that we developed in different interdisciplinary collaborations for different production domains.

14.3   A Journey Through Different Industrial User Interfaces

Next, we describe various demonstrators ranging from the shop floor to a strategic level of production, where Industrial User Interfaces have been designed, developed, and tested. For each demonstrator, we describe the context, research questions as well as the applied methodological approach. In addition, central findings as well as advantages and challenges are presented. Due to the breadth of the presented studies, the methodological approach and the results can only be touched upon. The corresponding referenced articles present the work in greater detail.

Our journey through the demonstrators and methodological approaches is structured as follows: We start on the shop floor level in Sect. 14.3.1 with an assistance system to support manual manufacturing tasks, taking carbon-fiber reinforced polymers (CFRP) as an example. Section 14.3.2 presents a study on design and contextual factors in human-robot-collaboration. Next, in Sect. 14.3.3 we ask what happens if autonomous robots are used in environments that require moral judgements. For the following sections, we leave the shop floor and focus on operational issues from supply chain operations and quality management, thus typical office activities. In Sect. 14.3.4 we present a methodological approach with focus on understanding information processing in the context of handling production

---

[1] https://www.iop.rwth-aachen.de/

Table 14.1 Demonstrators, approaches, and key metrics presented in the following sections

(Section) Title	Approach	Key metrics
(3.1) AR-based feed-forward to improve CFRP product quality	Lab based experiment	Speed and accuracy
(3.2) Understanding Motives and Barriers to Human-Robot Collaboration	Scenario based survey	Trust and social acceptance
(3.3) What happens when autonomous agents face moral judgements?	Interactive decision game	Trust, Reliance, Influence of Individual differences
(3.4) Basic Information Processing when handling Production Data	Lab based experiment	Speed and Accuracy, Influence of Individual Differences
(3.5) Studying Supply Chain phenomena	Interactive decision game	Performance, Influence of individual differences
(3.6) Supply Chains with Complexity—The Quality Management Game	Interactive decision game	Performance, Influence of interface and individual differences

data. Section 14.3.5 introduces a justification and a framework to study supply chain phenomena. Lastly, we conclude with a set of studies on decision support systems in quality assurance at a managerial level in Sect. 14.3.6. Table 14.1 illustrates the different stations of our journey.

14.3.1 AR-Based Feed-Forward to Improve CFRP Product Quality

Our first example investigates how feed-forward (process guidance to achieve a desired target state) can improve speed and/or accuracy of manufacturing tasks, taking Carbon-fiber-reinforced polymers (CFRP) as a use case. CFRP have many applications for manufacturing lightweight components in the automotive, energy, and aerospace sectors due to high strength-to-weight ratios and high stiffness. A common process for CFRP manufacturing is stacking multiple layers of carbon fiber cloth into a mold and then adding the matrix (such as a polymer resin) to bind both materials into the final composite. However, this process is rarely automated because the carbon fiber mats are flexible and difficult to grasp by robotic arms, or the geometry of the component and thus the mold requires sensitivity in processing to avoid wrinkles [37]. A challenge in the manual processing of carbon fiber mats is that these must be stacked with the correct orientation of the carbon fibers to ensure maximum stability (e.g., quasi-isotropic with $0°$, $+60°$, or $-60°$ offset).

Question Can assistive feed-forward systems improve the performance in manual assembly tasks?

Fig. 14.1 Illustration of the system and the task, where several layers of CFRP fabrics had to be placed (image from [38])

Approach In one project, we conceptualized and prototyped an interactive feed-forward system to assist workers to correctly place the carbon fiber layers into a mold [38]. The system used a Microsoft Kinect to track the current orientation of the current carbon fiber layers and provided auditory or visual feed-forward towards the correct placement of the different carbon fiber mats. Figure 14.1 shows the interface. In a comparative study, we then compared how auditory, visual, or combined feed-forward compared against no feed forward regarding speed and accuracy.

Key Results A major finding is that the assistance system significantly *reduced* the speed, but at the same time significantly increased the accuracy of task completion. Processing quality is essential in the manufacture of CFRP components, as quality inspection can only be performed much later in the process (after the matrix has hardened). Accordingly, the shift of the speed-accuracy tradeoff is a positive result.

Advantages and Challenges This approach facilitated prototypical users to experience different types of the feed-forward system and the influence on working speed and accuracy could be measured. However, some challenges still must be overcome before the system can be used productively: Processing of carbon or fiberglass fabric is difficult as the material is flexible without its matrix and may throw wrinkles depending on the component geometry. So far, this could only be addressed insufficiently in the demonstrator. Thus, follow-up studies with an assistance system integrated into real production and with more complex component geometries are necessary.

14.3.2 Understanding Motives and Barriers to Human-Robot Collaboration

A second topic we have addressed is the upcoming field of human-robot interaction and collaboration (HRI/HRC) [39]. HRC gets increasingly important for floor operations in industry [37] but also in other domains such as healthcare [40]. However, most existing studies addressed HRC in specific deployment scenarios and acceptance-related factors were not studied across different deployment domains. Essential questions in the field of HRC are whether and how design parameters favor the interaction with and acceptance of such systems but also how user characteristics influence acceptance and use. Recently, [41] presented a taxonomy for Human-Robot Interaction that formalized the design space of HRC. Dimensions are, for example, field of application, the role of the human, degree of autonomy, and the robots' morphology (anthropomorphic, zoomorphic, or technical appearance). It is notable that acceptance factors are often studied in detail for individual application scenarios but rarely across different application areas, morphologies, or other aspects of the design space.

Question What are characteristics people attribute to robots in human-robot collaboration scenarios depending on usage context *and* robot morphology?

Approach We used a survey with a vignette study (i.e. evaluation of scenarios systematically created based on predictor variables) to understand if perception of and trust in collaborative robots is driven by usage context and design of the robot [22]. To understand the influence of usage context, we presented robots in the contrasting environments of care and production. As design factor, we presented on the one hand a robot whose appearance was rather functional and on the other hand a robot that had human features (anthropomorph). We measured the effect of the factors *context* and *morphology* on different outcome variables, such as *Intention to Collaborate* (based on the Technology Acceptance Model [42], trust in the robot, as well as a set of properties attributed to the robots. Figure 14.2 illustrates the (simplified) research model and a key result on acceptance across both application fields.

Key Results First, the study indicates that the perceived trust in the robots as affected by the combination of both context and appearance. The results indicate that the willingness to cooperate with robots is higher in the production domain than in the health care domain. Likewise, they expected HRC more likely to happen in the production domain than in the

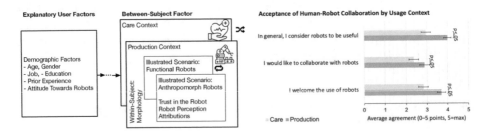

Fig. 14.2 Illustration of the research model and one of the findings that context shapes acceptance of human-robot collaboration (details in [22]).

care context and HRC was perceived as more desirable in production than in care. Lastly, the study showed that the participants' attributions towards the robots based on design (functional vs. anthropomorph) differed more strongly in the care domain (with anthropomorphic robots being evaluated as more positive), whereas the differences faded in the production domain.

Advantages and Challenges Although the study was conducted as a vignette study (and thus without real interaction between participants and robots), some known findings could be confirmed, and new ones efficiently identified and described. Of course, further studies with tangible robots will have to corroborate the findings. Nevertheless, it is precisely the comparison of different usage contexts with tangible robots that becomes difficult.

14.3.3 What Happens When Autonomous Agents Face Moral Judgements?

The last section presented a study that looked at the requirements and barriers for human-robot collaboration in the context of moral decision making. Even in the context of industrial use of autonomous robots and human-robot collaboration, there are very many decision situations in which moral decisions become relevant.

Awad et al. addressed this question in the context of autonomous driving and had different parameterized trolly dilemmas evaluated in a conjoint study (i.e., participants decided whether the car should change lanes to cause more or fewer casualties). Essentially, the study showed that utilitarian behavior is usually preferred [43].

Question We addressed a similar question and studied if the pattern is observable across multiple usage contexts as environmental factors (i.e., mobility, healthcare, and production), how different types of decision aid, as well as reliability and risk influence decision making in situations with moral dilemmas [44].

Approach In a series of decision tasks participants could choose between automation and manual execution and a trade-off between higher costs or possibly faulty automation, which can lead to property damage or personal injury. Their decision was supported by either a human agent, a statistical analysis, or an AI). Again, we used a game-based approach and participants could accumulate profits over the course of the game. Figure 14.3 shows a typical decision task and the game's interface.

Key Results The results suggest that peoples' decision making is influenced by context and risk. Reliance on automation was highest in the production context and lowest in the healthcare domain. Apparently, the contexts are perceived as having different sensitivities. Participants were most likely to rely on automation when there was no risk or only property damage to worry about. As soon as personal injury was to be feared, the willingness to

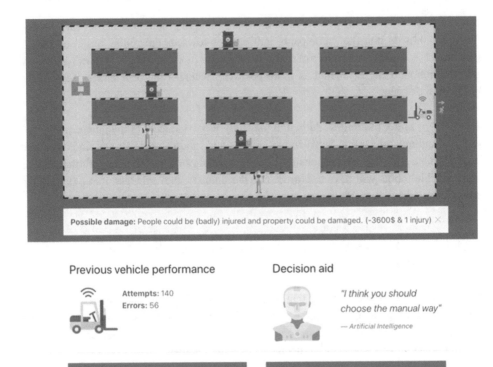

Fig. 14.3 Screenshot of the game with moral decision-making situations

automate decreased considerably. This is surprising in that personal damage was not penalized in the game logic.

Also, reliability of the system strongly influences the trust in automation. However, different types of decision support (human agent, statistical analysis, or AI) did not influence decision making. We were able to show that the subjects in the experiment reacted sensitive to most of the experimental conditions. Above all, the subjects measured the physical integrity of life in the game as more important than profit; they sacrificed profit to protect people in the game.

Advantages and Challenges This approach facilitates the analysis of moral decision making in production and other context without having to risk actual damages or harm. Despite the virtual and game-based environment, the participants responded in compliance with the hypothesis to the experimental conditions, such as context or reliability.

Our findings thus show that game-based approaches are suitable for investigating morally sensitive situations in the context of innovative interfaces such as human-robot collaboration. This is accompanied by advantages for the implementation of these interfaces in an industrial context. In the early stages of development and before implementation, interactions can be trained, and possible risks can be made visible without having to change processes. As challenges we see the need for a fundamental ethical and legal evaluation of human-robot-interaction. Our approach can only make the user perspective visible. However, this should not mean that this assessment should and can be the only orientation. In concrete terms, we also see further challenges for the design and evaluation in relation the human-robot-interface: It is still open whether concrete design parameters could enhance or support moral sensitive decision making from the human perspective. Furthermore, it should be examined whether and how the evaluation of moral decisions changes in real-world use compared to the game-based approach. A profound comparison would put the game-based approach on a more stable ground and support a more robust transfer from the virtual world to the shop floor.

14.3.4 Understanding Information Processing When Handling Production Data

While the previous demonstrator exemplified the challenges of manual production and how workers can be supported, an increasing amount of work nowadays deals with production planning and operation. Here, an operator typically interacts with a production system by reading information presented on a computer screen, processing and integrating this information into their mental model of the production system, and deciding on possible reactions, which are then fed back to the system via mouse and keyboard. The rapidly increasing amount of data and its analysis by means of algorithms, AI, and other digital means is massively changing the tools and interfaces of people in the field of production planning. A central question in this context is to understand how and under which conditions this interface can be designed. A fundamental understanding of the complex usage situation is essential. Clearly, the task itself, the ergonomic properties of the interface, as well as individual differences of the users will influence the decision speed and quality.

Questions How do people process the tasks' information? How does task complexity affect decision outcome and speed, if interface usability and decision aids can support decision making? And what happens if these automated decision aids fail?

Approach To investigate the relationships within this factor space, we conceptualized a set of experiments with task that relate to operative decisions in supply chain management. The task resembles a typical task in inventory management and procurement.

For incoming orders, the operator needs to check if sufficient resources are available or can be produced (see e.g., Fig. 14.5). The data is presented in tabular form and for each row, each standing for an individual product, multiple aspects need to be considered (the referenced papers provide more detail on the task and presentation). Note that this study only analyses the operators' response to a given task (stimulus-response), without adding further complexity to the task, such as (hidden) feedback loops or non-linear relationships that must be discovered and handled (as in the following two sections).

In the first experiment, we manipulated task complexity and interface usability [16]. We found that the influence of task complexity on decision accuracy and speed was significantly bigger than that of interface usability. The later had a small but still significant positive effect. However, a further analysis showed that the originally small effect of interface usability increases if the tasks get more complex.

Hence, interface usability may not appear as particularly important at first sight, but its benefits play out in more complicated tasks.

A further finding is that the individual perceptual speed of participants also made a difference: Not only were people with lower perceptual speed slower in the task (an expectable finding) but their disadvantage also grew with increasing task complexity: With increasing task complexity, successful operation got increasingly difficult for people with lower perceptual speed.

A second experiment studied if decision making can be supported by decision aids and how decision making is affected if these decision aids err [45]. The main finding was that a correctly functioning decision aid increases speed and accuracy; corroborating findings from automation research.

However, if the decision aid it not working correctly, operators still comply with the—now misleading—decision aid, yielding much lower accuracy. This effect, known as *Automation Complacency* [46], occurred particularly in tasks with higher complexity. Thus, task complexity is—again—important, as it is a mediator for complacency. Broken decision aids are easier to ignore when the tasks are simple but harder to ignore when task complexity increases.

Key Results First and foremost, if automated decision aids are integrated in IUI, their reliability is of paramount importance. If the decision aids err, the operators' have the tendency to automation complacency. Second, task complexity should never be neglected. In both studies, complexity contributed to interactions with other factors: In more complex tasks usability became important and automation complacency became problematic. Lastly, the results from the first study suggest that well designed Industrial User Interfaces may mitigate automation complacency, even if the tasks are complex.

Advantages and Challenges The two key drawbacks of this approach are communicability and transferability. On the one hand, project partners must be convinced of the utility of this approach. On the other hand, when transferring the approach to complex working environments, it must be checked which of the identified cause-effect relationships has which influence under real conditions.

14.3.5 Studying Basic Supply Chain Phenomena

Beyond conventional studies on ergonomic topics, the complexity of the underlying production system plays an enormous role in task performance and Industrial User Interfaces must take this into account.

A common example of the challenges raised by the underlying production system is the *Bull Whip effect* in supply chains and supply chain management [47, 48]. First described by Forrester, this term describes a phenomenon observable in multi-tier supply chains: If multiple participants along a demand-driven value chain do not coordinate (which they often don't because they are economically independent entities and are reluctant to disclose to share order information and thus information about their financial stability), small fluctuations on the demand side can build up along the supply chain and lead to substantial instabilities at the end of the chain.

Question Can we identify factors influencing operator's behavior in supply chains?

Approach To sensitize employees to this phenomenon, the "Beer Distribution Game" is often used as an abstraction of the challenges in supply chain management and thus made experienceable: Players take one position in a linear supply chain with four tiers (from a point of sale to a producing factory). Based in incoming orders, players pass their own orders to their successor in the chain and receive the products with two game iterations delay. Fulfilled orders earn profits, unfulfilled ones incur penalty costs. The customer's demand at the beginning of the supply chain is modelled by a simple function: At the beginning t_0, customer demand is $D_{0...3}$ units per week (usually crates of beer, hence the name); after 4 weeks, customer demand suddenly doubles to $D_{>4} = 8$ and stays there until the end of the game after 20 iterations.

To address our research questions, we developed a web-based supply chain game that resembles the Beer Distribution Game [49] from above. Players are at a (random) position in the chain and an artificial intelligence simulates the other agents. Based on the incoming order and the current inventory level, the players determine their order quantity. The basis for all decisions (e.g., incoming order, inventory, ...) and the player's decisions are saved, linked to surveys before (demographics, personality) and after playing the game (evaluation), and analyzed iteration by iteration or as an aggregate over the whole game.

Key Results First, after the singular change in customer demand, a dramatic increase in order volumes along the supply chain was observed. Thus, and as Fig. 14.4 illustrates, we could replicate Forrester's Bull Whip effect in our game environment. Second, we could also link player performance to personality traits collected in questionnaires before and after the game. In detail, we found performance differences based on gender, technology self-efficacy, and the need for safety. In the game, women and participants with a higher need for safety were marginally worse. A high self-efficacy in interacting with technology,

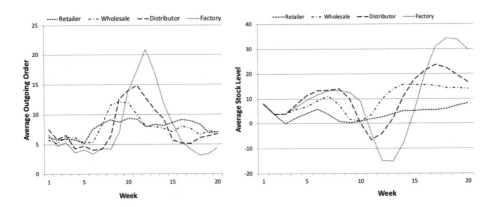

Fig. 14.4 The game metrics show that the Bull Whip effect is clearly present and that it gets extremer the farer up a player is in the supply chain (details in [49])

on the other hand, had a clearly positive effect. In addition, we found that performance increased across multiple rounds of the game. Thus, successful handling changes in the supply chain was learnable.

Advantages and Challenges Recruiting participants is usually cumbersome. Due to the interactive and partly challenging game component within the study, it was relatively easy to recruit participants. As typical effects could be confirmed, this approach does not seem to affect research validity. However, although interesting results could be realized, the demonstrator has a limited problem complexity. Thus, the expressiveness and transferability of the findings is limited. We therefore did not work on improving the player's interface to interact with this artificially simple supply chain model or to provide them with decision support, but conceptualized a significantly more realistic, complex, and relevant system model presented in the following.

14.3.6 Supply Chains with Added Complexity: The Quality Management Game

The previous section showed that behavior in complex production systems can be studied through business simulation games. To increase the transferability of the findings into practice, we significantly increased the complexity of the decision in this study on the one hand. And on the other one, we addressed a broader set of research questions (such as the utility of automated decision aids). Within four partial studies we addressed different research questions. The questions include how different system configurations (difficulties) affect decision behavior and performance, if the interface can contribute to better performance, how reliability of decision aids influences performance and trust, and if trust can be restored after critical incidents.

Questions Do user notice changes in the production? Can IUIs improve user performance? Do users rely on DSS? Can trust in DSS' can be restored after errors?

Approach To face the complex character of decision making in industrial contexts, we developed a much more sophisticated simulation model that also took aspects from quality management into account (an in-depth presentation of system dynamics model and its validation is given in [50]). Within a game-based approach our participants needed to inspect over 20 company metrics from a "company dashboard" and balance investment on three different dimensions (see Fig. 14.5): First, the incoming orders as in the Beer Distribution Game from above. Second, investments in the own production quality, as low production quality yields broken products and thus penalties. Third, investments in the incoming good inspection, as faulty supply parts would yield faulty products and thus likewise penalties. Complexity arises, for example, from the fact that the accuracy of certain input variables is influenced by the decisions made in previous iterations. If the investment in incoming goods inspection is neglected, its quality measures will no longer be correct, yielding defect parts and penalties. Players are now challenged to find the cause of this issue. However, they often attribute this to their own production quality and thus increase investments at the wrong place. Figure 14.5 shows a screenshot of the game's interface, the various input parameters on the company dashboard, as well as the three investment parameters that needed to be carefully balanced. With this IUI approach we have run in total four user studies all addressing different aspects of the interplay of system, user, and interface:

In a **first study**, we experimentally controlled the simulation environment and changed parameters of the production system after a few game iterations under experimental

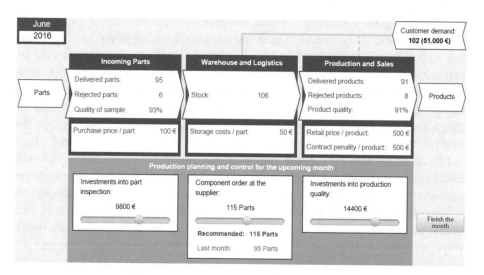

Fig. 14.5 Screenshot of the *Quality Management* game [21] for studying decision making in complex systems and how decision aids can support human operators

conditions. We analyzed how participants managed the company, if sudden changes at the side of the incoming goods (sudden quality drop), or at the side of the own production (sudden drop of the own production quality) occur [51]. As the research landscape on decision support in supply chain management is relatively sparse [52], we wanted to study if changes to the IUI by means of better decision support can improve performance. Within a **second study** we therefore revised the company dashboard and added hints to key performance indicators (i.e. inventory, customer complaints/penalties, and measured product quality) that visualize the relative change compared to the previous iteration [53]. Within this study we compared the original against the revised interface as randomized within subject factor. In a **third study**, we varied the reliability of the support system (see also Sect. 14.3.4). By chance, the support system was either correct or defect. In the correct case, the system provided good but not outstanding support throughout a round of the game. In the defect case, the system started correct but the quality degraded strongly after a few iterations and did not recover. As the previous study showed that defect decision aids diminish trust in the system. As outlined earlier, trust calibration is essential for successful cooperation between operators and automated decision aids, and missing trust leads to under-reliance and under-use. Thus, trust needs to be restorable, after a defect decision aid has been detected and repaired. Consequently, we conducted a **fourth study** to investigate if trust can be restored after a defect and which approaches are most suitable.

Key Results According to the complex research design and the four different partial studies in this section the findings are presented in the order of research questions presented above. A summary of key findings is presented below:

The results of the **first study** with focus on production changes show that no changes yield best performance (baseline condition) and that drops at both sides (incoming and own production) were most difficult to detect and compensate and therefore yielded worst performance. Further, the results suggest that changes at the suppliers' side were more likely to be detected than changes in the own simulated company. Participants were less likely to believe that changes happened in their own realm. The **second study** with focus on the performance has shown, that a small change to the IUI yielded significantly higher company profits and higher product quality. Strikingly, a survey that assessed the participants' perception of the old and the revised industrial user interface regarding perceived support, difficulty, and adequacy of the presented information showed little to no differences between the original and the revised interface. Consequently, the study showed that decision aids in Industrial User Interfaces can have a strong positive impact on performance that yielded significant higher profits for the simulated company, although the participants perception of the interface was not affected. In **study three**, of course, a correct and reliable support system improved the average performance of the participants, trust in the system, perceived usefulness and ease of use, and overall use intention. A defect led to significantly lower profits, and lower trust, usefulness, perceived ease of use, and use intention. An important finding that justifies intensive work on IUIs is a negative correlation between the perceived usability of a system and the automation complacency (i.e., the tendency to follow decision aids blindly even if they are wrong). In other words, people who found the system easy to use were more able to ignore the erroneous suggestions and could thus make better decisions. A finding that is consistent with information processing theories, such as Chandler and Sweller's Cognitive Load Theory [54]. Consequently, when IUIs are designed right, even malfunctioning decision support systems can be spotted and compensated. **Study four**, focusing on the re-establishment of trust revealed that although our findings regarding the best methods to restore trust were inconclusive, we found that trust was restorable. When the error and restoration of functionality was clearly communicated, trust and usage of the aids increased again, although not to the full extend [23].

Key Results The underlying simulation model was derived from a complex multi-tier supply chain simulation and the company dashboard provided an interface to data from and means of interaction with the simulation model. Thus, the game-based approach facilitated studying a broad range of different factors with a single demonstrator by manipulating parameters from the simulation or the user interface and linking these to user factors, performance, and evaluations. Key results are that changes in the simulation model (e.g., sudden drops in quality) had a strong impact on performance and were often noticed too late. Also, automated decision aids could improve overall performance and well-designed user interfaces help to compensate erroneous decision aids.

Advantages and Challenges This approach facilitates the empirical and experimental investigation how participants interact with complex, dynamic, and non-linear interactions in supply chain and quality management. The use of such complex simulation games for studies requires that participants develop the necessary knowledge and expertise for successful use in a short period of time. We ensured that through various iterations on the instruction that eventually included written text, videos, and simple shows. This may be a big investment to study human behavior in supply chains, the influence of decision support systems, and trust in automation. However, the experimental environments may also be used for other purposes, such as academic teaching.

Challenges in the context of the presented approach are twofold. First, we must realize that even with increased complexity the complexity of real applications can still not be achieved. Therefore, we recommend conducting studies with more specific applications in the future. Second, the studies performed with this demonstrator can only cover short temporary slots. The impact of a sustained failure of a system must be addressed in the context of extended studies of designs.

14.4 Destination and Conclusion of Our Journey

In the previous section we presented Industrial User Interfaces and their evaluations from several use cases from the research cluster Internet of Production. The interfaces and approaches presented range from shop floor operations ("blue collar work", see Sects.

14.3.1, 14.3.2, and 14.3.3) to higher level operative and strategic questions ("white collar work"), such as supply chain and quality management (see Sect. 14.3.5 and 14.3.6). Throughout the studies, we found that different factors influenced interaction with and evaluation of the interfaces on the one hand, and the overall task performance on the other. For example, the right type of feedback in the assistance system in carbon-fiber manufacturing yielded higher product quality (see Sect. 14.3.1), whereas higher task complexity led to lower performance in material disposition tasks (see Sect. 14.3.3.4) and higher technical self-efficacy yielded lower supply chain disruptions in the supply chain game (see Sect. 14.3.5).

To better systematize the space of possible factors influencing the evaluation of and performance with Industrial User Interfaces, we conceptualized a research framework for these and future studies on IUIs: Our *System × Interface × User* (SIU) framework differentiates the relevant aspects on the basis of the three areas **production system** or process (Do factors from the underlying production system or process influence the outcome?), **interface** (What is the impact of the user interface?) [55], and **user** (How do individual user factors relate to the outcome?). As a quantifiable result of the influence of these three components, we include the measurable **outcome(s)** of the performed human-system-interaction which can, for example, be task performance, accuracy, speed, achieved profit, worker or customer satisfaction, or perceived autonomy and well-being etc. (see Fig. 14.6). In the next sections we present the single components of the framework. Furthermore, we present implications for studying user interaction with IUIs according to the framework (see Sect. 14.4.2).

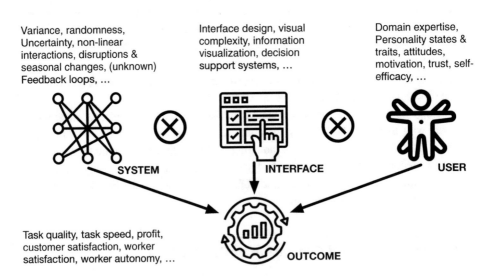

Fig. 14.6 Illustration of the *System × Interface × User Framework* (SIU) with exemplary factors from each domain (adapted from [55])

14.4.1 The SIU Framework

To systemize research on Industrial User Interfaces, we specified the SIU framework. It should be noted that ISO 9241/110, for example, also takes the operators' goals as well as the work environment into account [28], which match potential outcome metrics. In contrast to the ISO standard, however, our model makes the context and underlying production engineering factors explicitly visible. Likewise, the ISO norm focuses on the human-centered quality as a target variable, whereas our model may further include other metrics such as the performance of the overall production system. Furthermore, performance is made explicit via the presentation of parameters like, e.g., sold products or achieved quality. Thus, the goal of our framework is by no means to challenge the existing norms and approaches, but rather to draw attention to the complex interplay of the various influencing aspects relevant for a holistic perspective.

For instance, in our studies on information processing when handling production data (see Sect. 14.3.4), we found that task complexity (from the system domain) and interface usability (from the interface domain) interacted. Hence, smaller or no influences of factors from one domain cannot easily be neglected when designing future IUIs, as they may grow in importance when combined with factors from others. Within the next subsections we specify our understanding of the four dimensions:

System Under the term system we understand the underlying production process or production system. Focusing on system related aspects requires a precise consideration of deployment conditions. Depending on industry domain, product portfolio, or company size the requirements for system are extremely different and do often not allow a simple transfer of principles. Influencing factors on the system or process are, for example, disruptions caused by unexpected delays, quality or quantity deviations from suppliers, and inaccuracies generated by a production machine. For the design of future IUIs, it is essential that the space of possible system factors is adequately described and then systematically explored.

Interface In the context of IUIs a software interface builds the neuralgic intersection between human and systems which is decisive for the performance of an IUIs. We see three different aspects that are of importance when thinking about designing IUIs: First, IUIs differ in their requirements regarding the representation of information density. Second, the fields of application range from shop floor to offices. In each usage setting special local requirements like lighting conditions, or noise must be considered. Third, a wide range of interfaces is principally available to present information, e.g., VR glasses, small screen devices, ambient intelligence, stationary computers. In the future, a prudent selection of the appropriate interface will be of immense importance.

Current studies usually describe pilot phases of new interfaces. Domain-specific strengths and weaknesses analyses will show where the best fit will be in the future. All level that are corresponding to the interface are highly related to usability issues. A suitable usability and information design supports the human in the loop by supporting the HCI and providing information to the operators. In general, it is essential that an interface is not better or worse per se but must fit both to the task and processes at hand as well as the people interacting with the system.

User Putting the user in the focus even more challenging aspects come up that might be of importance in the context of IUI application and use. Beside factors like information processing speed, mental models, perceived cognitive load, educational level, technical expertise, and perceived locus of control also diversity aspects (e.g., age, gender, or cultural background) must be assessed. In this context, it is important to consider whether any associations are due to stable (personality traits) or temporary user characteristics (personality states).

Outcome and Optimization Dimension The combination of the three aspects system, interface, user leads to some form of human-system interaction, resulting in an observable interaction and outcome.

But what is suitable as an outcome variable? The choice of suitable outcome variables certainly depends on the individual use cases, the goals of the activity, and the companies' values. On the one hand, technical parameters such as decision-making quality, task performance, or the contribution to value creation are suitable; on the other hand, parameters such as workers' satisfaction, user experience, or the workers' experience of autonomy are becoming increasingly important in a labor market that is competing for skilled workers. Of course, it is also possible to consider the combination of several variables and to weight them individually according to company goals and values.

14.4.2 Application of the Framework

Each domain of the framework comes along with its own implications for user interaction with IUIs. It is important to keep in mind that the individual factors are difficult to consider in isolation, as they often arise and take effect in the interplay or interaction of factors across the categories. This fact challenges research in the field of IUIs to a large extent, as novel user interfaces cannot be evaluated in a sequential "*divide and conquer*" methodology. If a factor is identified as irrelevant in one study, this does not mean that it is irrelevant in general. For example, one of our studies showed that usability plays—on average—a negligible role for task performance, but that usability *in combination* with task complexity makes the difference between better or worse Industrial User Interfaces (see Sect. 14.3.4). The aspects mentioned here are not meant to be a complete but is intended primarily to help structure your own tasks. In the context of specific applications, other factors will certainly turn out to be important, or specific combinations will turn out to be significant.

A challenge around the design and evaluation of user interfaces in production environments is finding a match between tasks, usage contexts, and people on the one side, and tools and methods to evaluate these adequately on the other. Especially since the different evaluation methods and techniques have their own individual advantages and disadvantages depending on task, context of use and users but also the maturity of the interfaces [56]. For example, the presented experiments on production information processing presented in Sect. 14.3.4 yielded the insight that a persons' processing speed is crucial, especially for tasks with higher complexity. However, it is still unclear to what extend this holds true for real world work environments, where more complexity needs to be considered and colleagues or customers interrupt the work.

Accordingly, and as is so often the case, the research methods employed must be carefully tailored to the object of study, the maturity of the demonstrators, and the specific questions being addressed. In some cases, it does make sense to carefully explore the specific relationship of single system, user, and interface parameters in clean "laboratory experiments" (where the influence of other external factors is precisely controlled). In other more complex field cases, the influences of the factors will blur and will no longer be so clearly distinguishable. However, this also reflects real life work environments, where tasks are not carried out under clean, experimental conditions, but under the noise of machine tools, insufficient lightning, or office noise.

In the next section we will discuss challenges for the future in the form of an outlook. The focus will be on which fields of action future research in user interactions with Industrial User Interfaces can have.

14.5 Outlook and Future Journeys to Go

In this article we have presented different Industrial User Interfaces and methodological approaches for evaluating them. All interfaces originate from the Cluster of Excellence Internet of Production and are thus linked to the grand vision of Smart Factories, Industry 4.0 as well as real requirements from production research and practice. The presented SUI framework is certainly not the single solution for addressing the upcoming tasks at hand but might support understanding and facing the problem space when studying user interactions with IUIs. Especially, as we have found that an isolated consideration of aspects from a single domain (may it be factors from the system or underlying process, the interface, or the users) will certainly yield blind spots and therefore only partial solutions that cannot easily be combined. Therefore, the interconnection of the three domains must be considered when designing and evaluating future IUIs.

Certainly, we have only addressed a fraction of the possible applications of IUIs, and we have only presented a selection of methodological approaches to improve the fit between operators and production systems through adequately designed IUIs. Soon, more research in the field of Industrial User Interfaces, user interaction, and new evaluation approaches will be increasingly needed as the transformation to Industry 4.0 is just in its infancy and the challenges of the world with increasing volatility, uncertainty, complexity, and ambiguity (VUCA) [57] as well as the demographic change with its changing workforce will increasingly affect manufacturing companies. Derived from our studies and against the backdrop of increasing dynamics in the production industry, we see a wide range of challenges and implications that need to be addressed in future research:

The first challenge for research in the field of IUI is the increasing amount of available production data. In all areas of production (from simulation and design, over shop-floor operations to the usage cycle of products), more data is being captured, saved, and analyzed. Data increases in volume (created data growth exponentially), velocity (data is generated faster and often in real-time), and variety (heterogeneous data from process simulations to customer feedback) [58]. Since people will continue to be responsible for processes in the years and decades to come, we need more understanding of the tasks, the way people differentiate and process information from (big) data analysis. A key question in this context will be how to exploit this information by transforming it into actionable knowledge for workers and management of manufacturing companies by suitable aggregation, decision support, and visualization. This leads to the question how suitable Industrial User Interfaces can support workers from shop floor to management. In this context information visualization and the design of adequate decision support measures is of importance. From our point of view this can only be achieved using user-centered and participatory design approaches that address the interplay between the production system and the task, the interface, and its users.

Second, beyond capturing data from design, production, and use-cycles as digital shadows or digital twins, we expect that digital representations of the factory or office workers will also be created [59]. On the one hand, this will offer exiting new possibilities by, for example, being able to schedule operators based on their competencies and capabilities, by adapting the processes, task speed, or the Industrial User Interfaces to the individuals' needs [60], or on the other hand, there is the question of how to store and handle this sensitive personal data, which data can and should be stored and for how long, or who may access and use this data.

Third, the search for the sweet spot between automation of production process and harnessing peoples' expertise and thus facilitating hybrid operation and decision making will require a lot of attention. While some process will be increasingly automated others will still rely on human interaction or human oversight. Thus, we see trust calibration as continuing major research field in the context of IUI. Although the theoretical underpinnings are well known for decades [20, 61–63], trust needs to be carefully calibrated for each automated system, interface, and user individually. If not, we risk that the benefits of automation—that will get increasingly smarter due to the promises of Industry 4.0—are threatened by over- or under trust. If people do not trust a system, it will not be able to realize its optimization potential. If they trust the system too much, this leads to errors in production or production control which may be discovered too late or not at all. Further, a specific challenge will be solving the *automation conundrum* [64]: The more tasks get automated and the better automation gets the lower will the situational awareness be and the more difficult supervision, intervention, or manual control will be. We believe that these questions can only be solved through rigorous user-centered and participatory design. Meaning that automation is not imposed from above but is developed and introduced together with the employees. Beyond conventional metrics, such as performance and safety, humane and people-centered aspects, such as self-determination, perceived competence, relatedness, and experience of autonomy should also be taken into account [65] to increase the fit between the automated systems and its operators.

Fourth, focusing on company-related aspects in the context of IUI, we see different aspects that affect research in the field of IUIs: In some production domains, long life cycles of machines pose a challenge for rapid digitization and thus also for the emergence of new IUIs. For this reason, it will be necessary to think about the possibility and design of retrofitting machines. For designing interfaces this means that bridging technologies must be taken into consideration. In addition to the design of human-machine interfaces, the adaptation of workflows and other organizational conditions will also gain in importance in the future. These strategic and organizational aspects will have a direct and/or indirect effect on users of IUIs. For this reason, we recommend to include the influence of the context and conditions of use as a variable according to the SIU framework and findings from technology acceptance research [42, 66]. If we focus on the operators, issues such as an ageing workforce and impending shortages of skilled workers, technology acceptance but also aspects of education and training arise. This broad range of topics requires a focus on the user with his or her characteristics and professional skills. A fundamental understanding of factors and their interaction allows, for example, to offer target group-specific teaching and training formats.

Lastly, artificial intelligence is currently one of the major topics in Industry 4.0 and elsewhere. Here, transparency and explainability of the models is one of the most prominent issues [67]. The question how to design AI to allow comprehensive and traceable decisions made by AI is not only relevant regarding ethical and legal aspects, but also for people operating with AI based information transported via IUI. Factors like perceived control, transparency, and trust are of importance for acceptance.

In summary, it is important for us to emphasize that there is still a long way and lots of work to in the context of Industrial User Interfaces. Diverse processes, branches, and people require both novel design approaches and empirical methods to study interactions with Industrial User Interfaces.

Acknowledgments Funded by the Deutsche Forschungsgemeinschaft (DFG, German Research Foundation) under Germany's Excellence Strategy – EXC-2023 Internet of Production – 390621612. With thank the anonymous reviewers for their valuable feedback and Nina Braun for editing the manuscript.

References

1. Liao, Y., Deschamps, F., de Loures, E. F. R., & Ramos, L. F. P. (2017). Past, present and future of Industry 4.0: A systematic literature review and research agenda proposal. *International Journal of Production Research, 55*, 3609–3629. https://doi.org/10.1080/00207543.2017.1308576

2. Bruner, J. (2013). *Industrial Internet - The machines are talking*. Reilly Media.

3. Jeschke, S., Brecher, C., Meisen, T., Özdemir, D., & Eschert, T. (2017). Industrial Internet of things and cyber manufacturing systems. In *Industrial Internet of things* (pp. 3–19). Springer. https://doi.org/10.1007/978-3-319-42559-7_1

4. Sisinni, E., Saifullah, A., Han, S., Jennehag, U., & Gidlund, M. (2018). Industrial internet of things: Challenges, opportunities, and directions. *IEEE Transactions on Industrial Informatics, 14*, 4724–4734. https://doi.org/10.1109/TII.2018.2852491

5. Brauner, P., Dalibor, M., Jarke, M., Kunze, I., Koren, I., Lakemeyer, G., Liebenberg, M., Michael, J., Pennekamp, J., Quix, C., Rumpe, B., Van Aalst, W., Der Wehrle, K., Wortmann, A., & Ziefle, M. (2021). A computer science perspective on digital transformation in production. *ACM Transactions on Internet of Things (TIOT), 3*, 1–32.

6. Pennekamp, J., Glebke, R., Henze, M., Meisen, T., Quix, C., Hai, R., Gleim, L., Niemietz, P., Rudack, M., Knape, S., Epple, A., Trauth, D., Vroomen, U., Bergs, T., Brecher, C., Buhrig-Polaczek, A., Jarke, M., & Wehrle, K. (2019). Towards an infrastructure enabling the Internet of production. In *2019 IEEE International Conference on Industrial Cyber Physical Systems (ICPS)* (pp. 31–37). IEEE. https://doi.org/10.1109/ICPHYS.2019.8780276

7. Tao, F., Zhang, H., Liu, A., & Nee, A. Y. C. (2019). Digital twin in industry: State-of-the-art. *IEEE Transactions on Industrial Informatics, 15*, 2405–2415. https://doi.org/10.1109/TII.2018.2873186

8. Kadir, B. A., Broberg, O., & da Conceição, C. S. (2019). Current research and future perspectives on human factors and ergonomics in industry 4.0. *Computers & Industrial Engineering, 137*, 106004. https://doi.org/10.1016/j.cie.2019.106004

9. Frazzon, E. M., Hartmann, J., Makuschewitz, T., & Scholz-Reiter, B. (2013). Towards socio-cyber-physical systems in production networks. *Procedia CIRP, 7*, 49–54. https://doi.org/10.1016/j.procir.2013.05.009

10. Hermann, M., Pentek, T., & Otto, B. (2016). Design principles for Industrie 4.0 scenarios. In *2016 49th Hawaii International Conference on System Sciences (HICSS)* (pp. 3928–3937). IEEE. https://doi.org/10.1109/HICSS.2016.488

11. Kagermann, H. (2015). Change through digitization—Value creation in the age of Industry 4.0. In *Management of permanent change* (pp. 23–45). Springer.

12. Santoni de Sio, F., & Mecacci, G. (2021). Four responsibility gaps with artificial intelligence: Why they matter and how to address them. *Philosophy & Technology*. https://doi.org/10.1007/s13347-021-00450-x

13. Baethge-Kinsky, V. (2020). Digitized industrial work: Requirements, opportunities, and problems of competence development. *Frontiers in Sociology, 5*. https://doi.org/10.3389/fsoc.2020.00033

14. Hirsch-Kreinsen, H. (2016). Digitization of industrial work: Development paths and prospects. *Journal for Labour Market Research., 49*, 1–14. https://doi.org/10.1007/s12651-016-0200-6

15. Paelke, V., & Röcker, C. (2015). User interfaces for cyber-physical systems: Challenges and possible approaches. In *International Conference on Design, User Experience, and Usability (DUXU)* (pp. 75–85). Springer. https://doi.org/10.1007/978-3-319-20886-2_8

16. Mittelstädt, V., Brauner, P., Blum, M., & Ziefle, M. (2015). On the visual design of ERP systems – The role of information complexity, presentation and human factors. In *6th International Conference on Applied Human Factors and Ergonomics (AHFE 2015) and the Affiliated Conferences, AHFE 2015* (pp. 270–277). IEEE. https://doi.org/10.1016/j.promfg.2015.07.207

17. Wilkesmann, M., & Wilkesmann, U. (2018). Industry 4.0 – Organizing routines or innovations? *VINE Journal of Information and Knowledge Management Systems, 48*, 238–254. https://doi.org/10.1108/VJIKMS-04-2017-0019

18. Karabey, B. (2012). Big data and privacy issues. In *International symposium on information management in a changing world* (p. 3). Springer.

19. Ziefle, M., Halbey, J., & Kowalewski, S. (2016). Users' willingness to share data on the Internet: Perceived benefits and caveats. In *Proceedings of the International Conference on Internet of Things and Big Data* (pp. 255–265). SCITEPRESS - Science and and Technology Publications. https://doi.org/10.5220/0005897402550265

20. Hoff, K. A., & Bashir, M. (2015). Trust in automation: Integrating empirical evidence on factors that influence trust. *Human Factors, 57*, 407–434. https://doi.org/10.1177/0018720814547570

21. Brauner, P., Philipsen, R., Calero Valdez, A., Ziefle, M., & Philipsen, R. (2019). What happens when decision support systems fail?—The importance of usability on performance in erroneous systems. *Behaviour & Information Technology, 38*, 1225–1242. https://doi.org/10.1080/0144929X.2019.1581258

22. Biermann, H., Brauner, P., & Ziefle, M. (2021). How context and design shape human-robot trust and attributions. *Paladyn Journal of Behavioural Robotics., 12*, 74–86. https://doi.org/10.1515/pjbr-2021-0008

23. Philipsen, R., Brauner, P., Valdez, A. C., & Ziefle, M. (2018). Evaluating strategies to restore trust in decision support systems in cross-company cooperation. In W. Karwowski, S. Trzcielinski, B. Mrugalska, M. Di Nicolantonio, & E. Rossi (Eds.), *Advances in manufacturing, production management and process control. AHFE 2018. Advances in intelligent systems and computing* (pp. 115–126). Springer. https://doi.org/10.1007/978-3-319-94196-7_11

24. Courage, C., & Baxter, K. (2005). *Understanding your users: A practical guide to user requirements methods*. Tools & Techniques. Morgan Kaufmann Publishers.

25. Shneiderman, B., & Plaisant, C. (2004). *Designing the user interface: Strategies for effective human-computer interaction* (4th ed.). Pearson Addison Wesley.

26. Dix, A., Finlay, J., Abowd, G. D., & Beale, R. (Eds.). (2003). *Human computer interaction*. Pearson.

27. DIN/EN/ISO 9241–11. (1998). *Ergonomic requirements for office work with visual display terminals (VDTs): Guidance on usability specifications and measures*.

28. DIN EN ISO 9241-110. (2006). Ergonomie der Mensch-System-Interaktion – Teil 110: Grundsätze der Dialoggestaltung (ISO 9241-110:2006). *Deutsche Fassung EN ISO, 2006*, 9241.

29. International Organization for Standardization: ISO 9241-210:2010. (2019). *Ergonomics of human-system interaction - Part 210: Human-Centred Design for Interactive Systems*. ISO.

30. Ardanza, A., Moreno, A., Segura, Á., de la Cruz, M., & Aguinaga, D. (2019). Sustainable and flexible industrial human machine interfaces to support adaptable applications in the Industry 4.0 paradigm. *International Journal of Production Research., 57*, 4045–4059. https://doi.org/10.1080/00207543.2019.1572932

31. Fox, W. M. (1995). Sociotechnical system principles and guidelines: Past and present. *The Journal of Applied Behavioral Science., 31*, 91–105. https://doi.org/10.1177/0021886395311009

32. Trist, E. L., & Bamforth, K. W. (1951). Some social and psychological consequences of the Longwall method of coal-getting. *Human Relations, 4*, 3–38. https://doi.org/10.1177/001872675100400101

33. Liebenberg, M., & Jarke, M. (2020). Information systems engineering with digital shadows: Concept and case studies. In *International Conference on Advanced Information Systems Engineering* (pp. 70–84). Springer. https://doi.org/10.1007/978-3-030-49435-3_5

34. Jarke, M., Schuh, G., Brecher, C., Brockmann, M., & Porte, J.-P. (2018). Digital shadows in the internet of production. *ERCIM News, 115*, 26–28.
35. Schuh, G., Prote, J.-P., Gützlaff, A., Thomas, K., Sauermann, F., & Rodemann, N. (2019). Internet of production: Rethinking production management. In *Production at the leading edge of technology* (pp. 533–542). Springer. https://doi.org/10.1007/978-3-662-60417-5_53
36. Brauner, P., Brillowski, F., Dammers, H., Königs, P., Kordtomeikel, F., Petruck, H., Schaar, A. K., Schmitz, S., Steuer-Dankert, L., Mertens, A., Gries, T., Leicht-Scholten, C., Nagel, S., Nitsch, V., Schuh, G., & Ziefle, M. (2020). A research framework for human aspects in the internet of production – An intra-company perspective. In *Advances in intellifgent systems and computing* (pp. 3–17). Springer. https://doi.org/10.1007/978-3-030-51981-0_1
37. Villani, V., Pini, F., Leali, F., & Secchi, C. (2018). Survey on human–robot collaboration in industrial settings: Safety, intuitive interfaces and applications. *Mechatronics, 55*, 248–266. https://doi.org/10.1016/j.mechatronics.2018.02.009
38. Brauner, P., Bremen, L., Ziefle, M., Altorf, L., Rast, M., & Rossmann, J. (2014). Impact of different feedback conditions on worker's performance in carbon-fiber reinforced plastic manufacturing. In T. Ahram, W. Karwowski, & T. Marek (Eds.), *Proceedings of the 15th International Conference on The Human Aspects of Advanced Manufacturing (HAAMAHA): Manufacturing enterprises in a digital world* (pp. 5087–5097). CRC Press.
39. Chandrasekaran, B., & Conrad, J. M. (2015). Human-robot collaboration: A survey. *Conference Proceedings - IEEE SOUTHEASTCON*. https://doi.org/10.1109/SECON.2015.7132964
40. Heinz, M., Martin, P., Margrett, J. A., Yearns, M., Franke, W., Yang, H.-I., Wong, J., & Chang, C. K. (2013). Perceptions of technology among older adults. *Journal of Gerontological Nursing., 39*, 42–51. https://doi.org/10.3928/00989134-20121204-04
41. Onnasch, L., & Roesler, E. (2020). A taxonomy to structure and analyze human–robot interaction. *International Journal of Social Robotics*. https://doi.org/10.1007/s12369-020-00666-5
42. Davis, F. D. (1993). User acceptance of information technology: System characteristics, user perceptions and behavioral impacts. *International Journal of Man-Machine Studies, 38*, 475–487. https://doi.org/10.1006/imms.1993.1022
43. Awad, E., Dsouza, S., Kim, R., Schulz, J., Henrich, J., Shariff, A., Bonnefon, J.-F., & Rahwan, I. (2018). The moral machine experiment. *Nature, 1*. https://doi.org/10.1038/s41586-018-0637-6
44. Liehner, G. L., Brauner, P., Schaar, A. K., & Ziefle, M. (2021). Delegation of moral tasks to automated agents: The impact of risk and context on trusting a machine to perform a task. *IEEE Transactions on Technology and Society., 1–14*. https://doi.org/10.1109/TTS.2021.3118355
45. Brauner, P., Calero Valdez, A., Philipsen, R., & Ziefle, M. (2016). Defective still deflective – How correctness of decision support systems influences user's performance in production environments. In F. F.-H. Nah & C.-H. Tan (Eds.), *HCI in Business, Government, and Organizations (HCIGO), held as part of HCI international 2016* (pp. 16–27). Springer. https://doi.org/10.1007/978-3-319-39399-5_2
46. Goddard, K., Roudsari, A., & Wyatt, J. C. (2012). Automation bias: A systematic review of frequency, effect mediators, and mitigators. *Journal of the American Medical Informatics Association, 19*, 121–127. https://doi.org/10.1136/amiajnl-2011-000089
47. Forrester, J. W. (1961). *Industrial dynamics*. Wiley.
48. Lee, H. L., Padmanabhan, V., & Whang, S. (1997). Information distortion in a supply chain: The bullwhip effect. *Management Science, 43*, 546–558.
49. Brauner, P., Runge, S., Groten, M., Schuh, G., & Ziefle, M. (2013). Human factors in supply chain management – Decision making in complex logistic scenarios. In S. Yamamoto (Ed.), *Proceedings of the 15th HCI International 2013, Part III, LNCS 8018* (pp. 423–432). Springer.

50. Brauner, P., Philipsen, R., Fels, A., Fuhrmann, M., Ngo, H., Stiller, S., Schmitt, R., & Ziefle, M. (2016). A game-based approach to meet the challenges of decision processes in ramp-up management. *Quality Management Journal, 23,* 55–69. https://doi.org/10.1080/10686967.2016. 11918462

51. Philipsen, R., Brauner, P., Stiller, S., Ziefle, M., & Schmitt, R. (2014). The role of human factors in production networks and quality management. In *First International Conference, HCIB 2014, held as part of HCI International 2014, Heraklion, Crete, Greece, June 22–27, 2014* (Proceedings, LNCS 8527) (pp. 80–91). Springer. https://doi.org/10.1007/978-3-319-07293-7_8

52. Teniwut, W. A., & Hasyim, C. L. (2020). Decision support system in supply chain: A systematic literature review. *Uncertain Supply Chain Management, 8,* 131–148. https://doi.org/10.5267/j. uscm.2019.7.009

53. Philipsen, R., Brauner, P., Stiller, S., Ziefle, M., & Schmitt, R. (2014). Understanding and supporting decision makers in quality management of production networks. In *Advances in the ergonomics in manufacturing. Managing the enterprise of the future 2014: Proceedings of the 5th International Conference on Applied Human Factors and Ergonomics, AHFE 2014* (pp. 94–105). CRC Press.

54. Chandler, P., & Sweller, J. (1991). Cognitive load theory and the format of instruction. *Cognition and Instruction, 8,* 293–332. https://doi.org/10.1207/s1532690xci0804_2

55. Brauner, P., Valdez, A. C., Philipsen, R., & Ziefle, M. (2016). On studying human factors in complex cyber-physical systems. In *Workshop human factors in information visualization and decision support systems part of the mensch und computer 2016.* Gesellschaft für Informatik. https://doi.org/10.18420/muc2016-ws11-0002

56. Calero Valdez, A., Brauner, P., Schaar, A. K., Holzinger, A., & Ziefle, M. (2015). Reducing complexity with simplicity - Usability methods for Industry 4.0. In *19th Triennial Congress of the International Ergonomics Association (IEA 2015).* IEA. https://doi.org/10.13140/RG.2.1.4253. 6809

57. Bennett, N., & Lemoine, G. J. (2014). What a difference a word makes: Understanding threats to performance in a VUCA world. *Business Horizons, 57,* 311–317. https://doi.org/10.1016/j. bushor.2014.01.001

58. McAfee, A., & Brynjolfsson, E. (2012). Big data: The management revolution. *Harvard Business Review., 90,* 61–68. https://doi.org/10.1007/s12599-013-0249-5

59. Mertens, A., Putz, S., Brauner, P., Brillowski, F., Buczak, N., Dammers, H., Van Dyck, M., Kong, I., Konigs, P., Kordtomeikel, F., Rodemann, N., Schaar, A. K., Steuer-Dankert, L., Wlecke, S., Gries, T., Leicht-Scholten, C., Nagel, S. K., Piller, F. T., Schuh, G., et al. (2021). Human digital shadow: Data-based modeling of users and usage in the Internet of production. In *14th International Conference on Human System Interaction (HSI)* (pp. 1–8). IEEE. https://doi. org/10.1109/HSI52170.2021.9538729

60. Miraz, M. H., Ali, M., & Excell, P. S. (2021). Adaptive user interfaces and universal usability through plasticity of user interface design. *Computer Science Review., 40,* 100363. https://doi. org/10.1016/j.cosrev.2021.100363

61. Lee, J. D., & See, K. A. (2004). Trust in automation: Designing for appropriate reliance. *Human Factors, 46,* 50–80. https://doi.org/10.1518/hfes.46.1.50.30392

62. Masalonis, A. J., & Parasuraman, R. (1999). Trust as a construct for evaluation of automated aids: Past and future theory and research. *Proceedings of the Human Factors and Ergonomics Society Annual Meeting., 43,* 184–187. https://doi.org/10.1177/154193129904300312

63. Parasuraman, R., & Riley, V. (1997). Humans and automation: Use, misuse, disuse, abuse. *Human Factors: The Journal of the Human Factors and Ergonomics Society, 39,* 230–253. https://doi.org/10.1518/001872097778543886

64. Endsley, M. R. (2017). From here to autonomy: Lessons learned from human-automation research. *Human Factors, 59*, 5–27. https://doi.org/10.1177/0018720816681350
65. Deci, E. L., & Ryan, R. M. (2000). The "what" and "why" of goal pursuits: Human needs and the self-determination of behavior. *Psychological Inquiry, 11*, 227–268.
66. Molino, M., Cortese, C. G., & Ghislieri, C. (2020). The promotion of technology acceptance and work engagement in industry 4.0: From personal resources to information and training. *International Journal of Environmental Research and Public Health, 17*, 2438. https://doi.org/10.3390/ijerph17072438
67. Barredo Arrieta, A., Díaz-Rodríguez, N., Del Ser, J., Bennetot, A., Tabik, S., Barbado, A., Garcia, S., Gil-Lopez, S., Molina, D., Benjamins, R., Chatila, R., & Herrera, F. (2020). Explainable explainable Artificial Intelligence (XAI): Concepts, taxonomies, opportunities and challenges toward responsible AI. *Information Fusion, 58*, 82–115. https://doi.org/10.1016/j.inffus.2019.12.012

Printed in the United States
by Baker & Taylor Publisher Services